Special Issue in Honor of Dr. Marie-Hélène Metz-Boutigue 75th Birthday: "Recent Advances in Multifunctional Antimicrobial Peptides as Preclinical Therapeutic Studies and Clinical Future Applications"

Special Issue in Honor of Dr. Marie-Hélène Metz-Boutigue 75th Birthday: "Recent Advances in Multifunctional Antimicrobial Peptides as Preclinical Therapeutic Studies and Clinical Future Applications"

Editors

Scavello Francesco
Amiche Mohamed
Jean-Eric Ghia

MDPI

Basel • Beijing • Wuhan • Barcelona • Belgrade • Novi Sad • Cluj • Manchester

Editors
Scavello Francesco
IRCCS Humanitas Research Hospital
Rozzano, Italy

Amiche Mohamed
Institute of Biology Paris Seine CNRS Sorbonne University
Paris, France

Jean-Eric Ghia
Departments of Immunology and Internal Medicine, Section of Gastroenterology, Max Rady College of Medicine, Rady Faculty of Health Sciences, University of Manitoba
Winnipeg, Canada

Editorial Office
MDPI
St. Alban-Anlage 66
4052 Basel, Switzerland

This is a reprint of articles from the Special Issue published online in the open access journal *Pharmaceutics* (ISSN 1999-4923) (available at: https://www.mdpi.com/journal/pharmaceutics/special_issues/honor_AMP).

For citation purposes, cite each article independently as indicated on the article page online and as indicated below:

Lastname, A.A.; Lastname, B.B. Article Title. *Journal Name* **Year**, *Volume Number*, Page Range.

ISBN 978-3-0365-9307-4 (Hbk)
ISBN 978-3-0365-9306-7 (PDF)
doi.org/10.3390/books978-3-0365-9306-7

Contents

About the Editors

Scavello Francesco

Francesco Scavello obtained a double PhD in Life Sciences and Physics-Physical Chemistry from University of Calabria and University of Strasbourg, dissecting the role of Chromogranin A-derived antimicrobial peptides in cardiac inflammation and nosocomial infections. During his first post-doc period at IRCCS Centro Cardiologico (Milan), he studied the role of Soluble Receptor for Advanced Glycation End-products in cardiac aging and Inflammaging. At the moment, Dr. Scavello covers a post-doc position in the Laboratory of Experimental Immunopathology at Humanitas Research Hospital. He studies the role of the long Pentraxins (PTXs) in the response to COVID-19 and bacterial or fungal infections, focusing also on the validation of PTX3 as an early biomarker for COVID-19 acute and chronic complications, such as secondary infections and cardiovascular complications. His research interests also include the crosstalk between the immune system, cancer and cardiovascular diseases. In particular, these studies aim to elucidate the role of negative regulators of inflammatory responses, such as IL-1R8, in the context of chemotherapy treatments, both in terms of cancer growth and adverse cardiovascular events.

Amiche Mohamed

After a DEA in molecular and cellular pharmacology (University of Paris VI, 1987), Dr. Amiche obtained a PhD from the same university in 1990, focusing on the structural and pharmacological study of dermorphin and dermenkephalin, opioid peptides isolated from amphibian skin. A year later, he was recruited as a 1st class research fellow by the CNRS and assigned to the Institute Jacques Monod (UMR9922 CNRS). Between 1993 and 1994, Dr. Amiche joined the Protein Engineering Department of the Proteins at the "Commissariat à l'Energie Atomique, Paris-Saclay" for his Post-Doctoral internship. In 2000, he obtained his HDR in natural and life sciences from the University of Paris VI. His research interests lie at the interface between chemistry and biology. His work involves exploring animal biodiversity and discovering new therapeutic peptides. Dr. Amiche use the amphibian skin model for isolation, structural and pharmacological characterization of peptides involved in pain, antimicrobial and anticancer activity.

Jean-Eric Ghia

Dr. Ghia is a Professor with a remarkable academic journey spanning several prestigious institutions. His academic pursuits began with a Ph.D. in Neurosciences in 2003 at the University of Strasbourg, where he focused on the physiological role of Chromogranin A-derived peptides. After earning his doctorate, Dr. Ghia gained valuable experience as a postdoctoral fellow at McMaster University under the mentorship of Prof. Stephen M Collins, exploring the relationship between depressive-like behaviour and colitis. In 2010, he joined the University of Manitoba. Currently, he leads the Gastrointestinal Basic Biology Research department at the Manitoba Inflammatory Bowel Diseases (IBD) Clinical Research Center. He oversees a multidisciplinary team dedicated to advancing IBD therapeutics through cutting-edge research. His Neurogastroenterology lab is a local hub for innovative colitis prevention strategies targeting chromogranin-A and the brain-gut axis. Dr. Ghia also provides resources for other laboratories, teaches at the University of Manitoba, and is the co-editor of a textbook in Gastroenterology. Since 2015, he has made significant contributions to science communication as a co-producer, columnist, host, scientific director, co-author, and curator for over 200 radio segments, magazines, exhibits, and web series in Canada. He currently manages publications and archives for the Canadian Association of Gastroenterology and is the director of the microdiploma in scientific communication at the University of Manitoba. Dr. Ghia has received numerous accolades, including the Ken Hughes Award, the Canadian Association of Gastroenterology New Investigator Award, the University of Manitoba Presidential Outreach Award, and the Canadian Association of Gastroenterology Excellence in Education award. He was also honoured with the 2022 Queen Elisabeth II Platinum Jubilee Medal and was inducted by the French Government as a Knight of the Order of Academic Palms.

Preface

This Special Issue is in honor of Dr. Marie-Hélène Metz-Boutigue, acknowledging her influential scientific contributions to research on Antimicrobial Peptides (AMPs).

During her career, Dr. Metz-Boutigue has published many papers in the field of the biochemical characterization of proteins and peptides, identifying new antimicrobial agents with a large spectrum against different pathogens. In the last few years, she has also studied the integration of AMPs in new functionalized antimicrobial biomaterials to prevent nosocomial infections and inflammation.

Due to her great expertise in self-endogenous AMPs belonging to the humoral immune system, she collaborated with several scientists in many publications concerning their involvement in cardiovascular diseases, inflammation, the nervous system, intestinal homeostasis and concerning the properties of these agents in immunomodulation.

This Special Issue aimed to gather and highlight significant advancements for AMPs as an alternative to antibiotics for increasing resistant strains and recent discoveries for future therapeutic approaches. Since the AMPs are multifunctional peptides with antitumor and immunomodulating effects, original research articles and reviews in these fields were particularly welcomed. Manuscripts regarding antimicrobial biomaterials including AMPs were also of interest.

Scavello Francesco, Amiche Mohamed, and Jean-Eric Ghia
Editors

pharmaceutics

Editorial

The Editorial Position on 'Recent Advances in Multifunctional Antimicrobial Peptides as Preclinical Therapeutic Studies and Clinical Future Applications'

Francesco Scavello [1,*], Mohamed Amiche [2,*] and Jean-Eric Ghia [3,4,*]

1 IRCCS Humanitas Research Hospital, Rozzano, 20089 Milan, Italy
2 Laboratoire de Biogenèse des Signaux Peptidiques (BioSiPe), Institut de Biologie Paris-Seine, Sorbonne Université-CNRS, 75252 Paris, France
3 Department of Immunology, Rady Faculty of Health Sciences, University of Manitoba, Winnipeg, MB R3E 0T5, Canada
4 Section of Gastroenterology, Department of Internal Medicine, Rady Faculty of Health Sciences, University of Manitoba, Winnipeg, MB R3E 0T5, Canada
* Correspondence: francesco.scavello@humanitasresearch.it (F.S.); mohamed.amiche@upmc.fr (M.A.); jean-eric.ghia@umanitoba.ca (J.-E.G.)

Citation: Scavello, F.; Amiche, M.; Ghia, J.-E. The Editorial Position on 'Recent Advances in Multifunctional Antimicrobial Peptides as Preclinical Therapeutic Studies and Clinical Future Applications'. *Pharmaceutics* **2023**, *15*, 2383. https://doi.org/10.3390/pharmaceutics15102383

Received: 13 September 2023
Accepted: 16 September 2023
Published: 26 September 2023

Antibiotic resistance has recently been recognized as an alarming issue and one of the leading causes of death worldwide [1]. In the last few years, the frequent use of antibiotics in COVID-19 patients has exacerbated antimicrobial resistance [2]. Antimicrobial peptides (AMPs) are an alternative to antibiotic therapy, and they have a great potential to respond to multidrug-resistant pathogens [3]. They offer several advantages compared to classical antibiotics such as broad-spectrum activity and a lower tendency to develop resistance mechanisms and modulators of the host immune response [3]. Some of these molecules are named Multifunctional AMPs (MF-AMPs) due to their ability to influence several biological processes and their comprehensive antimicrobial function.

This Special Issue in *Pharmaceutics*, "Recent Advances in Multifunctional Antimicrobial Peptides as Preclinical Therapeutic Studies and Clinical Future Applications", is in honor of Dr. Marie-Hélène Metz-Boutigue, acknowledging her influential scientific contributions to research on AMPs and highlighting the recent advances in MF-AMPs research. In biomedical research, there are few pioneers whose groundbreaking contributions leave an indelible mark on their field. Among these stands Dr. Metz-Boutigue, a visionary scientist whose work in the field of AMPs has revolutionized our understanding of some of them. Her collaborative efforts with esteemed researchers have ushered in a new era of potential treatments for infectious diseases. Dr. Metz-Boutigue's interest in AMPs dates back to her early career when she recognized the immense therapeutic potential of these peptides. Her journey was marked by relentless dedication, determination, and a passion for unravelling the complexities of the mechanism of action of these peptides. Through her prolific research, Dr. Metz-Boutigue deepened our knowledge of AMPs and discovered novel classes of these peptides with potent antimicrobial activity. Her work paved the way for developing innovative therapeutic strategies against various infectious agents, including bacteria, fungi, and even viruses.

One of the significant aspects of Dr. Metz-Boutigue's remarkable success has been her collaborative spirit, as she has fostered partnerships with brilliant researchers worldwide. Among her notable collaborators are Prof. Schneider, renowned for his expertise in intensive care medicine and infection, and together they discovered the role of several AMPs as biomarkers in critically ill patients [4]; Dr. Prévost, an expert in microbiology, who, together with Dr. Metz-Boutigue, identified new AMPs against resistant *S. aureus* strains [5]; and Dr. Goumon, an expert in neuroscience, with whom Dr. Metz-Boutigue discovered several AMPs derived from the neuroendocrine system [6]. Of note are her collaborations with Dr. Jollès and Dr. Strub, specialists in protein biochemistry and mass spectrometry, respectively, through which they characterized the biochemistry and processing of several proteins producing AMPs [7,8]. Concerning the findings demonstrating the multifunctional profile of AMPs, we mentioned the collaborations with Prof. Angel and Prof. Ghia, prominent gut physiologists, who, together with Dr. Metz-Boutigue, characterized the role of some AMPs in colonic motility [9]. Additionally of note are her collaborations with Prof. Corti and Prof. Helle, experts in biochemistry, who together have demonstrated the role of several AMPs derived from Chromogranins (Cgs) in different biological processes [10]. Concerning Cgs-derived peptides, Dr. Metz-Boutigue collaborated with Prof. Mahata, with whom she discovered the role of Catestatin as an antimicrobial agent [11], and with Dr. Marban-Doran and Prof. Rohr, with whom she identified new Chromogranin A (CgA)-derived AMPs and their role in the context of in nosocomial infections [12,13]. Also relevant are the studies conducted in collaboration with Prof. Aunis, a prominent neuroscientist, showing the role of neuropeptides in innate immunity [14], as well as her work in collaboration with Prof. Tota and Prof. Angelone, who are renowned for their expertise in cardiac physiology, demonstrating the role of AMPs in cardiac physiopathology [10]. In the last few years, Dr. Metz-Boutigue's collaborations with Prof. Haikel, a specialist in oral biology, should be mentioned, which enabled the application study of AMPs in dental biomaterials and infections [15]. Finally, her fruitful collaborations with Dr. Lavalle and Prof. Schaaf, experts in biomaterials, enabled the development of functionalized biomaterials with antimicrobial and anti-inflammatory properties [16,17]. All these collaborations have synergized the researchers' diverse knowledge and methodologies, resulting in transformative research outcomes. Beyond her academic achievements, the collaborative work of Dr. Metz-Boutigue has not only catalyzed the development of international partnerships in the battle against infectious diseases but has also helped the next generation of scientists in this field of research by undoubtedly reshaping the landscape of AMPs and suggesting the concept of MF-AMPs, notably for Cgs and their derived AMPs. As we look toward the future, we eagerly await the progress of the next generation in this field of research.

In this Special Issue, several manuscripts reported the ability of MF-AMPs as an alternative or adjuvant to classical antibiotics to overcome antibiotic resistance. In general, Hassan et al. recapitulate the positive aspects of these molecules, such as a broad spectrum of antimicrobial activities, high pathogen specificity, and low toxicity for the host. In addition, they summarized the sources, structure, molecular mechanisms of action, and therapeutic potential of AMP, evaluating the advantages over classical antibiotics such as a low propensity for the development of resistance, endogenous origins, and anti-biofilm activity. However, they reported several limitations, such as poor absorption and distribution, a short half-life, low permeability, and some toxic effects [18].

Regarding the resistance problem, Kuhn and Di found that several pathogens were resistant to last-report antibiotic peptides classically used for multidrug-resistant bacterial treatment, such as Colistin. They demonstrated the importance of mutation timing in the progression of the *Klebsiella pneumoniae* Colistin-resistant strain, identifying a time-specific mutation of Colistin-resistant genes during this pathogen evolution. In particular, the authors observed the onset of an early resistance with mutations relevant to capsule production, cell membrane integrity, and energy metabolism in response to Colistin treatment, followed by mutations influencing LPS structure, strongly impacting Colistin-resistant progression [19].

Also, different studies are presented that dealt with improving antimicrobial activities and reducing their adverse effects using structure–function relationship studies, engineered variations, chemical modifications, and nanoparticle incorporation [20–22]. To identify the structural requirements of AMP activity directed against Gram-negative bacteria, Madi-Moussa and collaborators heterologously produced a series of variants of Lacticaseicin 30 including the N-terminal (amino acids 1 to 39) and central and C-terminal (amino acids 40 to 111) domains in *E. coli* and assayed their antibacterial activities. In addition, ten mutations were introduced to obtain several engineered variants. In comparison to the original AMP, they observed no differences for the N-terminal peptide and variants E32G, T33P, and

D57G preserving the antibacterial effects, but demonstrated a reduced antimicrobial activity against Gram-negative bacteria for the C-terminal peptide and the E6G, T7P, T52P, A74P, Y78S, Y93S, and A97P variants [20]. Finally, the authors synthesized an N terminal domain without the first 20 amino acids in the first helix, showing that this truncated peptide did not have antimicrobial activity [20]. Another approach used by Etayash et al. involved chemically modifying AMP IDR1018 by covalently linking it with a short-chain PEG (PEG6) and a glucose moiety (N-acetyl glucosamine: GlcNAc) and assessing the impact of this chemical modification on the peptide's antimicrobial and biological properties. The pegylation and glycosylation significantly reduced the aggregation, hemolysis, and cytotoxicity effect of IDR1018 [21]. Of interest, PEG6-IDR1018 maintained the immunomodulatory activity of IDR1018 while the Glc-IDR1018 significantly increased the anti-inflammatory effects of this peptide (production of MCP-1 and IL1-RA *vs.* lipopolysaccharide-induced IL-1β) [21]. On the other hand, these chemical modifications partially reduced the antimicrobial and anti-biofilm activity [21]. To deliver the antimicrobial Esc peptides specifically into the lung, Cappiello and coworkers developed engineered poly(lactide-co-glycolide) (PLGA) nanoparticles (NPs). They demonstrated that the peptides alone and the nanoparticle version did not affect lung epithelium integrity. No changes in pulmonary global gene expression profile were observed after the treatments [22]. The Esc diastereomer showed the highest antimicrobial activity without lung inflammation; however, the nanoparticulate system increased the pulmonary safety and delivery of Esc peptides, prolonging the biostability of these peptides in the mouse bronchoalveolar lavage [22].

In this Special Issue, we included several studies that also reported the possible applications of MF-AMPs in different human and clinical contexts. Some of these applications concern their antimicrobial nature, while others extend to new research fields. Jati and coworkers suggested that the pleiotropic CgA-derived peptide Catestatin is a perfect example of an MF-AMP, reporting its potential therapeutics in antimicrobial and immunological applications. They summarized this peptide's antibacterial, antifungal, and anti-yeast activity and its N-terminal domain of 15 amino acids called Cateslytin. The author reported the chemically modified isoform for this smaller peptide, with the total D-amino acids changing, and its increased antimicrobial effects against various bacterial and fungal strains [23]. Finally, they dealt with the human variants of Catestatin (Gly364Ser and Pro370Leu) and the evolutionary conservation of this peptide in mammals, pointing out the potential role of this peptide as a therapy for antibiotic-resistant superbugs and its role in innate immunity and the regulation of gut microbiota [23]. Also, Rizzetto and collaborators recapitulated the experience of their research group over 20 years in a narrative review describing the antibacterial and antifungal properties of several AMPs such as BMAP-28, Citropin 1.1, Temporin A, Distinctin, Magainin II, LL-37, Tritrpticin, Colistin, Distinctin, Magainin II, Cecropin A, and many others [24]. This review strongly supports the benefits of these molecules, suggesting AMPs as a valuable alternative for treating infectious diseases related to multiresistant microorganisms and overcoming the clinical issue of resistance against the commonly used antibiotics or antifungals.

One of the most relevant applications for AMPs in biomedical devices is in the field of biomaterials due to the anti-biofilm properties of these molecules [24]. As reported in many in vitro and in vivo studies, these molecules are elective factors for oral infection and biomaterials to treat early carious lesions, promote cell adhesion, and enhance the adhesion strength of dental implants while preventing infection at the surgery site [25]. However, the small amount of data regarding animal models or clinical trials is a limitation for the future use of this biomaterial containing AMPs [25]. In this Special Issue, Cunha and coworkers developed a clinical trial to evaluate in vivo the antimicrobial properties and dental efficacy of a nisin-biogel in dogs [26]. This study reported that the nisin-biogel group was associated with a significantly reduced periodontal pocket depth and dental plaque index after 90 days from the treatment compared to control animals. Also, a non-significant reduction in the gingivitis index was observed, while no influence on total bacterial counts and adverse effects was detected. Finally, the author suggested the nisin–biogel system as a valuable biomaterial for future human dentistry study. In addition, the crucial role of AMPs in the reduction of endodontic inflammation and infection was demonstrated by Morio et al. They, evaluating different datasets of ultraviolet (UV)-induced molecules from other tissues, such as endodontic tissue, identified 32 UV-induced molecules containing nine antimicrobial peptides (Cathelicidin and several Defensins), ten cytokines, six growth factors, three enzymes, two transmembrane receptors, and two transcription regulators. Many of these molecules are related to the wound-healing signaling pathway and communication between immune cells and antimicrobial response. They concluded that, also via the UV irradiation technique, AMPs production requires part of the activate innate immune response with a final aim to reduce infection but also assist tissue healing [27].

Another important application for MF-AMPs in human health is their roles as the biomarkers of several diseases both related and unrelated to infections [28–31]. In this Special Issue, Scnheider and collaborators summarized the biomarker profile of CgA and its derived antimicrobial peptides in the context of critically ill patients [4]. They reported the role of CgA as a biomarker of outcome in non-selected critically ill patients. Then, they focused on Vasostatin-I, an N-terminal fragment of CgA, and its ability as an early prognostic biomarker for these patients associated with age and lactate. Also, for trauma patients, CgA levels can predict the onset of nosocomial infections, and their CgA-derived peptides can act as inhibitors of inflammation and antimicrobial peptides. Finally, they also demonstrated that a 4% concentration of non-oxidized albumin infusion reverses the multimerization of these peptides, restoring their anti-inflammatory and antimicrobial properties [4]. They concluded that CgA and its peptides could be considered relevant biomarkers in critically ill patients. In addition, Scavello and coworkers reported the role of the multifunctional agents of CgA-derived AMPs, describing these as endogenous modulators of inflammation [32]. Concerning MF-AMPs and cancer, Dermaseptins, a class of α-helical-shaped polycationic peptides isolated from the skin secretions of several leaf frogs, have been identified as examples of exogenous molecules characterized by their anticancer activity [32].

Finally, Neuropilin-1 and Integrins have been recently described by Corti et al. as receptors with a high affinity for CgA-derived peptides [33]. In particular, the multifunctional properties of this protein and its fragments reported in angiogenesis, wound healing, and tumor growth are mediated by a specific interaction and binding of Neuropilin-1 and Integrin avb6 [33]. Moreover, Siegel and collaborators described an immunogenicity test for dendritic and CD4-positive T cells where different peptides and proteins have been used to stimulate both dendritic and CD4-positive cells [34], suggesting that MF-AMPs may be used in the future in this context to assess immunogenicity as well.

In summary, this research topic has brought to light new findings regarding MF-AMPs and their possible applications, as well as future studies, for human health.

Author Contributions: Writing—original draft preparation, F.S.; writing—review and editing, M.A. and J.-E.G. All authors have read and agreed to the published version of the manuscript.

Funding: This research received no external funding.

Acknowledgments: The authors thank Marie-Hélène Metz-Boutigue for the stimulating discussions and her full support for all the above research on identifying new MF-AMPs, such as the CgA-derived peptides.

Conflicts of Interest: The authors declare no conflict of interest.

References

1. Antimicrobial Resistance Collaborators. Global burden of bacterial antimicrobial resistance in 2019: A systematic analysis. *Lancet* **2022**, *399*, 629–655. [CrossRef] [PubMed]
2. Langford, B.J.; So, M.; Simeonova, M.; Leung, V.; Lo, J.; Kan, T.; Raybardhan, S.; Sapin, M.E.; Mponponsuo, K.; Farrell, A.; et al. Antimicrobial resistance in patients with COVID-19: A systematic review and meta-analysis. *Lancet Microbe* **2023**, *4*, e179–e191. [CrossRef] [PubMed]

3. Magana, M.; Pushpanathan, M.; Santos, A.L.; Leanse, L.; Fernandez, M.; Ioannidis, A.; Giulianotti, M.A.; Apidianakis, Y.; Bradfute, S.; Ferguson, A.L.; et al. The value of antimicrobial peptides in the age of resistance. *Lancet Infect. Dis.* **2020**, *20*, e216–e230. [CrossRef] [PubMed]
4. Schneider, F.; Clère-Jehl, R.; Scavello, F.; Lavigne, T.; Corti, A.; Angelone, T.; Haïkel, Y.; Lavalle, P. Chromogranin A and Its Fragments in the Critically Ill: An Expanding Domain of Interest for Better Care. *Pharmaceutics* **2022**, *14*, 2178. [CrossRef]
5. Aslam, R.; Marban, C.; Corazzol, C.; Jehl, F.; Delalande, F.; Van Dorsselaer, A.; Prévost, G.; Haïkel, Y.; Taddei, C.; Schneider, F.; et al. Cateslytin, a chromogranin A derived peptide is active against Staphylococcus aureus and resistant to degradation by its proteases. *PLoS ONE* **2013**, *8*, e68993. [CrossRef]
6. Metz-Boutigue, M.H.; Goumon, Y.; Strub, J.M.; Lugardon, K.; Aunis, D. Antimicrobial chromogranins and proenkephalin-A-derived peptides: Antibacterial and antifungal activities of chromogranins and proenkephalin-A-derived peptides. *Ann. N. Y. Acad. Sci.* **2003**, *992*, 168–178. [CrossRef]
7. Metz-Boutigue, M.H.; Jollès, J.; Mazurier, J.; Spik, G.; Montreuil, J.; Jollès, P. An 88 amino acid long C-terminal sequence of human lactotransferrin. *FEBS Lett.* **1982**, *142*, 107–110. [CrossRef]
8. Metz-Boutigue, M.H.; Goumon, Y.; Lugardon, K.; Strub, J.M.; Aunis, D. Antibacterial peptides are present in chromaffin cell secretory granules. *Cell. Mol. Neurobiol.* **1998**, *18*, 249–266. [CrossRef]
9. Ghia, J.E.; Pradaud, I.; Crenner, F.; Metz-Boutigue, M.H.; Aunis, D.; Angel, F. Effect of acetic acid or trypsin application on rat colonic motility in vitro and modulation by two synthetic fragments of chromogranin A. *Regul. Pept.* **2005**, *124*, 27–35. [CrossRef]
10. Helle, K.B.; Corti, A.; Metz-Boutigue, M.H.; Tota, B. The endocrine role for chromogranin A: A prohormone for peptides with regulatory properties. *Cell. Mol. Life Sci.* **2007**, *64*, 2863–2886. [CrossRef]
11. Briolat, J.; Wu, S.D.; Mahata, S.K.; Gonthier, B.; Bagnard, D.; Chasserot-Golaz, S.; Helle, K.B.; Aunis, D.; Metz-Boutigue, M.H. New antimicrobial activity for the catecholamine release-inhibitory peptide from chromogranin A. *Cell. Mol. Life Sci.* **2005**, *62*, 377–385. [CrossRef] [PubMed]
12. Zaet, A.; Dartevelle, P.; Daouad, F.; Ehlinger, C.; Quilès, F.; Francius, G.; Boehler, C.; Bergthold, C.; Frisch, B.; Prévost, G.; et al. D-Cateslytin, a new antimicrobial peptide with therapeutic potential. *Sci. Rep.* **2017**, *7*, 15199. [CrossRef] [PubMed]
13. Schneider, F.; Marban, C.; Ajob, G.; Helle, S.; Guillot, M.; Launoy, A.; Maestraggi, Q.; Scavello, F.; Rohr, O.; Metz-Boutigue, M.H. In Trauma Patients, the Occurrence of Early-Onset Nosocomial Infections is Associated With Increased Plasma Concentrations of Chromogranin A. *Shock* **2018**, *49*, 522–528. [CrossRef]
14. Metz-Boutigue, M.H.; Kieffer, A.E.; Goumon, Y.; Aunis, D. Innate immunity: Involvement of new neuropeptides. *Trends Microbiol.* **2003**, *11*, 585–592. [CrossRef] [PubMed]
15. Scavello, F.; Kharouf, N.; Lavalle, P.; Haikel, Y.; Schneider, F.; Metz-Boutigue, M.H. The antimicrobial peptides secreted by the chromaffin cells of the adrenal medulla link the neuroendocrine and immune systems: From basic to clinical studies. *Front. Immunol.* **2022**, *13*, 977175. [CrossRef]
16. Mateescu, M.; Baixe, S.; Garnier, T.; Jierry, L.; Ball, V.; Haikel, Y.; Metz-Boutigue, M.H.; Nardin, M.; Schaaf, P.; Etienne, O.; et al. Antibacterial Peptide-Based Gel for Prevention of Medical Implanted-Device Infection. *PLoS ONE* **2015**, *10*, e0145143. [CrossRef]
17. Özçelik, H.; Vrana, N.E.; Gudima, A.; Riabov, V.; Gratchev, A.; Haikel, Y.; Metz-Boutigue, M.H.; Carradò, A.; Faerber, J.; Roland, T.; et al. Harnessing the multifunctionality in nature: A bioactive agent release system with self-antimicrobial and immunomodulatory properties. *Adv. Healthc. Mater.* **2015**, *4*, 2026–2036. [CrossRef]
18. Hassan, M.; Flanagan, T.W.; Kharouf, N.; Bertsch, C.; Mancino, D.; Haikel, Y. Antimicrobial Proteins: Structure, Molecular Action, and Therapeutic Potential. *Pharmaceutics* **2022**, *15*, 72. [CrossRef]
19. Kuhn, J.M.; Di, Y.P. Determination of Mutational Timing of Colistin-Resistance Genes through Klebsiella pneumoniae Evolution. *Pharmaceutics* **2023**, *15*, 270. [CrossRef]
20. Madi-Moussa, D.; Deracinois, B.; Teiar, R.; Li, Y.; Mihasan, M.; Flahaut, C.; Rebuffat, S.; Coucheney, F.; Drider, D. Structure of Lacticaseicin 30 and Its Engineered Variants Revealed an Interplay between the N-Terminal and C-Terminal Regions in the Activity against Gram-Negative Bacteria. *Pharmaceutics* **2022**, *14*, 1921. [CrossRef]
21. Etayash, H.; Yip, F.; Hancock, R.E.W. Impacts of PEGylation and Glycosylation on the Biological Properties of Host Defense Peptide IDR1018. *Pharmaceutics* **2023**, *15*, 1391. [CrossRef] [PubMed]
22. Cappiello, F.; Casciaro, B.; Loffredo, M.R.; Puglisi, E.; Lin, Q.; Yang, D.; Conte, G.; d'Angelo, I.; Ungaro, F.; Ferrera, L.; et al. Pulmonary Safety Profile of Esc Peptides and Esc-Peptide-Loaded Poly(lactide-co-glycolide) Nanoparticles: A Promising Therapeutic Approach for Local Treatment of Lung Infectious Diseases. *Pharmaceutics* **2022**, *14*, 2297. [CrossRef] [PubMed]
23. Jati, S.; Mahata, S.; Das, S.; Chatterjee, S.; Mahata, S.K. Catestatin: Antimicrobial Functions and Potential Therapeutics. *Pharmaceutics* **2023**, *15*, 1550. [CrossRef]
24. Rizzetto, G.; Gambini, D.; Maurizi, A.; Candelora, M.; Molinelli, E.; Cirioni, O.; Brescini, L.; Giacometti, A.; Offidani, A.; Simonetti, O. Our Experience over 20 Years: Antimicrobial Peptides against Gram Positives, Gram Negatives, and Fungi. *Pharmaceutics* **2022**, *15*, 40. [CrossRef] [PubMed]
25. Hardan, L.; Chedid, J.C.A.; Bourgi, R.; Cuevas-Suárez, C.E.; Lukomska-Szymanska, M.; Tosco, V.; Monjarás-Ávila, A.J.; Jabra, M.; Salloum-Yared, F.; Kharouf, N.; et al. Peptides in Dentistry: A Scoping Review. *Bioengineering* **2023**, *10*, 214. [CrossRef]
26. Cunha, E.; Carreira, L.M.; Nunes, T.; Videira, M.; Tavares, L.; Veiga, A.S.; Oliveira, M. In Vivo Evaluation of the Efficacy of a Nisin-Biogel as a New Approach for Canine Periodontal Disease Control. *Pharmaceutics* **2022**, *14*, 2716. [CrossRef]

27. Morio, K.A.; Sternowski, R.H.; Zeng, E.; Brogden, K.A. Antimicrobial Peptides and Biomarkers Induced by Ultraviolet Irradiation Have the Potential to Reduce Endodontic Inflammation and Facilitate Tissue Healing. *Pharmaceutics* **2022**, *14*, 1979. [CrossRef]
28. Álvarez, Á.H.; Martínez Velázquez, M.; Prado Montes de Oca, E. Human β-defensin 1 update: Potential clinical applications of the restless warrior. *Int. J. Biochem. Cell. Biol.* **2018**, *104*, 133–137. [CrossRef]
29. Schneider, F.; Le Borgne, P.; Herbrecht, J.E.; Danion, F.; Solis, M.; Hellé, S.; Betscha, C.; Clere-Jehl, R.; Lefebvre, F.; Castelain, V.; et al. Assessment of plasma Catestatin in COVID-19 reveals a hitherto unknown inflammatory activity with impact on morbidity-mortality. *Front. Immunol.* **2022**, *13*, 985472. [CrossRef]
30. Mani, I.; Alexopoulou, A.; Vasilieva, L.; Hadziyannis, E.; Agiasotelli, D.; Bei, M.; Alexopoulos, T.; Dourakis, S.P. Human beta-defensin-1 is a highly predictive marker of mortality in patients with acute-on-chronic liver failure. *Liver Int.* **2019**, *39*, 299–306. [CrossRef]
31. Zhou, Q.; Pan, L.L.; Xue, R.; Ni, G.; Duan, Y.; Bai, Y.; Shi, C.; Ren, Z.; Wu, C.; Li, G.; et al. The anti-microbial peptide LL-37/CRAMP levels are associated with acute heart failure and can attenuate cardiac dysfunction in multiple preclinical models of heart failure. *Theranostics* **2020**, *10*, 6167–6181. [CrossRef] [PubMed]
32. Scavello, F.; Amiche, M.; Ghia, J.E. Recent Advances in Multifunctional Antimicrobial Peptides as Immunomodulatory and Anti-cancer Therapy: Chromogranin A-Derived Peptides and Dermaseptins as Endogenous versus Exogenous Actors. *Pharmaceutics* **2022**, *14*, 2014. [CrossRef] [PubMed]
33. Corti, A.; Anderluzzi, G.; Curnis, F. Neuropilin-1 and Integrins as Receptors for Chromogranin A-Derived Peptides. *Pharmaceutics* **2022**, *14*, 2555. [CrossRef] [PubMed]
34. Siegel, M.; Steiner, G.; Franssen, L.C.; Carratu, F.; Herron, J.; Hartman, K.; Looney, C.M.; Ducret, A.; Bray-French, K.; Rohr, O.; et al. Validation of a Dendritic Cell and CD4+ T Cell Restimulation Assay Contributing to the Immunogenicity Risk Evaluation of Biotherapeutics. *Pharmaceutics* **2022**, *14*, 2672. [CrossRef]

pharmaceutics

MDPI

Review

Antimicrobial Proteins: Structure, Molecular Action, and Therapeutic Potential

Mohamed Hassan [1,2,3,*], Thomas W. Flanagan [4], Naji Kharouf [1,2], Christelle Bertsch [1,2], Davide Mancino [1,2] and Youssef Haikel [1,2]

1 Department of Endodontics, Faculty of Dental Medicine, Strasbourg University, 67000 Strasbourg, France
2 Department of Biomaterials and Bioengineering, INSERM UMR_S 1121, Biomaterials and Bioengineering, 67000 Strasbourg, France
3 Research Laboratory of Surgery-Oncology, Department of Surgery, Tulane University School of Medicine, New Orleans, LA 70112, USA
4 Department of Pharmacology and Experimental Therapeutics, LSU Health Sciences Center, New Orleans, LA 70112, USA
* Correspondence: dr.hassan@gmx.de; Tel.: +1-504-339-2671

Abstract: Second- and third-line treatments of patients with antibiotic-resistant infections can have serious side effects, such as organ failure with prolonged care and recovery. As clinical practices such as cancer therapies, chronic disease treatment, and organ transplantation rely on the ability of available antibiotics to fight infection, the increased resistance of microbial pathogens presents a multifaceted, serious public health concern worldwide. The pipeline of traditional antibiotics is exhausted and unable to overcome the continuously developing multi-drug resistance. To that end, the widely observed limitation of clinically utilized antibiotics has prompted researchers to find a clinically relevant alternate antimicrobial strategy. In recent decades, the discovery of antimicrobial peptides (AMPs) as an excellent candidate to overcome antibiotic resistance has received further attention, particularly from scientists, health professionals, and the pharmaceutical industry. Effective AMPs are characterized by a broad spectrum of antimicrobial activities, high pathogen specificity, and low toxicity. In addition to their antimicrobial activity, AMPs have been found to be involved in a variety of biological functions, including immune regulation, angiogenesis, wound healing, and antitumor activity. This review provides a current overview of the structure, molecular action, and therapeutic potential of AMPs.

Keywords: antimicrobial peptide (AMP); multi-drug resistance (MDR); extracellular polymeric substances (EPSs); lipopolysaccharides (LPS); lipoteichoic acid (LTA)

Citation: Hassan, M.; Flanagan, T.W.; Kharouf, N.; Bertsch, C.; Mancino, D.; Haikel, Y. Antimicrobial Proteins: Structure, Molecular Action, and Therapeutic Potential. *Pharmaceutics* **2022**, *15*, 72. https://doi.org/10.3390/pharmaceutics15010072

Academic Editor: Giancarlo Morelli

Received: 19 October 2022
Revised: 16 December 2022
Accepted: 17 December 2022
Published: 26 December 2022

1. Introduction

The antimicrobial resistance (AMR) to available therapeutics is a serious healthcare problem that is mostly associated with the death of a specific portion of people worldwide. Moreover, current predictions indicate a significant increase in annual global death in the future [1].

AMR is a multifaceted health problem that represents a serious global threat. In addition, the traditional antibiotic pipeline is exhausted and unable to overcome the continuously developing multi-drug resistance. As consequence, the widely observed limitation of clinically utilized antibiotics has prompted researchers to find clinically relevant antimicrobial approaches. Accordingly, the discovery and development of alternative therapeutic approaches to overcome the widely reported AMR are urgently need. Antimicrobial peptides (AMPs) are group of small peptides, which are reported to play a crucial role in the host innate immunity against a broad spectrum of microorganisms, including bacteria (Gram-positive and Gram-negative), viruses, fungi, and parasites [1,2]. Accordingly, AMPs belong to the first-line defense of a host against invading pathogens. These AMPs are, in great part, amphipathic peptides with α-helical structures and β-sheets linked by disulfide bridges, extended loops, or cyclic configurations [2,3]. Thus, based on their broad-spectrum antimicrobial activity, killing potential, high selectivity, and low toxicity, AMPs have gained further interest from researchers and physicians as an alternative approach to the widely utilized antimicrobial agents, especially over the past two decades. The currently identified AMPs can be classified into different groups according to the information provided on the Data Repository of Antimicrobial Peptides (DRAMP) [http://dramp.cpu-bioinfor.org/ accessed on 1 August 2022]. AMPs are derived from six kingdoms (bacteria, archaea, protozoa, fungal, plants, and animals). As widely reported, AMPs are involved in a variety of biological functions, including immune regulation, angiogenesis, wound-healing, anti-inflammatory activities, and antitumor activity [3–6]. Furthermore, AMPs function as critical effectors in both innate and adaptive immunity. Thus, beyond their functional role as a link between innate and adaptive immunity, AMPs contribute to the resolution of inflammation [3,4]. The activation of the innate immune system by AMPs is recognized to be one of the key mechanisms that regulate AMP-mediated early clearance of infections. The immunomodulatory functions of AMPs are known to be very complex and involve various receptors, signaling pathways, and transcription factors [3,4]. AMPs can either directly or indirectly promote the recruitment of different immune cells, such as immature dendritic cells (iDCs), T lymphocytes, monocytes, eosinophils, and neutrophils, to the site of infection.

While the primary treatment of bacterial pathogens relies on established antibiotics, the development of multi-drug resistance (MDR) is ever evolving and changing [7,8]. In contrast to established antibiotics, AMPs have been approved for their therapeutic potential to overcome the multi-resistance of most microbial pathogens [5]. However, the clinical advantage of AMPs over currently available antibiotics resides in their mode of molecular action. Specifically, AMPs have been shown to kill microbial pathogens via a mechanism mediated by the destruction of plasma membranes and interference with intracellular components [2,7]. Despite the success of AMPs in clinical application, natural AMPs have doubtful properties that hinder their functional application [8]. However, the functional analysis of the chemical structure of AMPs may help to improve the molecular action and, subsequently, the therapeutic potential of AMPs. This review will discuss the structure, molecular action, and therapeutic potential of AMPs.

2. Sources and Structure of Antimicrobial Peptides

In addition to their distribution in all organisms, many if not all AMPs are evolutionarily conserved and derived from viral, bacterial, fungal, plant, and animal sources.

AMPs with viral sources include endolysins (lysins), virion-associated peptidoglycan hydrolases (VAPGHs), depolymerases, and holins, which are derived from bacteriophages [9].

AMPs with bacterial sources have been reported in several studies. For example, Bacillus strains have been shown to produce AMPs with promising inhibitory activity against *Shigella*, *Salmonella*, *E. coli*, and *Staphylococcus aureus* [10–13]. Furthermore, AMPs derived from Bacillus sp have been reported for their antimicrobial activity against *Staphylococcus aureus*, Alteromonas sp. strain CCSH174 and *Klebsiella pneumoni* [10,11]. Another example of bacterial AMPs includes those derived from *Propionibacterium jensenii* [14] and those isolated from Pseudomonas [12], which have been reported for their activity against *Shigella*, *Salmonella*, *E. coli*, and *Staphylococcus aureus* [15].

These AMPs with bacterial sources have been reported to be both ribosomally and non-ribosomally synthesized peptides.

Ribosomally synthesized bacterial AMPs are known as bacteriocins. Bacteriocins have been suggested as promising alternative approaches to the conventional small-molecule antibiotics. These bacteriocins can be divided into four classes. One of these classes is class I, which includes a group of AMPs that consist mainly of small peptides of 19–38 amino acids. The second class of the bacteriocins includes heat-stable AMPs, which are commonly synthesized as a prebacteriocin. The third class contains a group of large and heat-stable peptides, and the fourth class IV contains uniquely structured bacteriocins containing amino acids, lipids, or carbohydrates, in addition to being susceptible to lipolytic and glycolytic enzymes [9,16]. These AMPs are only active against bacteria, which are closely related to the producing strains but not their own producers [16].

Non-ribosomally synthesized AMPs of Gram-positive bacteria include cyclic lipopeptides, which are known as polymyxins, and linear peptides, which are known as tridecaptins. In contrast to Gram-positive-bacteria-derived AMPs, the majority of AMPs isolated from Gram-negative bacteria are common in *E. coli*, as well as in other species including Klebsiella spp. and *Pseudomonas* spp. [17]. These AMPs have limited activity against Gram-negative bacteria and can be classified into four classes, namely, colicins, colicin-like bacteriocins, microcins, and phage tail-like bacteriocins [17]. The classes of colicines are predominantly produced by *E. coli*. Although the class of colicin-like-bacteriocins is structurally and functionally similar to the colicins of *E. coli*, a number of other species, including *P. aeruginosa* and the Klebsiella genus, have been reported to produce colicin-like bacteriocins [18]. Furthermore, other bacteriocins such as microcins can be produced by Enterobacteriaceae, and they are active against phylogenetically close species [18]. The fourth class of bacteriocins includes Gram-negative-bacteria-derived AMPs, including phage tail-like bacteriocins [19]. This type of bacteriocin is characterized by its high molecular weight and cylindrical peptides [19].

Fungal AMPs are common AMPs, which are generally grouped into two main classes, fungal defensins and peptaibols [20].

Defensins are short, cysteine-rich peptides with different sources, including microorganisms, plants, and animals. Therefore, fungal-derived defensins are known as defensin-like peptides based on their high sequence and structural similarities. Although fungal AMPs are similar in their structure and peptide sequences, their activities against Gram-positive and/or Gram-negative bacteria and/or fungi differ [21].

Although there are different sources of AMPs, their numbers of amino acid residues range between 10 and 60 amino acids, and most of them are cationic with an average net charge of 3.32. In addition to cationic AMPs, there are also many anionic AMPs that contain several acidic amino acids, such as aspartic acid and glutamic acid.

Plant-derived AMPs are cysteine-rich peptides with broad-spectrum antimicrobial activity against bacteria, fungi, and viruses, and they possess immunomodulatory activities [22]. These AMPs are classified into various families based on their cysteine motifs, the arrangement of disulfide bridges, and sequence similarity. The most common members of plant-derived AMPs include α-hairpinin, defensins, hevein-like peptides, cyclic and linear knottin-type peptides, lipid transfer proteins, thionins, and snakins, in addition to unclassified cysteine-rich AMPs [22].

The AMPs with animal sources include invertebrate AMPs, fish and amphibian AMPs, reptile- and avian-derived peptides, and mammal-derived AMPs.

Invertebrate AMPs include those of insects, such as defensin and cecropin, mollusc AMPs (e.g., defensins), nematode AMPs (defensins), and horseshoe crabs (e.g., big defensins), in addition to invertebrate β-defensin and crustacean AMPs (e.g., crustins) [23]. Invertebrate AMPs are an integral component of humoral defense since the invertebrates lack an adoptive immune response when compared with those of the animal kingdom [23].

Vertebrates AMPs, particularly those of fish and amphibian origin, have been shown to play an essential role in defense responses to microorganisms. Although fish are a considerable source of several AMPs, such as cathelicidins, β-defensins, hepicidins, piscidins, and histone-derived peptides [24,25], amphibians are the largest source of AMPs among invertebrates. However, the most common amphibian AMPs include bombinins, buforin, cathelicidin, dermaseptins, esculentins, fallaxin, magainins, maximins, phylloseptins, phylloxin, plasticins, plasturins, pseudins, and ranateurins [25].

AMPs with reptile and avian sources belong to the members of the cathelicidin and defensin families [26]. Cathelicidins are small-sized AMPs secreted from macrophages and neutrophils upon their activation in response to infection. β-defensin was first discovered in reptiles as a 40-residue peptide isolated from leukocytes of the European pond turtle. Thus, based on its source, this type of AMP is known as turtle β-defensin 1 (TBD-1) [27]. Similarly, avian β-defensins include AvBD1-14 from the chicken, ostricacins from the ostrich (e.g., OSP-1 to OSP-4), and mallard duck β-defensins (AvBD2 and AvBD9), which are the common AMPs among avian family [27].

The most common mammalian AMPs belong to the members of the cathelicidin and defensins families. Mammalian cathelicidins are cationic peptides with an amphipathic structure in the form of α-helical, β-hairpin, or elongated conformations [26]. LL-37, the most well-studied cathelicidin, has an amphipathic structure, which can be modified into an aqueous solution to form an α-helix upon membrane interaction [26,28]. Mammalian defensins are classified into three sub-families: α, β, and θ. These subfamilies of defensins are synthesized first as prepropeptides, which share several features with mature peptides. These common features include cationic net charge (+1 to +11), short polypeptide sequences (18–45 amino acids), and three intramolecular disulfide bonds [27].

Based on their synthesis mechanisms, mammalian AMPs can be classified into either ribosomal-produced peptides [29] or non-ribosomal-produced peptides [30]. The synthesis of ribosomal AMPs occurs mainly in the cytoplasm of eukaryotic cells via the ribosome-dependent translation of genes encoding for AMPs, e.g., nisin [29,31]. By contrast, the synthesis of non-ribosomal AMPs is mediated by the peptide-synthesis-dependent mechanism in the cytosol of mammalian cells [25].

In contrast to ribosomal AMPs, the assembly of non-ribosomal AMPs contains not only the 20 common amino acids but also many rarer amino acids [32,33]. These different amino acids are synthesized by large enzymes, which are known as non-ribosomal peptide synthetases [33]. Non-ribosomal peptide synthetases are characterized by their ability to synthesize both cyclic and linear AMPs in the form of polypeptides, which give the AMPs their various molecular structures. [34]. A common example for both cyclic and linear AMPs is Gramicidin A (Figure 1), which appears as a small linear peptide with amphipathic and hydrophobic helices and a β-sheet secondary structure [35,36].

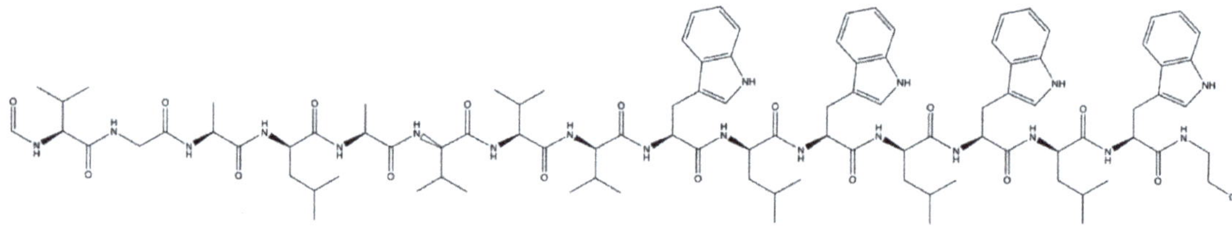

Figure 1. Chemical structure of antimicrobial peptide gramicidin A. Gramicidin A is a linear antimicrobial peptide and one of the three known gramicidins (A, B, and C). They are non-ribosomal peptides that consist of the following 15 L- and D-amino acids: formyl-L-X-Gly-L-Ala-D-Leu-L-Ala-D-Val-L-Val-D-Val-L-Trp-D-Leu-L-Y-D-Leu-L-Trp-D-Leu-L-Trp-ethanolamine. The difference between gramicidins A, B, and C is that the amino acid position Y is L-tryptophan in gramicidin A, L-phenylalanine in B, and L-tyrosine in C. The isoforms of the gramicidins A, B, and C are determined by the existence of L-valine or L-isoleucine at position X of anion acid and the origin [10].

Cyclic peptides, such as polymyxin B (Figure 2A) [37], bacitracin (Figure 2B) [38], and vancomycin (Figure 2C) [39], are characterized by their unique amino acid compositions that appear in the form of lipopeptides or macrocyclic peptides. By contrast, peptides such as α defensin appear as bundles of α-helical rods in lipid bilayers (Figure 2D) [40].

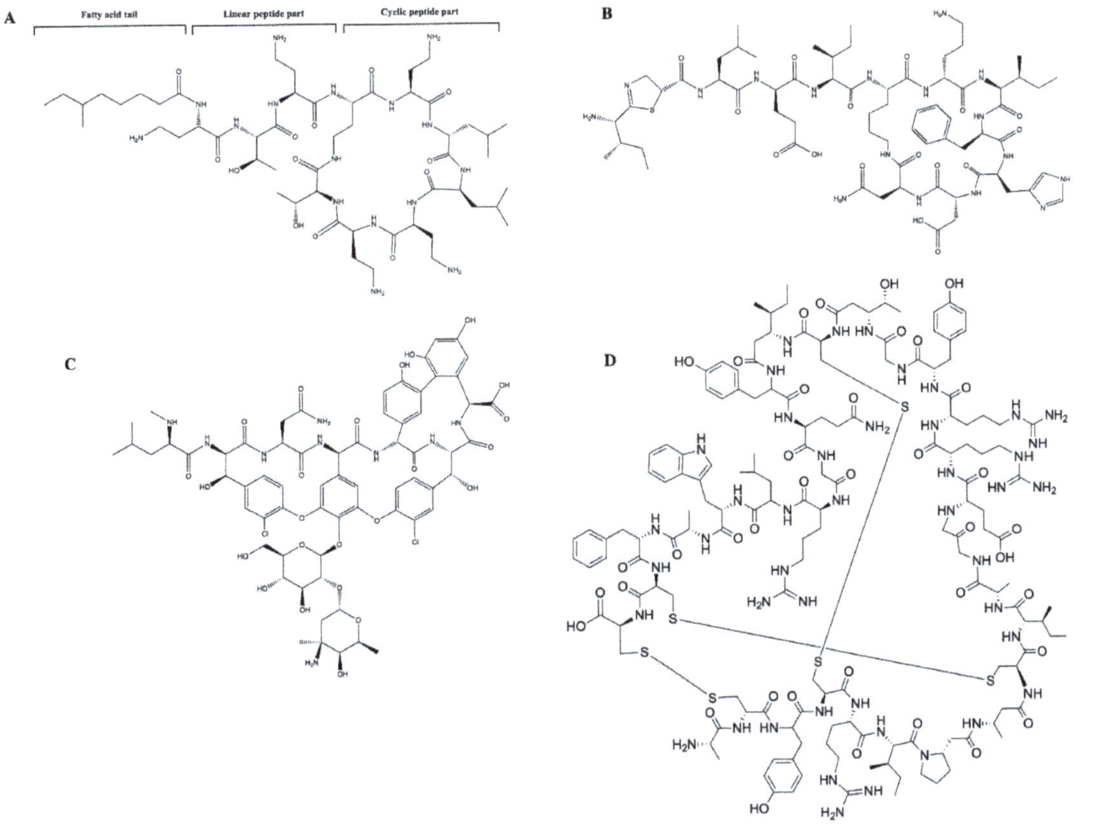

Figure 2. Chemical structure of cyclic antimicrobial peptides: (**A**) Structure of polymyxin B, including fatty acyl tail, linear peptide, and cyclic peptide (**B**) Chemical structure of bacitracin: bacitracin is an AMP that consists of D-aspartic acid, D-phenylalanine, D-ornithine, D-glutamic acid, and a ring of thiazoline containing amino acids. (**C**) Chemical structure of vancomycin: vancomycin is a branched tricyclic glycosylated non-ribosomal peptide. (**D**) Chemical structure of defensin: defensins comprise an N-terminal β-strand followed by an α-helix and two more β-strands. The β-strands form a triple-stranded antiparallel β-sheet that can be stabilized by disulphide bonds. Two of the disulphide bonds connect the α-helix and the central β-strand, while a third disulphide bond stabilizes the structure by linking the β-strand.

While the cationic amphipathic helix is common in the secondary structure, particularly among bacteriostatic peptides [41], α-helical peptides are either hydrophobic or anionic with less selectivity towards microbes [42]. Apart from their different net charges, helical peptides are characterized by their ability to form hexameric clusters that can traverse bilayer membranes and surround an aqueous pore [43]. Consequently, the mechanism by which AMPs kill bacteria is mediated via the formation of pores, which leads to the disintegration of pathogen cell membranes [44,45]. The biological functioning of AMPs therefore depends on their ability to undergo structural modifications that allow them to interact with the membrane and elements of the cellular matrix.

3. Molecular Mechanisms of AMP Action

AMPs are characterized by their diverse activities and modes of action. These characteristics are determined by the type of target organisms and the mechanisms via which the AMPs exert their antimicrobial activity. For example, AMPs with antiviral activity are mostly associated with viral assembly, adsorption, and entry processes, in addition to their ability to target both RNA and DNA viruses. Among these antiviral peptides are indolizidine and human α-defensin 1 [46]. These AMPs have been shown to eliminate viruses via their incorporation into the viral envelope, leading to the instability of the virus assembly, and they subsequently deliver the viral entry into the host cell [44]. By contrast, AMPs such as lactoferricin have been shown to inhibit viral adsorption by binding to the specific viral receptors on the target cells [46]. Further, AMPs such as NP-1, an alpha-defensin that is derived from rabbit neutrophils, has been shown to inhibit viral assembly and maturation by binding the intracellular components that are essential for the cellular translocation of the virus in the host cell [46].

The most investigated AMPs are those with antibacterial activity. This type of AMP is characterized by its ability to interact with anionic bacterial membranes, leading to the disruption of the lipid bilayer [40].

Based on their molecular action, peptides with antimicrobial activity can be classified into two types. One of these types includes membrane-disrupting peptides, while the other one includes non-membrane-targeting peptides [40]. Although the molecular action of the main types of AMPs is different, some bacterial AMPs exert their activity via both membrane- and non-membrane-dependent mechanisms. Most AMPs trigger bacterial membrane destruction via interaction between their positively charged peptide molecules and the negatively charged cell surface as well as through hydrophobic interactions between the peptide amphipathic domain and membrane phospholipids.

Cationic AMPs have been demonstrated to exert their antibacterial activities via interaction with negatively charged bacterial membranes. The electrostatic interaction between cationic AMPs and the anionic components of the plasma membrane results in an increase in membrane permeability, and the release of AMPs into the cytoplasmic membrane, which subsequently, causes the lysis of the plasma membrane and, finally, the death of the microbial pathogen. To that end, four models have been proposed to describe the mechanisms whereby AMPs trigger the destruction of the microbial membrane. These include the barrel-stave (Figure 3A), toroidal pore (Figure 3B), carpet (Figure 3C), and aggregate (Figure 3D) models. In the barrel-stave model, the increased number of peptides binding to the membrane triggers membrane aggregation and conformational transformation. Consequently, the shift of local phospholipid head groups leads to cell membrane instability.

The barrel-stave mechanism is mediated via the vertical aggregation of helices into the lipid bilayer. The insertion of the transmembrane peptide bundle is organized in the cell membrane as staves of a barrel so that their hydrophobic face region is aligned with the central lipid region of the lipid bilayer. In parallel, the hydrophilic peptide constituents form the inner pore region that is filled with water [36]. The stable channels, namely, the barrel-like pores, which are formed in the cell membrane, allow the outflow of the cytoplasm. As consequence, the severe damage of the cell membrane results in cell collapse and, finally cell, death [33].

The toroidal pore model is mechanistically similar to the barrel-stave model; however, the mode of insertion of AMPs into the membrane and the binding behavior of AMPs with lipid molecules are different. In the toroidal pore model, the insertion of peptides into the membrane results in a continuous bending of the lipid monolayer from top to bottom [43]. The central water core is wrinkled with the inserted peptides and lipid head groups. Upon the formation of toroidal pores, the polar regions of the peptides start to line up with the lipid polar head groups.

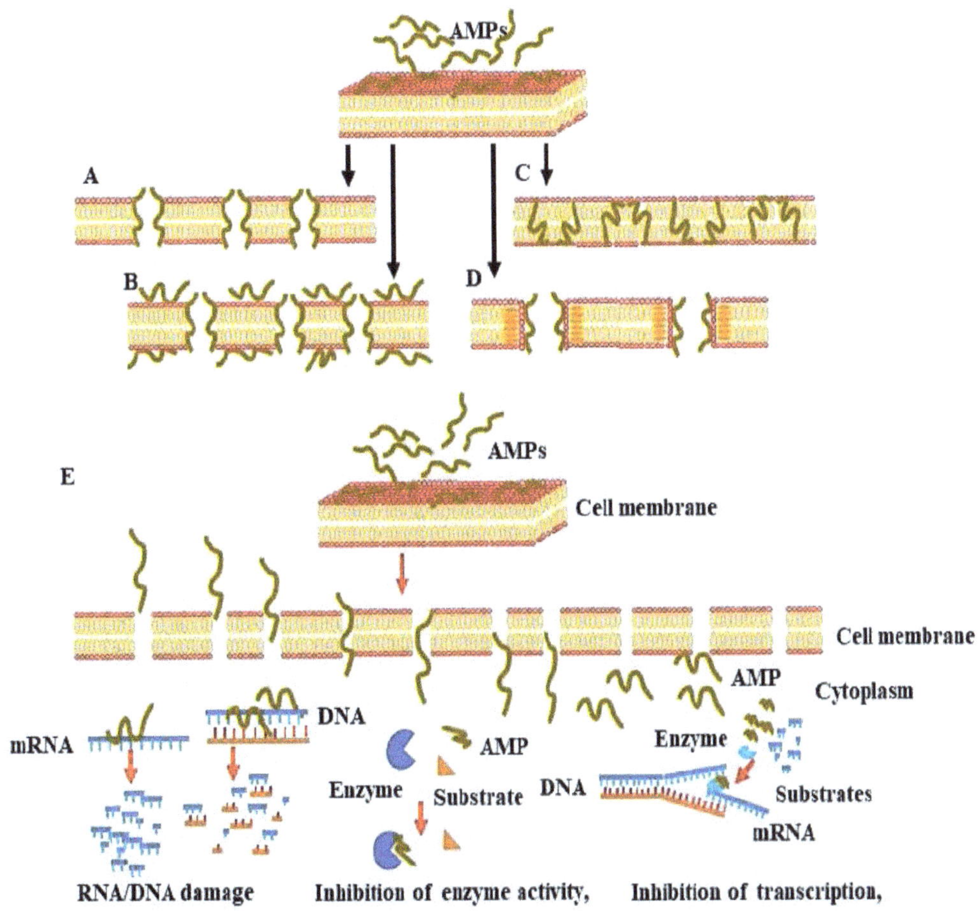

Figure 3. Proposed models for AMP-induced membrane permeability, membrane penetration, and interference with cellular components. AMPs exert their antimicrobial activity via interaction with negatively charged membranes to mediate and rapidly increase membrane permeability, cell membrane lysis, or the release of intracellular contents, leading to microbial cell death. There are four main models of membrane pore formation, namely, the barrel-stave model (**A**), toroidal pore model (**B**), carpet model (**C**) and aggregate model (**D**). (**E**) Mechanisms of the penetration of AMPs into the cytoplasm of the microbial cell and interference with intracellular components.

To mediate their antimicrobial activities, AMPs first undergo confirmational modifications so that they can penetrate the phospholipid membrane. Following the penetration of the phospholipid membrane, the hydrophobic regions of the AMPs combine with the internal hydrophobic regions of the phospholipid bilayer, exposing the hydrophilic regions to the outside and subsequently increasing the membrane permeability of the microbial cell, which ultimately results in microbial death. Upon their entry into the cytoplasm, AMPs start to interfere with the intracellular components, leading to the dysregulation of cellular function via the mechanism mediated by the enhancement of DNA/RNA damage, inhibition of enzyme activity, and suppression of the transcription/translation processes, which are necessary for cell wall synthesis. Although in both barrel-stave and toroidal pore models, the mode of AMP insertion into the membrane determines the action of AMPs, in the carpet model, the action of AMPs depends on the concentration levels and electrostatic effect of the AMPs, as well as the net charge of the anionic component [43]. The hypothesis of the carpet model relies mainly on the initial aggregation of the peptides on the membrane in the monomeric or oligomeric form that ultimately cover the membrane as a carpet. As a consequence, the hydrophobic regions start to interact with the cell membrane while the hydrophilic ends face the aqueous solution. Once the concentration threshold has been reached, the aggregation of the peptides starts to enhance membrane permeability and ultimately membrane disruption [34]. Finally, in the aggregate model, the binding of the AMPs to the anionic cytoplasmic membrane causes the peptides and lipids to form a peptide–lipid complex micelle that opens membrane channels and allows the release of ions and intracellular contents, which ultimately leads to cell death [46]. In all models, the molecular action by which AMPs trigger microbial death depends on both a conformational change in the AMPs and the peptide–lipid ratio of the AMPs and the microbial membrane [46]. The conformational change in α-helical AMPs following anionic lipid membrane binding transforms the disordered structure of the AMPs in the aqueous solution into an amphiphilic α-helical structure, which facilitates the interaction of the AMPs with the microbial membrane [46]. Of note, in contrast to α-helical AMPs, AMPs with β-sheets are unable to undergo major conformational transitions during the interaction with the microbial membrane [46] due to the β-sheet AMPs' stable disulfide bond bridges [16]. Peptide–lipid ratios likewise significantly impact conformational change and membrane lysis. At low peptide–lipid ratios, AMPs are located in parallel orientations on the surface of the plasma membrane [47], whereas at high peptide–lipid ratios, the AMPs becomes vertically oriented and are inserted into the hydrophobic center of the plasma membrane. This insertion of AMPs into the hydrophobic center of the plasma membrane increases the membrane permeability and subsequently enhances the release of both intracellular ions and metabolites that induce microbial cell death [48,49].

In addition to the destruction of the microbial membrane, AMPs have been reported to mediate their antimicrobial activity via intracellular-dependent mechanisms (Figure 3E). These include the induction of DNA/RNA damage, the inhibition of protein synthesis, enzyme activity, and the synthesis of a bacterial cell wall [50]. The above-mentioned AMP conformational changes and microbial membrane peptide–lipid ratios are unsurprisingly the main factors governing the ability of AMPs to pass through bacterial cell wall components, such as the lipopolysaccharides (LPSs) in the case of Gram-negative bacteria and lipoteichoic acid (LTA) and peptidoglycan in the case of Gram-positive bacteria [51]. AMP-mediated DNA/RNA damage has been found to be induced by the direct binding of AMPs to DNA or by the inhibition of DNA replication and transcription [48,49,52–59] AMPs such as Buforin II [52], a histone-derived antimicrobial peptide with a length of 21 amino acids, translocate across lipid membranes without affecting membrane permeability, and they trigger antimicrobial activity by binding to DNA/RNA [60]. Conversely, the AMP indolizidine, which displays antimicrobial activity against multi-drug resistance pathogens, has been shown to kill bacteria via the inhibition of DNA synthesis by penetrating membranes without inducing cell lysis. Anionic AMPs such as P2, isolated from Xenopus Leavis skin, was found to inhibit bacterial growth via interaction with microbial genomic DNA [61]. Other AMPs, such as PR-39, a proline/arginine-rich AMP isolated from the small intestine of pigs, has been reported to kill bacteria by penetrating the outer membranes of bacteria [62]. The entry of PR-39 into the cytoplasm was found to be associated with the inhibition of protein synthesis and acceleration of the ubiquitination of proteins, which are essential for DNA synthesis [63]. Proline-enriched AMPs have been reported to exert their antimicrobial activity through the interference of protein synthesis machinery by binding to ribosomes [63]. The N-terminal (1–25) and (1–31) residues of the non-lytic proline-rich AMP (PrAMP) Bac 5, for example, bind to the tunnel of ribosomes and prevent the translation process [64]. The proline-enriched AMP oncocin inhibits mRNA translation by binding the 70S ribosome, whereas apidaecin inhibits 50s ribosome assembly [65]. Api137, an apidaecin-derived peptide, binds to the ribosomes of *E. coli* and trap release factors 1 (RF1) or 2 (RF2) to trigger translation termination [66].

The inhibition of microbial pathogens' intracellular enzymes has also been reported to be a mechanism through which some AMPs exert their antimicrobial activity. Pyrrhocoricin binds to the bacterial heat shock protein DnaK and subsequently inhibits ATPase action [25]. Microcin J25, a ribosomal synthesized and post-translationally modified AMP, binds to the secondary channel of the RNA polymerase, blocking the entry of substrates through the channel [61]. LL-37 inhibits the activity of palmitoyl transferase PagP [57,58]. Pag P is an enzyme located in the outer membrane of Gram-negative bacteria and facilitates membrane permeability via activated lipid A acylation [67]. Finally, NP-6, isolated from Sichuan pepper seeds, inhibits *E. coli* β-galactosidase activity [67].

The inhibition of bacterial cell wall synthesis is a common therapeutic strategy to treat pathogenic bacteria infection. The anti-leishmanial drug candidate, human neutrophil peptide-1 (HNP-1), inhibits bacterial cell wall synthesis by penetrating the outer and inner membranes of *E. coli* and suppresses the synthesis of DNA, RNA, and proteins [68]. HNP1′s antimicrobial activity is mediated by its interaction with lipid II. HNP1 binds to a highly conserved non-peptide motif of peptidoglycan precursor (lipid II) and teichoic acid precursor (lipid III) [61], resulting in the inhibition of cell wall synthesis and subsequent lysis. HNP1 has excellent activity against a wide range of Gram-positive bacteria, including multi-drug resistant organisms, such as Methicillin-resistant *Staphylococcus aureus* (MRSA), Vancomycin Intermediate *S. aureus* (VISA), Vancomycin-resistant enterococci (VRE), Clostridium difficile, *Streptococcus pneumoniae*, and *Mycobacterium tuberculosis* [23]. Finally, teixobactin is a cyclic dipeptide containing an unusual amino acid, enduracididine [68]. This AMP is a member of a new class characterized by their specific action on unique targets in cell wall synthesis.

4. Therapeutic Potential of Antimicrobial Peptides

Human infections are typically polymicrobial and stem mainly from oral infections, surgical wounds, diabetic foot ulcers, cystic-fibrosis-related lung infections, urinary tract infections, and otitis media infections [69–71]. Therefore, the treatment of polymicrobial infection is more challenging when compared with monomicrobial infections. In contrast to traditional antibiotics, AMPs are characterized by their ability to target both monomicrobial and polymicrobial infections without the development of cross-resistance [72]. Thus, the advantage of AMPs over traditional antibiotics is their ability to act directly on the bacterial membrane when compared to their indirect action on the intracellular targets. Other advantages of AMPs over conventional antibiotics involve the actions mediated by their different characteristics, including their ability to function against both antibiotic-resistant and -sensitive microbial pathogens and their ability to target monomicrobial and polymicrobial infections without the development of cross-resistance [64,73]. However, the therapeutic success of AMPs in the treatment and prevention of bacterial infection may result from their ability to act directly on the bacterial membrane, rather than their indirect action on intracellular targets [74]. Furthermore, the ability of a single AMP to exert its antimicrobial activity via multiple mechanisms, and through different pathways [75], suggests the clinical relevance of AMPs in the treatment and prevention of microbial pathogens. However, the establishment of novel, clinically relevant therapeutic approaches that target multiple pathogens in mixed populations, thereby replacing traditional antibiotics, is tangible.

The production of endogenous AMPs by multicellular organisms constitutes a host defense mechanism against pathogenic microbes. Based on the broad spectrum of their antimicrobial activity, AMPs are promising therapeutic agents for infection control [25]. In addition to their antimicrobial activities against various pathogens, including bacteria (Gram-positive and Gram-negative bacteria), fungi, and viruses [76], many AMPs are effective against multi-drug resistant (MDR) bacteria and have low propensity for the development of resistance [77]. AMPs are also involved in the promotion and regulation of the innate immunity system [78]. Finally, the use of AMPs against biofilm formation has been widely reported over the last few decades [79]. Many AMPs kill cells in biofilms and inhibit biofilm formation via the interference with the abundant extracellular polymeric substances (EPSs) of microbial cells. These EPSs are known to be functionally responsible for the protection of microbial cells from the surrounding environment [80].

Despite the potential therapeutic benefits of AMPs when compared with existing antibiotics, AMPs have some limitations that hinder their development for clinical use [81]. Most natural AMPs are characterized by poor absorption, distribution, metabolism, and excretion, in addition to their short half-life and low permeability and solubility [82]. Moreover, AMPs have a high production cost and a degree of toxicity, particularly in oral administration. All these properties are considered as major hindrances for the development of novel AMP-based treatments. To overcome these AMP limitations that hamper clinical application, several studies are urgently needed to improve the functional properties of AMPs, such as their absorption, distribution, metabolism, excretion, cytotoxicity, and proteolytic stability. An improvement in the functional properties of AMPs may involve the alteration of the peptide composition and the modification of their post translation of AMPs.

To that end, several technical procedures have been proposed to improve the functional properties of AMPs. These include the modification of the chemical structure of AMPs via the introduction of unusual amino acids, such as D-form amino acids, or by the acetylation or amidation of the terminal regions of AMPs. As widely reported, the modification of the chemical structure of AMPs was found to improve the stability of their peptides and prevent their proteolytic degradation [83]. Similarly, the delivery of AMPs using liposome encapsulation was found to preserve the stability of AMPs and to reduce their toxicity [84].

AMPs are an essential component of the innate skin defense mechanisms and are considered to be a first-line barrier providing protection against microbial pathogens [85]. AMPs are closely associated the with innate skin immunity and are known to regulate immunity by interacting with various immune cells and linking innate and adaptive immune responses during infection. These AMPs include, β-defensins (BD) [86], cathelicidins (human hCAP18/LL37) [87], RNase 7 [88], and secretory leukocyte protease inhibitor (SLPI) [89]. Apart from their significant role in the regulation of innate skin immunity, AMPs such as defensins and cathelicidins have also been reported to play a key role in the regulation of the innate immunity of the lung [86,90]. To that end, both defensins and cathelicidins belong to a family of AMPs, which are mostly detected in the secretion of airways [87]. The exogenous administration of defensins and cathelicidins has been reported as an effective strategy in the prevention and treatment of infection. In particular, tachyplesin III, a β-sheet peptide isolated from the hemocytes of the horseshoe crab, has been evaluated for antimicrobial activity in lung polymicrobial co-infection pneumonia [89,91].

Naturally produced AMPs in the oral cavity play key roles in the maintenance of microbial homeostasis and oral cavity health stasis [91,92]. These AMPs are characterized by their antimicrobial activity against oral bacteria, which has been evaluated against oral infections, as widely reported in several studies [91,92]. D-Cateslytin (D-Ctl), an AMP derived from L-Cateslytin, has been observed to have therapeutic potential against bacterial infection in combination with several antimicrobials [92,93] and has been reported to be an antifungal agent in the treatment of oral infections associated with Candida albicans [94].

The most important advantages of AMPs over conventional therapeutics are attributed to the potential of AMPs to offer innovative and effective solutions to the treatment of mixed populations with polyinfections and to differentiate between pathogenic bacteria and protective normal flora. Therefore, the development and evaluation of AMPs with the ability to target multiple pathogens in mixed populations without the destruction of the protective normal flora represents an important public health issue.

5. Conclusions

AMPs are characterized by their broad spectrum of antimicrobial activities and are powerful regulators of innate immunity. AMPs have a strong cell-killing efficiency on microbial pathogens, particularly MDR bacteria. In addition, AMPs offer an alternative approach to overcome the antibiotic resistance of most microbial pathogens. Although AMPs may be able to overcome the limitations of current antibiotics due to their antimicrobial activity, their shortcomings include poor stability, toxicity, and unexplored adverse effects, which limit their clinical application. However, continued development and evaluation of functional AMPs may allow for the modification of natural AMPs, thereby facilitating the production of new AMPs with clinically desirable characteristics. Some AMPs have been approved for clinical application, while others remain under investigation in clinical trials. Therefore, the development and evaluation of AMPs with the ability to target multiple pathogens in mixed populations without the destruction of the protective normal flora represents an exciting antimicrobial therapeutic strategy.

Author Contributions: Conceptualization, M.H. and Y.H.; methodology, M.H.; software, M.H.; validation, M.H. and C.B.; resources, D.M.; data curation, T.W.F.; writing—original draft preparation, M.H.; writing—review and editing, N.K.; visualization, N.K. and C.B.; supervision, M.H.; project administration, M.H.; funding acquisition, Y.H. All authors have read and agreed to the published version of the manuscript.

Funding: This research received no external funding.

Institutional Review Board Statement: Not applicable.

Informed Consent Statement: Not applicable.

Data Availability Statement: The data that support the findings of this study are available from the corresponding author upon reasonable request.

Conflicts of Interest: The authors declare no conflict of interest.

References

1. Wang, J.; Dou, X.; Song, J.; Lyu, Y.; Zhu, X.; Xu, L.; Li, W.; Shan, A. Antimicrobial peptides: Promising alternatives in the post feeding antibiotic era. *Med. Res. Rev.* **2019**, *39*, 831–859. [CrossRef] [PubMed]
2. Lei, J.; Sun, L.; Huang, S.; Zhu, C.; Li, P.; He, J.; Mackey, V.; Coy, D.H.; He, Q. The antimicrobial peptides and their potential clinical applications. *Am. J. Transl. Res.* **2019**, *11*, 3919–3931. [PubMed]
3. Mookherjee, N.; Anderson, M.A.; Haagsman, H.P.; Davidson, D.J. Antimicrobial host defence peptides: Functions and clinical potential. *Nat. Rev. Drug Discov.* **2020**, *1*, 311–332. [CrossRef]
4. Dennison, S.R.; Harris, F.; Mura, M.; Phoenix, D.A. An Atlas of Anionic Antimicrobial Peptides from Amphibians. *Curr. Protein Pept. Sci.* **2018**, *19*, 823–838. [CrossRef]
5. Pfalzgraff, A.; Brandenburg, K.; Weindl, G. Antimicrobial Peptides and Their Therapeutic Potential for Bacterial Skin Infections and Wounds. *Front. Pharmacol.* **2018**, *9*, 281. [CrossRef] [PubMed]
6. Moretta, A.; Scieuzo, C.; Petrone, A.M.; Salvia, R.; Manniello, M.D.; Franco, A.; Lucchetti, D.; Vassallo, A.; Vogel, H.; Sgambato, A.; et al. Antimicrobial Peptides: A New Hope in Biomedical and Pharmaceutical Fields. *Front. Cell. Infect. Microbiol.* **2021**, *11*, 668632. [CrossRef]
7. Wangi, J.; Hao, X.; Lai, R.; Zhang, Z.Y. Antimicrobial peptides: New hope in the war against multidrug resistance. *Zool. Res.* **2019**, *40*, 488–505.
8. Bormann, N.; Koliszak, A.; Kasper, S.; Schoen, L.; Hilpert, K.; Volkmer, R.; Kikhney, J.; Wildemann, B.A. short artificial antimicrobial peptide shows potential to prevent or treat bone infections. *Sci. Rep.* **2017**, *1*, 1506. [CrossRef]
9. Wang, S.; Ye, Q.; Wang, K.; Zeng, X.; Huang, S.; Yu, H.; Ge, Q.; Qi, D.; Qiao, S. Enhancement of Macrophage Function by the Antimicrobial Peptide Sublancin Protects Mice from Methicillin-Resistant *Staphylococcus aureus*. *J. Immunol. Res.* **2019**, *2019*, 3979352. [CrossRef]
10. Kang, H.K.; Seo, C.H.; Luchian, T.; Park, Y. Pse-T2, an Antimicrobial Peptide with High-Level, Broad-Spectrum Antimicrobial Potency and Skin Biocompatibility against Multidrug-Resistant Pseudomonas aeruginosa Infection. *Antimicrob. Agents Chemother.* **2018**, *62*, e01493-18. [CrossRef]
11. Rodrigues, G.; Silva, G.G.O.; Buccini, D.F.; Duque, H.M.; Dias, S.C.; Franco, O.L. Bacterial Proteinaceous Compounds With Multiple Activities Toward Cancers and Microbial Infection. *Front. Microbiol.* **2019**, *10*, 1690. [CrossRef] [PubMed]
12. Brown, C.L.; Smith, K.; McCaughey, L.; Walker, D. Colicin-like bacteriocins as novel therapeutic agents for the treatment of chronic biofilm-mediated infection. *Biochem Soc. Trans.* **2012**, *40*, 1549–1552. [CrossRef] [PubMed]
13. Bhattacharjee, R.; Kumar, L.; Dhasmana, A.; Mitra, T.; Dey, A.; Malik, S.; Kim, B.; Gundamaraju, R. Governing HPV-related carcinoma using vaccines: Bottlenecks and breakthroughs. *Front. Oncol.* **2022**, *12*, 977933. [CrossRef]

14. Cheung, R.C.; Wong, J.H.; Pan, W.L.; Chan, Y.S.; Yin, C.M.; Dan, X.L.; Wang, H.X.; Fang, E.F.; Lam, S.K.; Ngai, P.H.; et al. Antifungal and antiviral products of marine organisms. *Appl. Microbiol. Biotechnol.* **2014**, *98*, 3475–3494. [CrossRef] [PubMed]
15. Liu, H.; Yang, N.; Teng, D.; Mao, R.; Hao, Y.; Ma, X.; Wang, J. Design and Pharmacodynamics of Recombinant Fungus Defensin NZL with Improved Activity against Staphylococcus hyicus In Vitro and In Vivo. *Int. J. Mol. Sci.* **2021**, *22*, 5435. [CrossRef]
16. Sharma, P.; Kaur, J.; Sharma, G.; Kashyap, P. Plant derived antimicrobial peptides: Mechanism of target, isolation techniques, sources and pharmaceutical applications. *J. Food Biochem.* **2022**, *46*, e14348. [CrossRef]
17. Egessa, R. Antimicrobial peptides from freshwater invertebrate species: Potential for future applications. *Mol. Biol. Rep.* **2022**, *49*, 9797–9811. [CrossRef] [PubMed]
18. Shi, J.; Camus, A.C. Hepcidins in amphibians and fishes: Antimicrobial peptides or iron-regulatory hormones? *Dev. Comp. Immunol.* **2006**, *30*, 746–755. [CrossRef]
19. Zasloff, M. Antimicrobial Peptides of Multicellular Organisms: My Perspective. *Adv. Exp. Med. Biol.* **2019**, *1117*, 3–6.
20. van Hoek, M.L.; Prickett, M.D.; Settlage, R.E.; Kang, L.; Michalak, P.; Vliet, K.A.; Bishop, B.M. The Komodo dragon (Varanus komodoensis) genome and identification of innate immunity genes and clusters. *BMC Genom.* **2019**, *20*, 684. [CrossRef]
21. Wu, J.; Zhou, X.; Chen, Q.; Chen, Z.; Zhang, J.; Yang, L.; Sun, Y.; Wang, G.; Dai, J.; Feng, T. Defensins as a promising class of tick antimicrobial peptides: A scoping review. *Infect. Dis. Poverty* **2022**, *11*, 71. [CrossRef] [PubMed]
22. Economou, N.J.; Cocklin, S.; Loll, P.J. High-resolution crystal structure reveals molecular details of target recognition by bacitracin. *Proc. Natl. Acad. Sci. USA* **2013**, *110*, 14207–14212. [CrossRef] [PubMed]
23. Hofmann, C.M.; Anderson, J.M.; Marchant, R.E. Targeted delivery of vancomycin to Staphylococcus epidermidis biofilms using a fibrinogen-derived peptide. *J. Biomed. Mater. Res. A* **2012**, *100*, 2517–2525. [CrossRef] [PubMed]
24. Eleftherianos, I.; Zhang, W.; Heryanto, C.; Mohamed, A.; Contreras, G.; Tettamanti, G.; Wink, M.; Bassal, T. Diversity of insect antimicrobial peptides and proteins—A functional perspective: A review. *Int. J. Biol. Macromol.* **2021**, *191*, 277–287. [CrossRef]
25. Dalphin, M.D.; Stangl, A.J.; Liu, Y.; Cavagnero, S. KLR-70: A Novel Cationic Inhibitor of the Bacterial Hsp70 Chaperone. *Biochemistry* **2020**, *59*, 1946–1960. [CrossRef]
26. Yokoo, H.; Hirano, M.; Ohoka, N.; Misawa, T.; Demizu, Y. Structure-activity relationship study of amphipathic antimicrobial peptides using helix-destabilizing sarcosine. *J. Pept. Sci.* **2021**, *27*, e3360. [CrossRef]
27. Tanaka, M.; Kato, T.; Oda, M. Conformational changes of α-helical peptides with different hydrophobic residues induced by metal-ion binding. *Biophys. Chem.* **2021**, *277*, 106661. [CrossRef]
28. Tecle, T.; Tripathi, S.; Hartshorn, K.L. Review: Defensins and cathelicidins in lung immunity. *Innate Immun.* **2010**, *16*, 151–159. [CrossRef]
29. Afanasyeva, E.F.; Syryamina, V.N.; De Zotti, M.; Formaggio, F.; Toniolo, C.; Dzuba, S.A. Peptide antibiotic trichogin in model membranes: Self-association and capture of fatty acids. *Biochim. Biophys. Acta Biomembr.* **2019**, *1861*, 524–531. [CrossRef]
30. Oliva, R.; Del Vecchio, P.; Grimaldi, A.; Notomista, E.; Cafaro, V.; Pane, K.; Schuabb, V.; Winter, R.; Petraccone, L. Membrane disintegration by the antimicrobial peptide (P)GKY20: Lipid segregation and domain formation. *Phys. Chem. Chem. Phys.* **2019**, *21*, 3989–3998. [CrossRef]
31. Kumar, P.; Kizhakkedathu, J.N.; Straus, S.K. Antimicrobial Peptides: Diversity, Mechanism of Action and Strategies to Improve the Activity and Biocompatibility In Vivo. *Biomolecules* **2018**, *8*, 4. [CrossRef] [PubMed]
32. Drozak, J.; Veiga-da-Cunha, M.; Kadziolka, B.; Van Schaftingen, E. Vertebrate Acyl CoA synthetase family member 4 (ACSF4-U26) is a β-alanine-activating enzyme homologous to bacterial non-ribosomal peptide synthetase. *FEBS J.* **2014**, *281*, 1585–1597. [CrossRef] [PubMed]
33. Lazzaro, B.P.; Zasloff, M.; Rolff, J. Antimicrobial peptides: Application informed by evolution. *Science* **2020**, *368*, eaau5480. [CrossRef]
34. Hong, J.; Lu, X.; Deng, Z.; Xiao, S.; Yuan, B.; Yang, K. How Melittin Inserts into Cell Membrane: Conformational Changes, Inter-Peptide Cooperation, and Disturbance on the Membrane. *Molecules* **2019**, *24*, 1775. [CrossRef] [PubMed]
35. Quemé-Peña, M.; Juhász, T.; Mihály, J.; Szigyártó, I.C.; Horváti, K.; Bősze, S.; Henczkó, J.; Pályi, B.; Németh, C.; Varga, Z.; et al. Manipulating Active Structure and Function of Cationic Antimicrobial Peptide CM15 with the Polysulfonated Drug Suramin: A Step Closer to in Vivo Complexity. *ChemBioChem* **2019**, *20*, 1578–1590. [CrossRef]
36. Lee, M.R.; Raman, N.; Ortiz-Bermúdez, P.; Lynn, D.M.; Palecek, S.P. 14-Helical β-Peptides Elicit Toxicity against C. albicans by Forming Pores in the Cell Membrane and Subsequently Disrupting Intracellular Organelles. *Cell. Chem. Biol.* **2019**, *26*, 289–299.e4. [CrossRef]
37. Gallardo-Godoy, A.; Hansford, K.A.; Muldoon, C.; Becker, B.; Elliott, A.G.; Huang, J.X.; Pelingon, R.; Butler, M.S.; Blaskovich, M.A.T.; Cooper, M.A. Structure-Function Studies of Polymyxin B Lipononapeptides. *Molecules* **2019**, *24*, 553. [CrossRef]
38. Xu, C.; Wang, A.; Marin, M.; Honnen, W.; Ramasamy, S.; Porter, E.; Subbian, S.; Pinter, A.; Melikyan, G.B.; Lu, W.; et al. Human Defensins Inhibit SARS-CoV-2 Infection by Blocking Viral Entry. *Viruses* **2021**, *26*, 1246. [CrossRef]
39. Böffert, R.; Businger, R.; Preiß, H.; Ehmann, D.; Truffault, V.; Simon, C.; Ruetalo, N.; Hamprecht, K.; Müller, P.; Wehkamp, J.; et al. The human α-defensin-derived peptide HD5(1-9) inhibits cellular attachment and entry of human cytomegalovirus. *Antivir. Res.* **2020**, *177*, 104779. [CrossRef]
40. Yan, Y.; Li, Y.; Zhang, Z.; Wang, X.; Niu, Y.; Zhang, S.; Xu, W.; Ren, C. Advances of peptides for antibacterial applications. *Colloids Surf. B Biointerfaces* **2021**, *202*, 111682. [CrossRef]

41. Strandberg, E.; Wadhwani, P.; Tremouilhac, P.; Dürr, U.H.; Ulrich, A.S. Solid-state NMR analysis of the PGLa peptide orientation in DMPC bilayers: Structural fidelity of 2H-labels versus high sensitivity of 19F-NMR. *Biophys. J.* **2006**, *90*, 1676–1686. [CrossRef] [PubMed]
42. Maisetta, G.; Batoni, G. Editorial: Interspecies Interactions: Effects on Virulence and Antimicrobial Susceptibility of Bacterial and Fungal Pathogens. *Front. Microbiol.* **2020**, *11*, 1922. [CrossRef] [PubMed]
43. Wang, Q.; Xu, Y.; Hu, J. Intracellular mechanism of antimicrobial peptide HJH-3 against. *RSC Adv.* **2022**, *12*, 14485–14491. [CrossRef] [PubMed]
44. Rajagopal, M.; Walker, S. Envelope Structures of Gram-Positive Bacteria. *Curr. Top. Microbiol. Immunol.* **2017**, *404*, 1–44. [PubMed]
45. Islam, M.Z.; Sharmin, S.; Moniruzzaman, M.; Yamazaki, M. Elementary processes for the entry of cell-penetrating peptides into lipid bilayer vesicles and bacterial cells. *Appl. Microbiol. Biotechnol.* **2018**, *102*, 3879–3892. [CrossRef]
46. Malanovic, N.; Lohner, K. Gram-positive bacterial cell envelopes: The impact on the activity of antimicrobial peptides. *Biochim. Biophys. Acta* **2016**, *1858*, 936–946. [CrossRef]
47. Paredes, S.D.; Kim, S.; Rooney, M.T.; Greenwood, A.I.; Hristova, K.; Cotten, M.L. Enhancing the membrane activity of Piscidin 1 through peptide metallation and the presence of oxidized lipid species: Implications for the unification of host defense mechanisms at lipid membranes. *Biochim. Biophys. Acta Biomembr.* **2020**, *1862*, 183236. [CrossRef]
48. Graf, M.; Mardirossian, M.; Nguyen, F.; Seefeldt, A.C.; Guichard, G.; Scocchi, M.; Innis, C.A.; Wilson, D.N. Proline-rich antimicrobial peptides targeting protein synthesis. *Nat. Prod. Rep.* **2017**, *34*, 702–711. [CrossRef]
49. Mardirossian, M.; Barrière, Q.; Timchenko, T.; Müller, C.; Pacor, S.; Mergaert, P.; Scocchi, M.; Wilson, D.N. Fragments of the Nonlytic Proline-Rich Antimicrobial Peptide Bac5 Kill Escherichia coli Cells by Inhibiting Protein Synthesis. *Antimicrob. Agents Chemother.* **2018**, *62*, e00534-18. [CrossRef]
50. Chen, X.; Li, L. Non-membrane mechanisms of antimicrobial peptide P7 against *Escherichia coli*. *Wei Sheng Wu Xue Bao* **2016**, *56*, 1737–1745.
51. Wu, C.; Biswas, S.; Garcia De Gonzalo, C.V.; van der Donk, W.A. Investigations into the Mechanism of Action of Sublancin. *ACS Infect. Dis.* **2019**, *5*, 454–459. [CrossRef]
52. Muñoz-Camargo, C.; Salazar, V.A.; Barrero-Guevara, L.; Camargo, S.; Mosquera, A.; Groot, H.; Boix, E. Unveiling the Multifaceted Mechanisms of Antibacterial Activity of Buforin II and Frenatin 2.3S Peptides from Skin Micro-Organs of the Orinoco Lime Treefrog (*Sphaenorhynchus lacteus*). *Int. J. Mol. Sci.* **2018**, *19*, 2170. [CrossRef] [PubMed]
53. Cardoso, M.H.; Meneguetti, B.T.; Costa, B.O.; Buccini, D.F.; Oshiro, K.G.N.; Preza, S.L.E.; Carvalho, C.M.E.; Migliolo, L.; Franco, O.L. Non-Lytic Antibacterial Peptides That Translocate Through Bacterial Membranes to Act on Intracellular Targets. *Int. J. Mol. Sci.* **2019**, *20*, 4877. [CrossRef] [PubMed]
54. Zhang, Y.; Liu, S.; Li, S.; Cheng, Y.; Nie, L.; Wang, G.; Lv, C.; Wei, W.; Cheng, C.; Hou, F.; et al. Novel short antimicrobial peptide isolated from *Xenopus laevis* skin. *J. Pept. Sci.* **2017**, *23*, 403–409. [CrossRef]
55. Florin, T.; Maracci, C.; Graf, M.; Karki, P.; Klepacki, D.; Berninghausen, O.; Beckmann, R.; Vázquez-Laslop, N.; Wilson, D.N.; Rodnina, M.V.; et al. An antimicrobial peptide that inhibits translation by trapping release factors on the ribosome. *Nat. Struct. Mol. Biol.* **2017**, *24*, 752–757. [CrossRef]
56. Yasir, M.; Dutta, D.; Kumar, N.; Willcox, M.D.P. Interaction of the surface bound antimicrobial peptides melimine and Mel4 with Staphylococcus aureus. *Biofouling* **2020**, *36*, 1019–1030. [CrossRef] [PubMed]
57. Yang, H.; Fu, J.; Zhao, Y.; Shi, H.; Hu, H.; Wang, H. Escherichia coli PagP Enzyme-Based De Novo Design and In Vitro Activity of Antibacterial Peptide LL-37. *Med. Sci. Monit.* **2017**, *23*, 2558–2564. [CrossRef] [PubMed]
58. Trent, M.S.; Pabich, W.; Raetz, C.R.; Miller, S.I. A PhoP/PhoQ-induced Lipase (PagL) that catalyzes 3-O-deacylation of lipid A precursors in membranes of Salmonella typhimurium. *J. Biol. Chem.* **2001**, *276*, 9083–9092. [CrossRef] [PubMed]
59. Braffman, N.R.; Piscotta, F.J.; Hauver, J.; Campbell, E.A.; Link, A.J.; Darst, S.A. Structural mechanism of transcription inhibition by lasso peptides microcin J25 and capistruin. *Proc. Natl. Acad. Sci. USA* **2019**, *116*, 1273–1278. [CrossRef]
60. Ling, L.L.; Schneider, T.; Peoples, A.J.; Spoering, A.L.; Engels, I.; Conlon, B.P.; Mueller, A.; Schäberle, T.F.; Hughes, D.E.; Epstein, S.; et al. A new antibiotic kills pathogens without detectable resistance. *Nature* **2015**, *517*, 455–459. [CrossRef]
61. De Leeuw, E.; Li, C.; Zeng, P.; Diepeveen-de Buin, M.; Lu, W.Y.; Breukink, E.; Lu, W. Functional interaction of human neutrophil peptide-1 with the cell wall precursor lipid II. *FEBS Lett.* **2010**, *584*, 1543–1548. [CrossRef] [PubMed]
62. Krishnakumari, V.; Binny, T.M.; Adicherla, H.; Nagaraj, R. *Escherichia coli* Lipopolysaccharide Modulates Biological Activities of Human-β-Defensin Analogues but Not Non-Ribosomally Synthesized Peptides. *ACS Omega* **2020**, *5*, 6366–6375. [CrossRef] [PubMed]
63. Woodburn, K.W.; Jaynes, J.M.; Clemens, L.E. Evaluation of the Antimicrobial Peptide, RP557, for the Broad-Spectrum Treatment of Wound Pathogens and Biofilm. *Front. Microbiol.* **2019**, *10*, 1688. [CrossRef] [PubMed]
64. Duplantier, A.J.; van Hoek, M.L. The Human Cathelicidin Antimicrobial Peptide LL-37 as a Potential Treatment for Polymicrobial Infected Wounds. *Front. Immunol.* **2013**, *4*, 143. [CrossRef]
65. Büchau, A.S.; Hassan, M.; Kukova, G.; Lewerenz, V.; Kellermann, S.; Würthner, J.U.; Wolf, R.; Walz, M.; Gallo, R.L.; Ruzicka, T. S100A15, an antimicrobial protein of the skin: Regulation by *E. coli* through Toll-like receptor 4. *J. Investig. Dermatol.* **2007**, *127*, 2596–2604. [CrossRef]
66. Boparai, J.K.; Sharma, P.K. Molecular Cloning, Functional and Biophysical Characterization of an Antimicrobial Peptide from Rhizosphere Soil. *Protein Pept. Lett.* **2021**, *28*, 1312–1322. [CrossRef]

67. Boll, J.M.; Tucker, A.T.; Klein, D.R.; Beltran, A.M.; Brodbelt, J.S.; Davies, B.W.; Trent, M.S. Reinforcing Lipid A Acylation on the Cell Surface of Acinetobacter baumannii Promotes Cationic Antimicrobial Peptide Resistance and Desiccation Survival. *mBio.* **2015**, *6*, e00478-15. [CrossRef]
68. Hou, X.; Feng, C.; Li, S.; Luo, Q.; Shen, G.; Wu, H.; Li, M.; Liu, X.; Chen, A.; Ye, M.; et al. Mechanism of antimicrobial peptide NP-6 from Sichuan pepper seeds against E. coli and effects of different environmental factors on its activity. *Appl. Microbiol. Biotechnol.* **2019**, *103*, 6593–6604. [CrossRef]
69. Jenssen, H.; Hamill, P.; Hancock, R.E. Peptide antimicrobial agents. *Clin. Microbiol. Rev.* **2006**, *19*, 491–511. [CrossRef]
70. Giuliani, A.; Pirri, G.; Nicoletto, S. Antimicrobial peptides: An overview of a promising class of therapeutics. *Cent. Eur. J. Biol.* **2007**, *2*, 1–33. [CrossRef]
71. Dowd, S.E.; Delton Hanson, J.; Rees, E.; Wolcott, R.D.; Zischau, A.M.; Sun, Y.; White, J.; Smith, D.M.; Kennedy, J.; Jones, C.E. Survey of fungi and yeast in polymicrobial infections in chronic wounds. *J. Wound Care* **2011**, *20*, 40–47. [CrossRef] [PubMed]
72. Magiorakos, A.P.; Srinivasan, A.; Carey, R.B.; Carmeli, Y.; Falagas, M.E.; Giske, C.G.; Harbarth, S.; Hindler, J.F.; Kahlmeter, G.; Olsson-Liljequist, B.; et al. Multidrug-resistant, extensively drug-resistant and pandrug-resistant bacteria: An international expert proposal for interim standard definitions for acquired resistance. *Clin. Microbiol. Infect.* **2012**, *18*, 268–281. [CrossRef] [PubMed]
73. Bayramov, D.; Li, Z.; Patel, E.; Izadjoo, M.; Kim, H.; Neff, J. A Novel Peptide-Based Antimicrobial Wound Treatment is Effective Against Biofilms of Multi-Drug Resistant Wound Pathogens. *Mil. Med.* **2018**, *183*, 481–486. [CrossRef] [PubMed]
74. Slezina, M.P.; Istomina, E.A.; Kulakovskaya, E.V.; Abashina, T.N.; Odintsova, T.I. Synthetic Oligopeptides Mimicking γ-Core Regions of Cysteine-Rich Peptides of Solanum lycopersicum Possess Antimicrobial Activity against Human and Plant Pathogens. *Curr. Issues Mol. Biol.* **2021**, *43*, 1226–1242. [CrossRef]
75. Li, J.; Fernández-Millán, P.; Boix, E. Synergism between Host Defence Peptides and Antibiotics Against Bacterial Infections. *Curr. Top. Med. Chem.* **2020**, *20*, 1238–1263. [CrossRef]
76. Almeida, B.; Nag, O.K.; Rogers, K.E.; Delehanty, J.B. Recent Progress in Bioconjugation Strategies for Liposome-Mediated Drug Delivery. *Molecules* **2020**, *25*, 5672. [CrossRef]
77. Kiatsurayanon, C.; Ogawa, H.; Niyonsaba, F. The Role of Host Defense Peptide Human β-defensins in the Maintenance of Skin Barriers. *Curr. Pharm. Des.* **2018**, *24*, 1092–1099. [CrossRef]
78. Masadeh, M.M.; Laila, S.A.; Haddad, R.; Alzoubi, K.; Alhaijaa, A.A.; Alrabadi, N. The Antimicrobial Effect Against Multi-Drug Resistant Bacteria of The SK4 Peptide: A Novel Hybrid Peptide of Cecropin-A and BMAP-27. *Curr. Pharm. Biotechnol.* **2022**. [CrossRef]
79. Hemmati, F.; Rezaee, M.A.; Ebrahimzadeh, S.; Yousefi, L.; Nouri, R.; Kafil, H.S.; Gholizadeh, P. Novel Strategies to Combat Bacterial Biofilms. *Mol. Biotechnol.* **2021**, *63*, 569–586. [CrossRef]
80. Batoni, G.; Maisetta, G.; Esin, S. Antimicrobial peptides and their interaction with biofilms of medically relevant bacteria. *Biochim. Biophys. Acta* **2016**, *1858*, 1044–1060. [CrossRef]
81. Rai, A.; Ferrão, R.; Palma, P.; Patricio, T.; Parreira, P.; Anes, E.; Tonda-Turo, C.; Martins, M.C.L.; Alves, N.; Ferreira, L. Antimicrobial peptide-based materials: Opportunities and challenges. *J. Mater. Chem. B* **2022**, *10*, 2384–2429. [CrossRef] [PubMed]
82. Qi, J.; Gao, R.; Liu, C.; Shan, B.; Gao, F.; He, J.; Yuan, M.; Xie, H.; Jin, S.; Ma, Y. Potential role of the antimicrobial peptide. *Infect. Drug Resist.* **2019**, *12*, 2865–2874. [CrossRef] [PubMed]
83. Dale, B.A.; Fredericks, L.P. Antimicrobial peptides in the oral environment: Expression and function in health and disease. *Curr. Issues Mol. Biol.* **2005**, *7*, 119–133. [PubMed]
84. Brancatisano, F.L.; Maisetta, G.; Barsotti, F.; Esin, S.; Miceli, M.; Gabriele, M.; Giuca, M.R.; Campa, M.; Batoni, G. Reduced human beta defensin 3 in individuals with periodontal disease. *J. Dent. Res.* **2011**, *90*, 241–245. [CrossRef]
85. Khurshid, Z.; Zafar, M.S.; Naseem, M.; Khan, R.S.; Najeeb, S. Human Oral Defensins Antimicrobial Peptides: A Future Promising Antimicrobial Drug. *Curr. Pharm. Des.* **2018**, *24*, 1130–1137. [CrossRef]
86. Wertz, P.W.; de Szalay, S. Innate Antimicrobial Defense of Skin and Oral Mucosa. *Antibiotics* **2020**, *9*, 159. [CrossRef]
87. Becknell, B.; Spencer, J.D.; Carpenter, A.R.; Chen, X.; Singh, A.; Ploeger, S.; Kline, J.; Ellsworth, P.; Li, B.; Proksch, E.; et al. Expression and antimicrobial function of beta-defensin 1 in the lower urinary tract. *PLoS ONE.* **2013**, *8*, e77714. [CrossRef]
88. Rodríguez-Carlos, A.; Trujillo, V.; Gonzalez-Curiel, I.; Marin-Luevano, S.; Torres-Juarez, F.; Santos-Mena, A.; Rivas-Santiago, C.; Enciso-Moreno, J.A.; Zaga-Clavellina, V.; Rivas-Santiago, B. Host Defense Peptide RNase 7 Is Down-regulated in the Skin of Diabetic Patients with or without Chronic Ulcers, and its Expression is Altered with Metformin. *Arch. Med. Res.* **2020**, *51*, 327–335. [CrossRef]
89. Zaiou, M.; Nizet, V.; Gallo, R.L. Antimicrobial and protease inhibitory functions of the human cathelicidin (hCAP18/LL-37) prosequence. *J. Investig. Dermatol.* **2003**, *120*, 810–816. [CrossRef]
90. Kawano, K.; Yoneya, T.; Miyata, T.; Yoshikawa, K.; Tokunaga, F.; Terada, Y.; Iwanaga, S. Antimicrobial peptide, tachyplesin I, isolated from hemocytes of the horseshoe crab (*Tachypleus tridentatus*). NMR determination of the beta-sheet structure. *J Biol Chem.* **1990**, *265*, 15365–15367. [CrossRef]
91. Zaet, A.; Dartevelle, P.; Daouad, F.; Ehlinger, C.; Quilès, F.; Francius, G.; Boehler, C.; Bergthold, C.; Frisch, B.; Prévost, G.; et al. D-Cateslytin, a new antimicrobial peptide with therapeutic potential. *Sci. Rep.* **2017**, *7*, 15199. [CrossRef] [PubMed]
92. Mancino, D.; Kharouf, N.; Scavello, F.; Hellé, S.; Salloum-Yared, F.; Mutschler, A.; Mathieu, E.; Lavalle, P.; Metz-Boutigue, M.-H.; Haïkel, Y. The Catestatin-Derived Peptides Are New Actors to Fight the Development of Oral Candidosis. *Int. J. Mol. Sci.* **2022**, *23*, 2066. [CrossRef] [PubMed]

93. Scavello, F.; Kharouf, N.; Lavalle, P.; Haikel, Y.; Schneider, F.; Metz-Boutigue, M.H. The antimicrobial peptides secreted by the chromaffin cells of the adrenal medulla link the neuroendocrine and immune systems: From basic to clinical studies. *Front. Immunol.* **2022**, *13*, 977175. [CrossRef] [PubMed]
94. Dartevelle, P.; Ehlinger, C.; Zaet, A.; Boehler, C.; Rabineau, M.; Westermann, B.; Strub, J.M.; Cianferani, S.; Haïkel, Y.; Metz-Boutigue, M.H.; et al. D-Cateslytin: A new antifungal agent for the treatment of oral Candida albicans associated infections. *Sci. Rep.* **2018**, *8*, 9235. [CrossRef] [PubMed]

pharmaceutics

Review

Our Experience over 20 Years: Antimicrobial Peptides against Gram Positives, Gram Negatives, and Fungi

Giulio Rizzetto [1,†], Daisy Gambini [1,†], Andrea Maurizi [1], Matteo Candelora [1], Elisa Molinelli [1], Oscar Cirioni [2], Lucia Brescini [2], Andrea Giacometti [2], Annamaria Offidani [1] and Oriana Simonetti [1,*]

1 Clinic of Dermatology, Department of Clinical and Molecular Sciences, Polytechnic University of Marche, 60126 Ancona, Italy
2 Clinic of Infectious Diseases, Department of Biomedical Sciences and Public Health, Polytechnic University of Marche, 60126 Ancona, Italy
* Correspondence: o.simonetti@staff.univpm.it; Tel.: +39-071-596-3494
† These authors contributed equally to this work.

Abstract: Antibiotic resistance is rapidly increasing, and new anti-infective therapies are urgently needed. In this regard, antimicrobial peptides (AMPs) may represent potential candidates for the treatment of infections caused by multiresistant microorganisms. In this narrative review, we reported the experience of our research group over 20 years. We described the AMPs we evaluated against Gram-positive, Gram-negative, and fungi. In conclusion, our experience shows that AMPs can be a key option for treating multiresistant infections and overcoming resistance mechanisms. The combination of AMPs allows antibiotics and antifungals that are no longer effective to exploit the synergistic effect by restoring their efficacy. A current limitation includes poor data on human patients, the cost of some AMPs, and their safety, which is why studies on humans are needed as soon as possible.

Keywords: anti-microbial peptides; antibiotic resistance; Gram-positive Gram-negative; fungi

Citation: Rizzetto, G.; Gambini, D.; Maurizi, A.; Candelora, M.; Molinelli, E.; Cirioni, O.; Brescini, L.; Giacometti, A.; Offidani, A.; Simonetti, O. Our Experience over 20 Years: Antimicrobial Peptides against Gram Positives, Gram Negatives, and Fungi. *Pharmaceutics* **2022**, *15*, 40. https://doi.org/10.3390/pharmaceutics15010040

Academic Editors: Scavello Francesco, Amiche Mohamed and Jean-Eric Ghia

Received: 26 October 2022
Revised: 4 December 2022
Accepted: 12 December 2022
Published: 22 December 2022

1. Introduction

Antibiotic resistance (AR) has for some decades been rapidly increasing and represents, nowadays, one of the world's greatest challenges affecting animal and human health. The emergence of AR can be attributed not only to inappropriate antimicrobial prescriptions in humans but also to the overuse of antibiotics in animal breeding and agriculture [1,2].

Therefore, the identification of new anti-infective therapies is urgently needed. Over the past years, numerous efforts have been made to identify new molecules or new methods that could overcome the growing microbial resistance. In this perspective, antimicrobial peptides (AMPs) may represent potential candidates for the treatment of infections caused by multiresistant microorganisms. Inspired by this research area, our group at the "Polytechnic University of Marche" decided to evaluate the effectiveness of selected new AMPs over the last 20 years.

AMPs are oligopeptides usually composed of 12–50 cationic and hydrophobic amino acids with a positive net charge, representing essential components of innate immunity [3–5]. AMPs display broad-spectrum activity against a wide variety of pathogens, such as yeast, fungi, viruses, and bacteria [6,7].

In particular, many AMPs kill pathogens by interacting with negatively charged bacterial cell membranes; this leads to a change in their electrochemical potential, which generates cell membrane damage. Although several AMPs may also kill pathogens indirectly by modulating host immune responses [8–11], AMPs can also act with other different mechanisms of action, such as inhibiting communication between pathogens. These aspects will be described in the specific paragraphs that follow. Additionally, some AMPs show synergistic interactions with conventional molecules, contributing to the decrease in the selection of antibiotic-resistant bacteria and allowing us to restore sensitivity of conventional treatments [12,13]. The aim of this article is to summarise the 20 year experience of our research. Initially, our team included researchers from the Polytechnic University of Marche affiliated with the Clinic of Infectious Diseases, the Clinic of Surgery, the INRCA Experimental Animal Models for Aging Unit, and the Department of Inorganic Chemistry at the Medical University of Gdansk (Poland). Our scientific production was initially focused on the antimicrobial properties of AMPs in medical device infection, biofilm, and sepsis. With the arrival of researchers from dermatology and the Institute of Pathological Anatomy and Histology, our interest shifted to bacterial and fungal skin diseases. Our main goal was to provide new treatments to overcome multiresistant infections, particularly those caused by methicillin-resistant *Staphylococcus aureus* (MRSA) in chronic and surgical skin wounds, including burns. We studied the effects of AMPs not only for treating resistant bacteria-related infections in the skin but also for wound healing.

Simultaneously, we also continued to evaluate the potential and spectrum of action of selected AMPs against fungi and Gram-positive and Gram-negative bacteria. In the beginning, the experimental model was in vitro on colonies of resistant microorganisms taken from patients and then switched, in the case of effective AMPs, to the in vivo animal model. As our research progressed, we decided to assess the potential impact on wound healing by considering histological features and immunohistochemical markers that could quantify the action of AMPs vs. conventional antibiotics, such as vascular endothelium growth factor (VEGF), matrix metalloproteinases (MMP) expression, growth factors, or their receptors.

In the following paragraphs, we summarised the main areas of research performed, indicating the methods used and the results obtained.

2. AMPs and Biofilm in Medical Devices

Biofilm is characterised by bacterial cells that adhere to a substratum, interface, or each other and are embedded in a matrix of substances produced by the cells themselves. Biofilm often tends to develop on medical devices, in particular long-term silicone catheters such as the central venous catheter (CVC). The biofilm protects bacteria from antimicrobial therapy, leading to frequent failure of conventional antibiotic therapy [14–16].

Due to this issue, in the last few years different novel drug technologies have been studied, including antimicrobial peptides.

2.1. BMAP-28

BMAP-28 is a 27 residue peptide. It has an amidated C-terminus, and it has been shown to have the ability to kill bacteria in vitro. Furthermore, in vivo studies have also demonstrated BMAP-28 efficacy in reducing mortality in different infections [17,18].

Cirioni et al. (2005) assessed the efficacy of BMAP-28 pre-coating in the treatment of *S. aureus* central venous catheter-associated infections using the antibiotic-lock technique. In vitro studies revealed that pretreatment with BMAP-28 and then the use of antibiotics reduced the Minimal Bactericidal Concentration (MBC) values against biofilm. In vivo studies demonstrated that catheters pre-treated with BMAP-28 or high-dose antibiotics have a lower bacterial load compared to catheters with standard-dose antibiotics or without BMAP-28 (from 10^7 to 10^3 CFU/mL and bacteremia from 10^3 to 10^1 CFU/mL). A further significant reduction in bacterial load, from 10^7 to 10^1 CFU/mL, was observed when catheters were impregnated with BMAP-28 and then treated with a higher dose of antibiotics [19].

Another experimental study was performed to evaluate the efficacy of BMAP-28. In particular, the efficacy of BMAP-28 alone and in combination with vancomycin was assessed in animal models of ureteral stent infection induced by *Enterococcus faecalis* and *Staphylococcus aureus*. In vivo studies revealed that BMAP-28 reduced bacterial load (from 8×10^6 to 5×10^4 against *S. aureus* and from 8.7×10^6 to 6.4×10^4 against *E. faecalis*) and enhanced the effect of vancomycin (no bacterial count). This result suggests that the BMAP-28-impregnated ureteral stent has lower rates of infection. In vitro studies support these results [20].

2.2. Citropin 1.1

Citropin 1.1 is a wide-spectrum amphibian antimicrobial cationic peptide produced by the glands of the green tree frog, *Litoria citropa* [21].

An experimental study conducted in 2006 evaluated the efficacy of citropin 1.1, minocycline, and rifampicin in the prevention of *S. aureus* central venous catheter (CVC)-associated infection using the antibiotic-lock technique. In vitro studies show that biofilms were strongly affected by the presence of citropin 1.1, which also acts synergistically with hydrophobic antibiotics. In vivo studies confirm the same results; in fact, citropin 1.1 alone not only reduced bacterial load on the CVC from 10^7 to 10^3 CFU/mL but also enhanced the effect of commonly used antibiotics, reducing bacterial load to 10^1 CFU/mL [22].

2.3. Temporin A

Temporin A is a hydrophobic, basic, antimicrobial peptide amide with antibiotic activity against a wide spectrum of microorganisms, including antibiotic-resistant Gram-positive cocci [23]. Temporin A is thought to act in conjunction with the formation of the ion channel in the bacterial cytoplasmic membrane [24].

Ghiselli et al. investigated the efficacy of temporin A as a prophylactic agent against methicillin sodium-susceptible (MS) and methicillin sodium-resistant (MR) *Staphylococcus epidermidis* vascular graft infection. In vitro studies revealed that both MR and MS were similarly susceptible to temporin A. In vivo studies support these results, showing that the use of a temporin A-soaked Dacron graft in vascular surgery can result in substantial bacterial growth inhibition (from 1.9×10^7 to 3.4×10^3 CFU/mL against *S. epidermidis* MS and from 3.9×10^7 to 6.1×10^3 CFU/mL against MR). Most of the antibiotic prophylactic treatments were helpful; nevertheless, only the association of a temporin A-soaked graft and intraperitoneal vancomycin hydrochloride inhibited bacterial growth for both the MR and MS strains [23].

Another study tested the efficacy of topical temporin A and RNAIII-inhibiting peptide (RIP) compared to rifampicin in preventing *S. epidermidis* and *S. aureus* graft infection in a rat pouch model [25].

RIP is a heptapeptide that has strong activity against *S. aureus* and *S. epidermidis*. Considering its mechanism of action, RIP inhibits cell-cell communication, also known as quorum sensing, preventing bacterial adhesion and virulence [26].

The results of this study showed that the use of temporin A-soaked and RIP-soaked Dacron grafts induced a significant bacterial growth inhibition. In fact, the combination of RIP and temporin A showed the lowest bacterial growth (negative quantitative cultures for VISE4 and from 6×10^7 to 6.9×10^1 CFU/mL for VISA4). More specifically, temporin A had a high antistaphylococcal activity, independent of the level of resistance shown by the isolates. RIP was more effective against staphylococcal strains when used alone than temporin A or rifampicin alone [25].

Giacometti et al. (2004) investigated the efficacy of temporin A soaking in combination with intraperitoneal linezolid in the prevention of vascular graft infection in a rat model of infection with *Staphylococcus epidermidis* with intermediate resistance to glycopeptides (GISE). The in vitro results show that temporin A and linezolid both have high activity against the GISE clinical strain. The in vivo study confirmed the strong activity against *S. epidermidis* of temporin A and linezolid, and it showed that the combination of temporin A with a parenteral antibiotic, such as linezolid (from 6.9×10^6 to 3.8×10^2 CFU/mL with linezolid and to 3.4×10^3 with temporin A), may become a valid opportunity for chemoprophylaxis in vascular surgery [27].

Temporin A, citropin 1.1, CA (1-7)M (2-9)NH2, and Pal-KGK-NH2 were also studied in 2019 for their synergistic activity against methicillin-resistant *Staphylococcus aureus* (MRSA) biofilms developed on polystyrene surfaces (PSS) and central venous catheters (CVC). The study highlighted that antimicrobial peptides have strong activity in inhibiting biofilm formation on both PSS (citropin 1.1 inhibited biofilm formation of all MRSA strains tested) and CVC (citropin 1.1 caused a biomass reduction for the reference strain with an OD_{570} of

0.152 compared with the control). The eradication of preformed biofilms, on the other hand, was more difficult and took 24 hours after contact between the AMP and biofilms. The combination of AMP had synergistic activity, leading to an improvement in the performance of all the peptides in the removal of biofilms [28].

2.4. Other Peptides

Polycationic peptides have been studied in recent years for their antimicrobial activity. Buforin II and ranalexin are polycationic peptides derived from amphibian tissues. Cerioni et al. investigated the efficacy of ranalexin and buforin II in the prevention of vascular prosthetic graft infection due to *Staphylococcus epidermidis* with intermediate glycopeptide resistance.

Both peptides demonstrate strong in vitro activity. In vivo studies demonstrated that buforin II and ranalexin (from 4.9×10^6 to 1.9×10^2 CFU/mL) had a stronger activity than vancomycin (from 4.9×10^6 to 6.2×10^3 CFU/mL) and teicoplanin (from 4.9×10^6 to 5.1×10^4 CFU/mL). In particular, buforin II was able to inhibit bacterial growth completely [29].

Another study compared the activity of the same polycationic peptides to that of rifampicin in the prevention of methicillin-susceptible and methicillin-resistant *Staphylococcus epidermidis* vascular prosthetic graft infections. This study found that polycationic activities against *Staphylococcus epidermidis* were comparable to rifampicin. The combinations of buforin II and ranalexin-coated grafts with cefazolin showed stronger activity against the methicillin-resistant strain (no evidence of infection) than that of the combination of rifampicin-coated grafts and cefazolin (10^2 bacterial growth) [30].

Pal-Lys-NH_2 and Pal-Lys-Lys are lipopeptides with bactericidal activity, and they are effective against different Gram-positive cocci [31].

A study investigated their action alone or in combination with vancomycin in the prevention of prosthesis biofilm in a subcutaneous rat pouch model of staphylococcal vascular graft infection. The results of this study showed that vancomycin (from 6.94 log to 3.65 log CFU/mL) and lipopeptides (from 6.94 log to 3.87 log CFU/mL for Pal-Lys-Lys NH_2 and from 6.94 log to 4.080 log CFU/mL for Pal-Lys-Lys) when used alone had similar activity. The combination of vancomycin with Pal-Lys-Lys-NH_2 had the strongest efficacy (from 6.94 log to 1 log CFU/mL). The in vitro study globally confirms the in vivo one [32].

Distinctin is an antimicrobial peptide with a heterodimeric structure. It has a lytic activity on unilamellar vesicles, suggesting their possible action on bacterial membranes [33].

In a study, the efficacy of distinctin was assessed in the treatment of *Staphylococcus aureus* CVC-associated infection, in particular in inhibiting the attachment of *S. aureus* to CVCs and in increasing its susceptibility to glycopeptides and carbapenems once it is adherent. The in vitro study showed a valid activity of distinctin on the biofilm and the ability to enhance the efficacy of antibiotics when used in combination. In vivo studies confirmed these results; furthermore, treatment with antibiotics and distinctin induced a significant reduction in bacterial loads on the CVC (from 10^6 to 10^1 CFU/mL) with no evidence of bacteriemia [34].

Protegrins are cysteine-rich AMPs and comprise 16–18 amino acids. IB-367 is a synthetic protegrin with bactericidal and fungicidal activity [35].

Ghiselli et al. evaluated the efficacy of a pre-treatment with IB-367 and its capacity for enhancing the efficacy of linezolid on Gram-positive biofilm in animal models of CVC infection. The study showed that IB-367 pre-treatment of CVC enhanced linezolid activity, causing a higher biofilm bacterial load reduction (from 10^6 to 10^1 CFU/mL) and absence of bacteriemia. In conclusion, IB-367 could be considered an interesting adjunctive agent to conventional antibiotics for the treatment of CVC and other medical devices [36].

Cirioni et al. investigated the efficacy of daptomycin and rifampicin alone and in combination in the prevention of vascular graft biofilm formation in a rat pouch model of Staphylococcal infection. Rifampicin is a glycopeptide antibiotic, while daptomycin is a lipopeptide. The results of this study showed that both rifampicin and daptomycin have good efficacy when used alone (from 7.4×10^6 to 3.3×10^2 CFU/mL for daptomycin and from 7.4×10^6 to 4.2×10^3 CFU/mL for rifampicin). When they are used in combination, their efficacy is higher than that of each single compound (from 7.4×10^6 to 10^1 CFU/mL). Moreover, their combination prevented the emergence of rifampicin resistance in adherent bacteria. These results were confirmed by in vitro studies [37].

Another study investigated the efficacy of levofloxacin, cefazolin, and teicoplanin in preventing vascular prosthetic graft infection induced by methicillin-susceptible and methicillin-resistant *Staphylococcus epidermidis*. The three compounds had similar efficacies, but levofloxacin (from 10^6 to 10^3 CFU/mL) showed slightly less efficacy than teicoplanin (from 10^6 to 10^2 CFU/mL) against the methicillin-resistant strain. Furthermore, the results highlighted that the most useful combination for the prevention of late-appearing vascular graft infections is rifampicin-levofloxacin (no infection detected). However, rifampicin-teicoplanin was also very effective (no infection was detected) [38].AMPs, in our experience, have a high efficacy in reducing bacterial load on the surface of medical devices; this efficacy is frequently comparable to that of the most commonly used antibiotics. Furthermore, when AMPs are used in combination with other antibiotics, they increase their efficacy, leading to no evidence of bacteriemia in most cases.

3. AMPs and Gram-Positive Sepsis

Sepsis represents a serious clinical problem given its severity, prevalence, and difficulty in treatment. Specifically, in 50% of cases, sepsis results from Gram-positive infections. The most frequently involved microorganisms are *Staphylococcus aureus* and *Streptococcus epidermidis*. Antibiotic therapy is not always effective, partly because of the increasing prevalence of antibiotic resistance. For this reason, new molecules such as antimicrobial peptides are increasingly being considered [39].

3.1. Distinctin

Distinctin is an amphipathic antimicrobial peptide that has a structure characterised by two different peptide chains connected by a disulfide bridge. It has been isolated from the skin of *Phyllomedusa distincta* and has shown good antimicrobial activity in vitro. The in vitro efficacy was also confirmed in vivo. In fact, this molecule demonstrated efficacy when administered intravenously in neutropenic mouse models infected with *Staphylococcus aureus*, either alone or in combination with other antibiotics. Notably, its efficacy was shown to be highest when administered together with glycopeptides in the absence of toxic events related to the administration of the peptide itself. Distinctin in combination with vancomycin and teicoplanin resulted in the lowest lethality rate in the aforementioned models [39].

3.2. Temporin A

Temporin A has demonstrated the ability to inhibit the production of TNF-alpha, IL-6, and NO by macrophages in mouse models and is active against antibiotic-resistant Gram-positive cocci. Specifically, it has shown efficacy against *Staphylococcus aureus* in mouse models and was particularly high 6 h after injection. The most effective antibiotic used in combination was imipenem (lethality rates of 25% for temporin A, 20% for imipenem, and 10% for temporin A + imipenem). Temporin A is able to facilitate the passage of imipenem across the bacterial membrane by destructuring it, as both act on peptidoglycan. In addition, temporin A appears to induce the migration of human monocytes, neutrophils, and macrophages [40].

3.3. BMAP-28

BMAP-28 has been shown to inhibit TNF-alpha release and NO production. In mouse models, it has shown similar lethality to antibiotics such as clarithromycin and imipenem against *Staphylococcus aureus*. In addition, it appears to have a superior ability to neutralise bacterial products released by Gram-positive bacteria, a positive factor in severe staphylococcal infections when used in combination with other antimicrobial agents [41].

3.4. Magainin II and Cecropin A

Magainin II and Cecropin A are two alpha-helical antimicrobial peptides that have demonstrated in vitro activity and in vivo efficacy against *Staphylococcus aureus* with intermediate resistance to glycopeptides in combination with vancomycin. In particular, the combination of magainin II and vancomycin has been shown to be particularly effective in reducing lethality in murine models of staphylococcal sepsis (lethality of 1/20 vs. 6/20 vancomycin vs. 10/20 magainin II vs. 12/20 cecropin A). These two peptides appear to be able to insert into the cytoplasmic membrane and activate murine bacterial hydrolases, resulting in peptidoglycan damage and cell lysis [42].

3.5. Aurein 1.2

This is an amphipathic, alpha-helical peptide composed of only 13 amino acids. It has demonstrated antimicrobial activity in vitro against Gram-positive cocci at concentrations ranging from 1 to 16 mg/litre and synergistic activity when administered in combination with clarithromycin and minocycline. In particular, aurein 1.2 exerts its action by making the membrane more permeable and less organised, allowing the entry of hydrophobic substrates [43].

Other peptides that demonstrated antimicrobial activity in vitro against Gram positive cocci include palmitol (pal)-lys-lys-NH_2 and pal-lys-lys, uperine 3.6, and citropin 1.1.

The lipopeptides showed in vitro efficacy mainly against enterococci and streptococci compared with Staphylococci and *Rhodococcus equi*. Their efficacy was higher when combined with beta-lactams and vancomycin; they also proved effective against both susceptible and multidrug-resistant germs.

Uperine 3.6 is a broad-spectrum antibiotic peptide isolated from the Australian amphibian *Uperoleia mjobergii*. It consists of only 17 amino acids, and for this reason, represents the smallest of the antibiotic peptides isolated from amphibians. Although most of the antibiotics tested were more effective than uperine 3.6, it was effective against both susceptible and multiresistant germs [44].

Citropin 1.1 is an antimicrobial peptide derived from the Australian frog *Litoria citropa*. It is a small peptide consisting of 16 amino acids produced by both the dorsal and submental glands of *Litoria citropa*. It has been shown to be effective in vitro against 12 nosocomial isolates of *Rhodococcus equi* at concentrations ranging from 2 to 8 mg/L. The peptide presented synergistic activity against this germ when combined with clarithromycin, doxycycline, and rifampicin. In addition, other in vitro studies have demonstrated its efficacy against staphylococci and streptococci at concentrations ranging from 1 to 16 mg/L. Synergy was demonstrated when citropin 1.1 was combined with hydrophobic antibiotics such as clarithromycin and doxyxline [45,46].

4. AMPs and Wound Infection

We report on four studies whose purpose was to evaluate the efficacy of new experimental AMPs in the management of infected surgical wounds in mouse models. The infection was established using MRSA, the most frequent aetiological agent in cSSTIs. In these studies, the role of antimicrobial peptides in infected wounds was evaluated mostly by considering bacterial growth (quantitative cultures of excised tissues) and healing parameters such as collagen organisation, degree of re-epithelialisation, granulation tissue formation, angiogenesis, and VEGF expression. The results were compared with data from control groups, such as animals with no infected wounds (treated and untreated), animals with infected but untreated wounds, and animals treated with other conventional antibiotics.

In 2013, Cirioni et al. studied the activity of the AMP IB-367 (a synthetic protegrin) as an immunomodulator and immune enhancer, evaluating whether pretreatment with this peptide resulted in an enhancement of antibiotic efficacy (daptomycin and teicoplanin) against MRSA wound infection in a mouse model. The main outcome measures were quantitative bacterial culture and analysis of natural killer (NK) cytotoxicity and leukocyte phenotype. Antibiotics alone showed a comparable antimicrobial efficacy, but their association with IB-367 significantly enhanced the antimicrobial activities. When compared to antibiotics alone (2 log reduction), IB-367 plus daptomycin showed a 4 log reduction in bacterial growth, with IB-367 plus daptomycin showing the highest efficacy (reduction in bacterial load of $2.7 \times 10^3 \pm 0.3 \times 10^3$ c.f.u. mL^{-1}). IB-367 action is also related to a modulation of the innate immune response, mainly represented by an increase in NK cell activity (but not NK cell number) and increasing the number of both CD11b and Gr-1 cells 3 days after MRSA challenge, over the levels observed in the respective controls [47].

LL-37 is the only human antimicrobial peptide that belongs to the cathelicidins. LL-37 showed a broad-spectrum against several different pathogens, such as Gram-positive and Gram-negative bacteria, viruses, and fungi [48]. Moreover, LL-37 revealed other biological activities, such as regulation of responses to inflammation and demonstration of an important activity in wound closure and angiogenesis [49].

Simonetti et al. (2021) evaluated the efficacy of synthetic cathelicidin LL-37 in MRSA-infected surgical wounds in mice, in comparison with teicoplanin treatment. The results of the study showed that LL-37 had a stronger effect than teicoplanin on the wound healing process in MRSA-infected mice, although with a lower effect on bacterial culture growth. LL-37 reduced bacterial numbers to $7.1 \times 10^5 \pm 0.6 \times 10^5$ CFU/g, and $6.9 \times 10^2 \pm 0.4 \times 10^2$ CFU/g when combined with topical LL-37, while i.p. teicoplanin produced a bacterial count of $7.4 \times 10^4 \pm 1.0 \times 10^4$ CFU/g and $3.0 \times 10^2 \pm 1.2 \times 10^2$ CFU/g when associated with topical teicoplanin. LL-37, after topical and parenteral administration, enhanced wound closure via stimulation of granulation tissue formation associated with a better angiogenesis process and better organised collagen deposition and reconstitution of the epithelium in comparison with the teicoplanin treatment group [50].

Another study evaluated the efficacy of distinctin, a heterodimeric peptide from the Amazonian frog *Phyllomedusa distincta*, in the management of cutaneous MRSA wound infections in an experimental mouse model. It was found that topical distinctin combined with parenteral teicoplanin inhibited bacterial growth to $3.4 \times 10^1 \pm 1.0 \times 10^1$ (levels comparable with those observed in uninfected animals), but the combination of topical and parenteral teicoplanin proved to be the most effective in reducing bacterial counts ($4.7 \times 10^1 \pm 1.6 \times 10^1$). Furthermore, when compared to mice treated only with teicoplanin, wounded areas of animals treated with distinctin were characterised by a more mature granulation tissue, a more organised and denser type of connective tissue, and a significant reduction in fibrinous exudation [51].

In 2007, Simonetti et al. conducted a study on temporin A, investigating the effect of its topical use in mouse models of MRSA-infected wounds. Temporin-A treatment combined with teicoplanin injection significantly reduced the bacterial load to $0.85 \times 10^1 \pm 0.1 \times 10^1$ CFU/mL. Histological examination showed that infected mice receiving temporin A-soaked Allevyn (with or without teicoplanin) had a higher degree of granulation tissue formation and collagen deposition compared to the other treated groups. A significant increase in serum VEGF expression was observed in mice receiving temporin A topically with or without intraperitoneal teicoplanin [52].

Wound Infection: Commercially Available AMPs and Perspectives

In this section, we mention studies that analyse the management of MRSA wound infections in mouse models with commercially available antimicrobial peptides such as teicoplanin, daptomycin, and dalbavancin.

Ghiselli et al. (2006) wanted to compare the efficacy of topical vs. systemic teicoplanin for the treatment of wound infection with *Staphylococcus aureus* in a mouse model. Data analysis showed that strong inhibition of bacterial growth was achieved in all groups treated with intraperitoneal teicoplanin. However, the highest inhibition of bacterial growth was obtained in the group that received teicoplanin-soaked Allevyn and intraperitoneal teicoplanin. Histological examination showed that each treatment modality was able to reduce the delay in wound repair, but the best results were obtained with teicoplanin-soaked Allevyn, with and without intraperitoneal teicoplanin, associated with a wound remodelling similar to that of not-infected mice (assessing microvessel density, VEGF expression, and granulation tissue formation in tissue sections) [53].

Daptomycin is a bactericidal lipopeptide antibiotic widely used to treat systemic infections caused by Gram-positive cocci [54]. In a study conducted by Simonetti et al. (2017), a mouse model was used to study the in vivo efficacy of daptomycin in the treatment of burn wound infections caused by *S. aureus* and evaluate the wound healing process through morphological and immunohistochemical analysis, compared to teicoplanin. The highest inhibition of infection in terms of bacterial load was achieved in the group that received daptomycin ($2.0 \times 10^3 \pm 0.4 \times 10^3$ CFU/g), which also showed better overall healing with epithelialisation and significantly higher collagen scores than the other groups, and these findings were also confirmed by immunohistochemical data on EGFR and FGF-2. The results of this in vivo study in an animal model showed that daptomycin demonstrated stronger antimicrobial activity than teicoplanin [55]. Moreover, daptomycin, in a previous study, showed synergy in its effect against MRSA when combined with other antibiotics such as tigecycline [56].

Dalbavancin is a semisynthetic lipoglycopeptide active against Gram-positive bacteria and has been approved for the treatment of acute bacterial skin and skin structure infections (ABSSSI) [57]. In a 2020 study conducted by Simonetti O. et al., the aim was to determine the effect of dalbavancin administration on wound healing compared to that of vancomycin and to elucidate if and how EGFR, MMP-1, MMP-9, and VEGF could be involved in its therapeutic mechanisms. A mouse model of MRSA skin infection was established, and mice were treated daily with vancomycin or weekly with dalbavancin at days 1 and 8. Both dalbavancin and vancomycin were effective in reducing the bacterial load ($8.71 \times 10^5 \pm 9.02 \times 10^5$ CFU/mL vs. $8.04 \times 10^6 \pm 7.96 \times 10^6$ CFU/mL, respectively). The wounds treated with dalbavancin showed well-organised granulation tissue with numerous blood vessels, although slightly less than that in the uninfected group. The immunohistochemical staining also showed elevated EGFR and VEGF expression in both treated groups (higher in dalbavancin-treated mice), decreased MMP-1 and MMP-9 levels in uninfected tissue, and in both treated tissues compared with untreated infected wounds. This study revealed faster healing with dalbavancin treatment than might be associated with higher EGFR and VEGF levels, with the lowest values of MMP-9 and MMP-1 expression [58].

5. AMPs and *Enterococcus faecalis* Infection

Enterococci are responsible for multiple nosocomial infections, and they have a high frequency of multidrug infections. The majority of enterococcal infections are caused by *Enterococcus faecalis*, which is often resistant to multiple antibiotics. Thus, it is very important to search for new antimicrobial compounds such as AMPs [59].

Giacometti et al. (2004) evaluated the in vitro activity of temporin A against *E. faecalis* nosocomial isolates, including vancomycin-resistant strains, and its in vitro activity combined with antibiotics. Temporin A demonstrated potent antibacterial activity against *E. faecalis*, including vancomycin-resistant strains, in vitro, especially when combined with amoxicillin/clavulanic acid and imipenem. In conclusion, this peptide could be used in the future as an adjuvant in the therapy for enterococcal infections [60].

Cirioni et al. conducted an experimental study to evaluate both the in vitro and in vivo interaction between the Laur-CKK-NH_2 lipopeptide and daptomycin using two *Enterococcus faecalis* strains with different patterns of susceptibilities. The in vitro experiments showed that the Laur-CKK-NH_2 dimer is able to prevent the emergence of daptomycin resistance. Additionally, for in vivo studies using a mouse model of enterococcal sepsis, the Laur-CKK-NH_2 dimer and daptomycin exhibited the highest efficacy in measuring lethality and bacteremia [61].

6. AMPs and Gram-Negative Bacteria

Infections sustained by multi-drug-resistant (MDR) Gram-negative bacteria represent one of the most important therapeutic challenges, considering that their resistance to antibiotics is expanding from extended-spectrum beta-lactamases and carbapenemases [62] to the *mcr* gene, which is responsible for colistin resistance [63]. This is why new molecules need to be evaluated in order to overcome AMRs. AMPs can also be a valuable aid in the treatment of Gram-negative infections.

6.1. Protegrin-1

Acinetobacter baumannii, in our experience, is a Gram-negative pathogen with a high risk of developing multiple antibiotic resistances, particularly in the hospital setting and in immunocompromised patients. Although it has been shown that the combination of colistin with daptomycin or teicoplanin can make antibacterial therapy effective in a mouse model [64], colistin may not be sufficient in cases of *A. baumannii* MDR. In an in-vitro model of cultures of *A. baumannii* MDR, also resistant to colistin, taken from surgical wounds, the minimum inhibitory concentration (MIC), 2 mcg/mL, and minimum bactericidal concentration (MBC), 8 mcg/mL, of Protegrin-1 (PG-1) were assessed. PG-1 is an 18-amino-acid beta-hairpin AMP belonging to the cathelicidin family, with excellent bactericidal action in monotherapy and excellent synergy with colistin. No resistance to PG-1 developed, but there was also no effect on biofilm. However, PG-1 is proposed as an interesting future perspective in gram-negative MDR infections [65].

6.2. Pexiganan

Another interesting AMP is pexiganan, a 22-amino acid synthetic magainin-based lysine-rich peptide that showed effective action against *A. baumannii* in a mouse model of sepsis. Both the groups treated with pexiganan (1 mg/kg intraperitoneal) and colistin (1 mg/kg intraperitoneal) showed good antibacterial efficacy, but the lowest bacterial count occurred in the pexiganan plus colistin combination group, which also achieved the highest survival rate (90%) [66]. This AMP may also help overcome MDR phenomena involving last-line antibiotics such as colistin.

In addition, pexiganan was shown to be synergistic with tigecycline in a mouse model infected with *Pseudomonas aeruginosa*, making a normally ineffective antibiotic effective against Gram-negative bacteria [67]. This offers new perspectives, considering the possible use of antibiotics that would not normally be effective against Gram-negative bacteria. In another study on a mouse model with urethral stents, the effect of pexiganan and imipenem at sub-MIC concentrations on the biofilm produced by *P. aeruginosa*, a slime producer was evaluated [68]. Studying biofilm production in vitro, the group of mice treated with pexiganan and imipenem showed a marked reduction in adhesion and biofilm expression compared to untreated controls (average reductions of 34 ± 8% and 27 ± 4%, respectively), highlighting a role for this AMP in the management of infections sustained by *P. aeruginosa* capable of producing biofilm.

6.3. Alpha-Helical AMPs

Magainin II and cecropin A, alpha-helical AMPs, were used in vitro and in vivo in a mouse model against *P. aeruginosa* MDR [69]. Magainin II and cecropin A showed a strong antimicrobial action, achieving a significant reduction in plasma endotoxin ($\leq$0.015 ± 0.0 EU/mL and $\leq$0.015 ± 0.0 EU/mL, respectively) and TNF-alpha concentrations (0.38 ± 0.02 ng/mL and 0.44 ± 0.03 ng/mL, respectively) compared to control (38.40 ± 2.89 EU/mL and 145.16 ± 18.32 ng/mL) and rifampicin-treated groups (29.45 ± 3.43 EU/mL and 98.0 ± 8.7 ng/mL). The latter, in monotherapy, showed no action against *P. aeruginosa*, as in other studies [70] while the combination of magainin II and cecropin A was proved significantly effective in reducing bacterial counts and mortality. This study highlights how the combination of AMPs and antibiotics that are normally ineffective against *P. aeruginosa* may be a novel solution for new therapeutic needs.

6.4. Tachyplesin III

Tachyplesin III, a potent disulphide-linked peptide, showed synergistic action in vitro with beta-lactams and colistin against *P. aeruginosa* MDR [71]. In a study by Cirioni et al. [72], the activity and in vivo efficacy of Tachyplesin III, colistin, and imipenem against a multiresistant *P. aeruginosa* strain, were investigated in a murine model of sepsis. Bacteremia levels were significantly lower in the combination therapy groups ($1.1 \times 10^1 \pm 0.1 \times 10^1$ CFU/mL, Tachyplesin III and imipenem), ($4.6 \times 10^1 \pm 0.5 \times 10^1$ CFU/mL colistin and imipenem) than in the single agent groups (control $5.8 \times 10^7 \pm 0.8 \times 10^7$ CFU/mL, Tachyplesin III $3.6 \times 10^3 \pm 0.6 \times 10^3$ CFU/mL), in particular Tachyplesin III with imipenem had the highest efficacy in terms of bacterial lethality, quantitative blood cultures, and plasma levels of lipopolysaccharide, tumour necrosis factor alpha, and interleukin-6. Once again, combination therapy with AMPs and traditional antibiotics proved to be a very useful option. Additionally, in a study with piperacillin/tazobactam (TZP), the authors [73] reported how mice treated with Tachyplesin III in combination with TZP demonstrated the greatest efficacy compared to monotherapy, implying that a urethral stent coated with Tachyplesin III can reduce *P. aeruginosa* bacterial growth by 1,000-fold.

Finally, the effects of Tachyplesin III and clarithromycin were studied in a mouse model of *Escherichia coli* sepsis. It was seen that Tachyplesin III (1 mg/kg intraperitoneally) alone resulted in greater antimicrobial action and a significant reduction in endotoxin and TNF-alpha plasma concentrations compared to the control and clarithromycin (50 mg/kg intraperitoneally) alone groups. The latter showed no antimicrobial activity but resulted in the reduction of endotoxins and TNF-alpha plasma concentrations. The combination group of Tachyplesin III and clarithromycin was seen to be the most effective in all parameters analysed [74].

6.5. LL-37 and Tritrpticin

LL-37, a human cathelicidin, showed an interesting anti-pseudomonas action in neutropenic patients. In a neutropenic mouse model, septic shock was induced by *P. aeruginosa*, and then the groups were randomised into those treated with placebo, imipenem, granulocyte CSF (G-CSF), LL-37 + G-CSF, or imipenem + G-CSF. Although all therapy groups were superior to the control, the LL-37 + G-CSF group was the most effective in preventing sepsis by significantly lowering neutrophil apoptosis in vitro. The authors [75] reported similar results were obtained by *Escherichia coli* 0111:B4 LPS and ATCC 25922 in the murine animal model of sepsis. The authors [76] used treatment groups which consisted of LL-37, polymyxin B, imipenem, or piperacillin vs. placebo. Despite the fact that all treatments reduced lethality, only LL-37 and polymyxin B showed a reduction in endotoxin ($\leq 0.015 \pm 0.0$ EU/mL and $\leq 0.015 \pm 0.0$ EU/mL, respectively, vs. 33.49 ± 4.69 ng/mL piperacillin) and TNF-α plasma levels (0.22 ± 0.01 ng/mL and 0.21 ± 0.01 ng/mL, respectively, vs. 131.12 ± 17.0 ng/mL piperacillin, see Figure 2 from [77]). Furthermore, there were no statistically significant differences in antimicrobial and antiendotoxin activities between LL-37 and polymyxyn B. Given its anti-inflammatory effect, LL-37 is an excellent candidate for the treatment of Gram-negative sepsis. Furthermore, LL-37 was combined with colistin against multidrug-resistant *Escherichia coli*, demonstrating good activity in reducing biofilm formation [77].

Another cathelicidin that has shown activity in vitro against *P. aeruginosa* MDR is tritrpticin, which completely inhibits the procoagulant activity of lipopolysaccharides and shows a synergistic effect with beta lactams [78].

6.6. IB-367

The efficacy of topical IB-367, a 17-amino acid synthetic protegrin, was evaluated in a mouse model [79] of a skin wound infected with *P. aeruginosa* or *E. coli*, both MDR, alone and in combination with colistin or imipenem (intraperitoneal). The group treated with IB-367 plus colistin showed the greatest inhibition of both bacterial strains, demonstrating excellent efficacy. In vitro, IB-367 inhibited both bacterial strains with a rapid killing time of 40 min. Therefore, IB-367 may be an excellent candidate for topical therapy of Gram-negative infected wounds in the future.

6.7. Citropin 1.1

Citropin 1.1 is an amphibian peptide studied alone and in combination with tazobactam-piperacillin (TZP) in a mouse model of *E. coli* sepsis. When compared to controls, all treatment groups—intraperitoneal 1 mg/kg cytropin 1.1, 120 mg/kg TZP, or 1 mg/kg cytropin 1.1 plus 60 mg/kg TZP—reduced lethality. The group with cytropin 1.1 alone showed a significant reduction in plasma endotoxins and inflammatory cytokines, while TZP exerted the opposite effect. The combination of cytropin 1.1 and TZP was most effective in reducing lethality, bacterial growth in blood and peritoneum, and oxidative stress indices in plasma. Citropin 1.1 is therefore an AMP with not only antimicrobial but also immunomodulatory properties and may be an interesting option in conditions of severe Gram-negative infection [80].

All of these AMPs are molecules that showed in vivo action against Gram-negative MDR bacteria, suggesting a possible use to overcome increasing antibiotic resistance, as proposed by other in vitro studies. [79–83] However, only some of these molecules, in our opinion, can be developed for use in humans. Particularly in Gram-negative skin infections, an interesting role could be played by IB-367 as the only topical agent to be combined with traditional therapy. Further studies in patients are needed to evaluate both the efficacy and safety of these molecules in humans.

6.8. Colistin

Colistin, previously studied in combination with pexiganam against Gram-negative bacterial infection [67], was also combined with teicoplanin or daptomycin in an experimental mouse model of multiresistant *Acinetobacter baumannii* infection. The permeabilising effect of colistin on the *A. baumannii* outer membrane allows glycopeptide and lipopeptide molecules to enter, which are normally excluded due to their size, resulting in a better patient outcome in severe infections caused by multiresistant microorganisms like *A. baumannii* (6.7. $\times 10^4 \pm 1.1 \times 10^4$ colistin alone, $5.0 \times 10^9 \pm 1.6 \times 10^9$ daptomycin alone, $7.3 \times 10^9 \pm 1.8 \times 10^9$ teicoplanin alone, $2.9 \times 10^2 \pm 0.4 \times 10^2$ colistin + daptomycin, and $3.1 \times 10^2 \pm 0.2 \times 10^2$ colistin + teicoplanin) [83].

7. AMPs and Fungi

Fungal infections represent one of the most frequent public health problems [84], also considering the progressive increase in resistance to traditional therapies and the side effects of some antimycotics that limit their use, especially in immunocompromised patients [85].

For this reason, it is necessary to evaluate new molecules to expand our therapeutic options. In our experience, we also evaluated the action of AMPs against fungal infections. Some AMPs show both antibacterial and antibiotic action and can therefore be excellent options in the treatment of mixed infections, allowing them to act on different mechanisms of action than current therapies, which mainly affect sterol biosynthesis [84–89].

7.1. IB-367

IB-367 is a protegrin with activity against Gram-negative bacteria as well as fungi. In an in vitro study, the efficacy of IB-367 alone and in combination with fluconazole, itraconazole, and terbinafine was evaluated against strains from patients infected with *Trichophyton mentagrophytes*, *T. rubrumand*, and *Microsporum canis*. In monotherapy, the lowest MIC was for terbinafine and itraconazole, but there was a synergy of 35% with IB-367/fluconazole, 30% with IB-367/ITRA, and 25% with IB-367/TERB [88]. This study suggests that IB-367 may be a molecule capable of increasing the efficacy of currently available antifungal therapies.

In addition, IB-367 showed in vitro a rapid fungicidal action against *Candida* spp., both sensitive and resistant to fluconazole. Synergistic action occurred in 41.6% of cases with fluconazole and 44% of cases with amphotericin B, without antagonism [89]. For these reasons, IB-367 is also a very promising molecule for treating candida infections.

7.2. Lipopeptide PAL-Lys-Lys-NH_2

The short lipopeptide palmitoyl PAL-Lys-Lys-NH_2 (PAL) was evaluated in vitro against *Candida* spp. alone and in combination with fluconazole, amphotericin B, and caspofungin. All drugs showed good activity against *Candida* strains; however amphotericin B and caspofungin had the lowest MIC. PAL showed relevant synergistic action with PAL/fluconazole (81.25%), PAL/amphotericin B (75%), and particularly with PAL/caspofungin (87.5%) [90]. In our opinion, the combination of PAL/caspofungin may be a new therapeutic option in cases of severe candida infection.

In cases of severe *Cryptococcus neoformans* infections, PAL was also effective in vitro, showing synergy in 21.4% of cases with amphotericin B, suggesting its possible use in infected patients to reduce the dosage and side effects of amphotericin B [91].

PAL was also studied in vitro against several clinical isolates of *dermatophytes* [92,93]. PAL and fluconazole showed a lower MIC and lower fungal biomass than gamma-terpinene, a component of tea tree oil. Finally, PAL was superior to fluconazole in reducing hyphal viability against both dermatophytes, suggesting its possible role in the treatment of these fungi as well.

7.3. Tachyplesin III

As seen against Gram-negatives, Tachyplesin III was also evaluated in mycotic dermatophyte infections in vitro. Terbinafine had a significantly lower MIC than Tachyplesin III ($p < 0.001$). The combination of the two therapies showed synergistic activity in 30% of cases, and no antagonism was recorded. Interestingly, both Tachyplesin III and terbinafine significantly reduced growth in *M. canis* ($p < 0.01$) [94]. This AMP could therefore be useful in combination with terbinafine to lower the dose of the antifungal while maintaining efficacy and safety.

7.4. C14-NleRR-NH_2 and C14-WRR-NH_2

Two lipopeptides (C14-NleRR-NH_2 and C14-WRR-NH_2) were studied to assess the antifungal activity against azole-resistant *Aspergillus fumigatus*. Both lipopeptides showed antifungal activity, with MICs ranging from 8 to 16 mg/L, and a dose-dependent effect was confirmed by both time-kill curves and XTT assays. Moreover, microscopy showed that hyphae growth was hampered at concentrations equal to or higher than MICs. Our results showed that both C14-NleRR-NH_2 and C14-WRR-NH_2 are effective against the resistant isolates tested, and this further prompts the research of lipopeptides as alternatives in antifungal therapy [95].

Finally, our experience also includes studies [96,97] evaluating the in vitro efficacy of AMPs on *Pneumocystis carinii* taken from patients with pneumonia. Cecropin P1, magainin II, indolicidin, and ranalexin alone and in combination with macrolides and dihydrofolate reductase inhibitors (DHFRs) were investigated. The four peptides suppressed the growth of *P. carinii* with no change in combination, with the exception of ranalexin, which showed synergistic activity with the macrolide [96]. Furthermore, an in vitro study on cell monolayers revealed that cecropin P1 and magainin II may be effective in inhibiting *P. carinii* growth at non-toxic cell monolayer concentrations [97].

8. Conclusions

In conclusion, our experience shows that AMPs can be a key option for treating infections sustained by multiresistant microorganisms and overcoming resistance mechanisms against currently used antibiotics or antifungals. The combination of AMPs allows antibiotics and antifungals that are no longer effective due to resistance to exploit the synergistic effect by restoring their efficacy. This will be crucial in the near future, considering the growing spread of antibiotic resistance. In conclusion, our experience shows that AMPs can be a key option for treating infections sustained by multiresistant microorganisms and overcoming resistance mechanisms against currently used antibiotics or antifungals. The combination of AMPs allows antibiotics and antifungals that are no longer effective due to resistance to exploit the synergistic effect by restoring their efficacy. Our research has shown that the peptides allow the penetration of antibiotic molecules inside the bacterial bodies, which would otherwise be primarily ineffective against those bacterial species, thus allowing an antibiotic action in some ways "unexpected" as in the case of some beta-lactams, macrolides, or tetracyclines when combined with peptides for the treatment of Gram-negative microorganism infections.

This will be crucial in the near future, considering the growing resistance.

Moreover, our studies and other more recent ones have highlighted the possibility of coating some devices with peptides (we did it by hand by immersing the device, e.g., Dacron, in a peptide solution), and now other researchers do it with covalent bonds. This possibility can be exploited, for example, for orthopedic prostheses, long-lasting catheters, etc. A current limitation is the lack of data on human patients, the high cost of some AMPs, and their safety (which is improving thanks to cytotoxicity studies on cell monolayers). In addition, the problem of the frequent peptides' short half-life must be considered. This issue will have to be addressed in the future by seeking solutions similar to those obtained with glycopeptides such as dalbavancin and oritavancin, glycolipopeptides with prolonged half-lives (250–350 h), allowing once-weekly (dalbavancin) administration or a unique single-dose regimen (oritavancin).

Author Contributions: Conceptualization, O.S. and O.C.; methodology, G.R.; validation, A.G. and A.O.; data curation, E.M. and L.B.; writing—original draft preparation, D.G., A.M., and M.C.; writing—review and editing, G.R.; supervision, O.S.; All authors have read and agreed to the published version of the manuscript.

Funding: This research received no external funding.

Institutional Review Board Statement: The animal study protocol was approved by the Institutional Animal Care Committee of the Ministery of Health and by the Animal Research Ethics Committee of IRCCS-INRCA (Istituto di Ricovero e Cura a Carattere Scientifico–Istituto nazionale di Riposo e Cura per Anziani) 767/2016 Pr 28/07/2016.

Informed Consent Statement: Not applicable.

Data Availability Statement: Not applicable.

Conflicts of Interest: The authors declare no conflict of interest.

References

1. World Health Organization (WHO). Antimicrobial Resistance. Available online: https://www.who.int/news-room/fact-sheets/detail/antimicrobial-resistance (accessed on 13 October 2022).
2. Healthcare Without Harm (HCWH). Antimicrobial Resistance in European Medical Schools. Available online: https://noharm-europe.org/sites/default/files/documents-files/6591/2020-11-18-Antimicrobial-resistance-in-European-medical-schools_WEB.pdf (accessed on 13 October 2022).
3. Van Hoek, M.L. Antimicrobial peptides in reptiles. *Pharmaceuticals* **2014**, *7*, 723–753. [CrossRef] [PubMed]
4. Nawrot, R.; Barylski, J.; Nowicki, G.; Broniarczyk, J.; Buchwald, W.; Goździcka-Józefiak, A. Plant antimicrobial peptides. *Folia Microbiol.* **2014**, *59*, 181–196. [CrossRef] [PubMed]
5. Lehrer, R.I.; Ganz, T. Antimicrobial peptides in mammalian and insect host defence. *Curr. Opin. Immunol.* **1999**, *11*, 23–27. [CrossRef] [PubMed]
6. Kostyanev, T.; Bonten, M.J.; O'Brien, S.; Steel, H.; Ross, S.; François, B.; Tacconelli, E.; Winterhalter, M.; Stavenger, R.A.; Karlén, A.; et al. The Innovative Medicines Initiative's New Drugs for Bad Bugs programme: European public-private partnerships for the development of new strategies to tackle antibiotic resistance. *J. Antimicrob. Chemother.* **2016**, *71*, 290–295. [CrossRef] [PubMed]
7. Carlet, J.; Collignon, P.; Goldmann, D.; Goossens, H.; Gyssens, I.C.; Harbarth, S.; Jarlier, V.; Levy, S.B.; N'Doye, B.; Pittet, D.; et al. Society's failure to protect a precious resource: Antibiotics. *Lancet* **2011**, *378*, 369–371. [CrossRef] [PubMed]

8. Steinstraesser, L.; Kraneburg, U.; Jacobsen, F.; Al-Benna, S. Host defense peptides and their antimicrobial-immunomodulatory duality. *Immunobiology* **2011**, *216*, 322–333. [CrossRef] [PubMed]
9. Hancock, R.E.W.; Nijnik, A.; Philpott, D.J. Modulating immunity as a therapy for bacterial infections. *Nat. Rev. Microbiol.* **2012**, *10*, 243–254. [CrossRef]
10. Hancock, R.E.; Sahl, H.G. Antimicrobial and host-defense peptides as new anti-infective therapeutic strategies. *Nat. Biotechnol.* **2006**, *24*, 1551–1557. [CrossRef]
11. Yeung, A.T.; Gellatly, S.L.; Hancock, R.E. Multifunctional cationic host defence peptides and their clinical applications. *Cell. Mol. Life Sci.* **2011**, *68*, 2161–2176. [CrossRef]
12. Zharkova, M.S.; Orlov, D.S.; Golubeva, O.Y.; Chakchir, O.B.; Eliseev, I.E.; Grinchuk, T.M.; Shamova, O.V. Application of Antimicrobial Peptides of the Innate Immune System in Combination With Conventional Antibiotics-A Novel Way to Combat Antibiotic Resistance? *Front. Cell Infect. Microbiol.* **2019**, *9*, 128. [CrossRef]
13. Li, J.; Fernández-Millán, P.; Boix, E. Synergism between Host Defence Peptides and Antibiotics Against Bacterial Infections. *Curr. Top. Med. Chem.* **2020**, *20*, 1238–1263. [CrossRef] [PubMed]
14. Simonetti, O.; Marasca, S.; Candelora, M.; Rizzetto, G.; Radi, G.; Molinelli, E.; Brescini, L.; Cirioni, O.; Offidani, A. Methicillin-resistant Staphylococcus aureus as a cause of chronic wound infections: Alternative strategies for management. *AIMS Microbiol.* **2022**, *8*, 125–137. [CrossRef] [PubMed]
15. Donlan, R.M.; Costerton, J.W. Biofilms: Survival mechanisms of clinically relevant microorganisms. *Clin. Microbiol. Rev.* **2002**, *15*, 167–193. [CrossRef] [PubMed]
16. Raad, I.; Hanna, H. Intravascular catheters impregnated with antimicrobial agents: A milestone in the prevention of bloodstream infections. *Support Care Cancer.* **1999**, *7*, 386–390. [CrossRef] [PubMed]
17. Benincasa, M.; Skerlavaj, B.; Gennaro, R.; Pellegrini, A.; Zanetti, M. In vitro and in vivo antimicrobial activity of two alpha-helical cathelicidin peptides and of their synthetic analogs. *Peptides* **2003**, *24*, 1723–1731. [CrossRef] [PubMed]
18. Skerlavaj, B.; Gennaro, R.; Bagella, L.; Merluzzi, L.; Risso, A.; Zanetti, M. Biological characterization of two novel cathelicidin-derived peptides and identification of structural requirements for their antimicrobial and cell lytic activities. *J. Biol. Chem.* **1996**, *271*, 28375–28381. [CrossRef] [PubMed]
19. Cirioni, O.; Giacometti, A.; Ghiselli, R.; Bergnach, C.; Orlando, F.; Mocchegiani, F.; Silvestri, C.; Licci, A.; Skerlavaj, B.; Zanetti, M.; et al. Pre-treatment of central venous catheters with the cathelicidin BMAP-28 enhances the efficacy of antistaphylococcal agents in the treatment of experimental catheter-related infection. *Peptides* **2006**, *27*, 2104–2110. [CrossRef] [PubMed]
20. Orlando, F.; Ghiselli, R.; Cirioni, O.; Minardi, D.; Tomasinsig, L.; Mocchegiani, F.; Silvestri, C.; Skerlavaj, B.; Riva, A.; Muzzonigro, G.; et al. BMAP-28 improves the efficacy of vancomycin in rat models of gram-positive cocci ureteral stent infection. *Peptides* **2008**, *29*, 1118–1123. [CrossRef] [PubMed]
21. Doyle, J.; Brinkworth, C.S.; Wegener, K.L.; Carver, J.A.; Llewellyn, L.E.; Olver, I.N.; Bowie, J.H.; Wabnitz, P.A.; Tyler, M.J. nNOS inhibition, antimicrobial and anticancer activity of the amphibian skin peptide, citropin 1.1 and synthetic modifications. The solution structure of a modified citropin 1.1. *Eur. J. Biochem.* **2003**, *270*, 1141–1153. [CrossRef]
22. Cirioni, O.; Giacometti, A.; Ghiselli, R.; Kamysz, W.; Orlando, F.; Mocchegiani, F.; Silvestri, C.; Licci, A.; Chiodi, L.; Lukasiak, J.; et al. Citropin 1.1-treated central venous catheters improve the efficacy of hydrophobic antibiotics in the treatment of experimental staphylococcal catheter-related infection. *Peptides* **2006**, *27*, 1210–1216. [CrossRef]
23. Ghiselli, R.; Giacometti, A.; Cirioni, O.; Mocchegiani, F.; Orlando, F.; Kamysz, W.; Del Prete, M.S.; Lukasiak, J.; Scalise, G.; Saba, V. Temporin A as a prophylactic agent against methicillin sodium-susceptible and methicillin sodium-resistant Staphylococcus epidermidis vascular graft infection. *J. Vasc. Surg.* **2002**, *36*, 1027–1030. [CrossRef]
24. Harjunpää, I.; Kuusela, P.; Smoluch, M.T.; Silberring, J.; Lankinen, H.; Wade, D. Comparison of synthesis and antibacterial activity of temporin A. *FEBS Lett.* **1999**, *449*, 187–190. [CrossRef] [PubMed]
25. Cirioni, O.; Giacometti, A.; Ghiselli, R.; Dell'Acqua, G.; Gov, Y.; Kamysz, W.; Lukasiak, J.; Mocchegiani, F.; Orlando, F.; D'Amato, G.; et al. Prophylactic efficacy of topical temporin A and RNAIII-inhibiting peptide in a subcutaneous rat Pouch model of graft infection attributable to staphylococci with intermediate resistance to glycopeptides. *Circulation* **2003**, *108*, 767–771. [CrossRef] [PubMed]
26. Balaban, N.; Gov, Y.; Bitler, A.; Boelaert, J.R. Prevention of Staphylococcus aureus biofilm on dialysis catheters and adherence to human cells. *Kidney Int.* **2003**, *63*, 340–345. [CrossRef] [PubMed]
27. Giacometti, A.; Cirioni, O.; Ghiselli, R.; Orlando, F.; D'Amato, G.; Kamysz, W.; Mocchegiani, F.; Sisti, V.; Silvestri, C.; Łukasiak, J.; et al. Temporin A soaking in combination with intraperitoneal linezolid prevents vascular graft infection in a subcutaneous rat pouch model of infection with Staphylococcus epidermidis with intermediate resistance to glycopeptides. *Antimicrob. Agents Chemother.* **2004**, *48*, 3162–3164. [CrossRef]
28. Ciandrini, E.; Morroni, G.; Cirioni, O.; Kamysz, W.; Kamysz, E.; Brescini, L.; Baffone, W.; Campana, R. Synergistic combinations of antimicrobial peptides against biofilms of methicillin-resistant Staphylococcus aureus (MRSA) on polystyrene and medical devices. *J. Glob. Antimicrob. Resist.* **2020**, *21*, 203–210. [CrossRef]
29. Giacometti, A.; Cirioni, O.; Ghiselli, R.; Goffi, L.; Mocchegiani, F.; Riva, A.; Scalise, G.; Saba, V. Efficacy of polycationic peptides in preventing vascular graft infection due to Staphylococcus epidermidis. *J. Antimicrob. Chemother.* **2000**, *46*, 751–756. [CrossRef]

30. Giacometti, A.; Cirioni, O.; Ghiselli, R.; Goffi, L.; Mocchegiani, F.; Riva, A.; Scalise, G.; Saba, V. Polycationic peptides as prophylactic agents against methicillin-susceptible or methicillin-resistant Staphylococcus epidermidis vascular graft infection. *Antimicrob. Agents Chemother.* **2000**, *44*, 3306–3309. [CrossRef]
31. Kamysz, W.; Silvestri, C.; Cirioni, O.; Giacometti, A.; Licci, A.; Della Vittoria, A.; Okroj, M.; Scalise, G. In vitro activities of the lipopeptides palmitoyl (Pal)-Lys-Lys-NH(2) and Pal-Lys-Lys alone and in combination with antimicrobial agents against multiresistant gram-positive cocci. *Antimicrob. Agents Chemother.* **2007**, *51*, 354–358. [CrossRef]
32. Cirioni, O.; Giacometti, A.; Ghiselli, R.; Kamysz, W.; Silvestri, C.; Orlando, F.; Mocchegiani, F.; Vittoria, A.D.; Kamysz, E.; Saba, V.; et al. The lipopeptides Pal-Lys-Lys-NH(2) and Pal-Lys-Lys soaking alone and in combination with intraperitoneal vancomycin prevent vascular graft biofilm in a subcutaneous rat pouch model of staphylococcal infection. *Peptides* **2007**, *28*, 1299–1303. [CrossRef]
33. Batista, C.V.; Scaloni, A.; Rigden, D.J.; Silva, L.R.; Rodrigues Romero, A.; Dukor, R.; Sebben, A.; Talamo, F.; Bloch, C. A novel heterodimeric antimicrobial peptide from the tree-frog Phyllomedusa distincta. *FEBS Lett.* **2001**, *494*, 85–89. [CrossRef] [PubMed]
34. Giacometti, A.; Cirioni, O.; Ghiselli, R.; Orlando, F.; Silvestri, C.; Renzone, G.; Testa, I.; Mocchegiani, F.; Della Vittoria, A.; Saba, V.; et al. Distinctin improves the efficacies of glycopeptides and betalactams against staphylococcal biofilm in an experimental model of central venous catheter infection. *J. Biomed. Mater. Res. A* **2007**, *81*, 233–239. [CrossRef] [PubMed]
35. Chen, J.; Falla, T.J.; Liu, H.; Hurst, M.A.; Fujii, C.A.; Mosca, D.A.; Embree, J.R.; Loury, D.J.; Radel, P.A.; Cheng Chang, C.; et al. Development of protegrins for the treatment and prevention of oral mucositis: Structure-activity relationships of synthetic protegrin analogues. *Biopolymers* **2000**, *55*, 88–98. [CrossRef] [PubMed]
36. Ghiselli, R.; Giacometti, A.; Cirioni, O.; Mocchegiani, F.; Silvestri, C.; Orlando, F.; Kamysz, W.; Licci, A.; Nadolski, P.; Della Vittoria, A.; et al. Pretreatment with the protegrin IB-367 affects Gram-positive biofilm and enhances the therapeutic efficacy of linezolid in animal models of central venous catheter infection. *JPEN J. Parenter. Enteral. Nutr.* **2007**, *31*, 463–468. [CrossRef] [PubMed]
37. Cirioni, O.; Mocchegiani, F.; Ghiselli, R.; Silvestri, C.; Gabrielli, E.; Marchionni, E.; Orlando, F.; Nicolini, D.; Risaliti, A.; Giacometti, A. Daptomycin and rifampin alone and in combination prevent vascular graft biofilm formation and emergence of antibiotic resistance in a subcutaneous rat pouch model of staphylococcal infection. *Eur. J. Vasc. Endovasc. Surg.* **2010**, *40*, 817–822. [CrossRef]
38. Giacometti, A.; Cirioni, O.; Ghiselli, R.; Mocchegiani, F.; Riva, A.; Saba, V.; Scalise, G. L'infezione delle protesi vascolari da Staphylococcus epidermidis: Efficacia di vari protocolli di profilassi perioperatoria in un modello animale [Vascular graft infection by Staphylococcus epidermidis: Efficacy of various perioperative prophylaxis protocols in an animal model]. *Infez. Med.* **2001**, *9*, 13–18.
39. Cirioni, O.; Ghiselli, R.; Orlando, F.; Silvestri, C.; De Luca, S.; Salzano, A.M.; Mocchegiani, F.; Saba, V.; Scalise, G.; Scaloni, A.; et al. Efficacy of the amphibian peptide distinctin in a neutropenic mouse model of staphylococcal sepsis. *Crit. Care Med.* **2008**, *36*, 2629–2633. [CrossRef]
40. Cirioni, O.; Giacometti, A.; Ghiselli, R.; Kamysz, W.; Orlando, F.; Mocchegiani, F.; Silvestri, C.; Licci, A.; Łukasiak, J.; Saba, V.; et al. Temporin A alone and in combination with imipenem reduces lethality in a mouse model of staphylococcal sepsis. *J. Infect. Dis.* **2005**, *192*, 1613–1620. [CrossRef]
41. Giacometti, A.; Cirioni, O.; Ghiselli, R.; Bergnach, C.; Orlando, F.; D'Amato, G.; Mocchegiani, F.; Silvestri, C.; Del Prete, M.S.; Skerlavaj, B.; et al. The antimicrobial peptide BMAP-28 reduces lethality in mouse models of staphylococcal sepsis. *Crit. Care Med.* **2004**, *32*, 2485–2490. [CrossRef]
42. Cirioni, O.; Silvestri, C.; Ghiselli, R.; Giacometti, A.; Orlando, F.; Mocchegiani, F.; Chiodi, L.; Vittoria, A.D.; Saba, V.; Scalise, G. Experimental study on the efficacy of combination of alpha-helical antimicrobial peptides and vancomycin against Staphylococcus aureus with intermediate resistance to glycopeptides. *Peptides* **2006**, *27*, 2600–2606. [CrossRef]
43. Giacometti, A.; Cirioni, O.; Riva, A.; Kamysz, W.; Silvestri, C.; Nadolski, P.; Della Vittoria, A.; Łukasiak, J.; Scalise, G. In vitro activity of aurein 1.2 alone and in combination with antibiotics against gram-positive nosocomial cocci. *Antimicrob. Agents Chemother.* **2007**, *51*, 1494–1496. [CrossRef] [PubMed]
44. Giacometti, A.; Cirioni, O.; Kamysz, W.; Silvestri, C.; Licci, A.; D'Amato, G.; Nadolski, P.; Riva, A.; Lukasiak, J.; Scalise, G. In vitro activity and killing effect of uperin 3.6 against gram-positive cocci isolated from immunocompromised patients. *Antimicrob. Agents Chemother.* **2005**, *49*, 3933–3936. [CrossRef] [PubMed]
45. Giacometti, A.; Cirioni, O.; Kamysz, W.; Silvestri, C.; Del Prete, M.S.; Licci, A.; D'Amato, G.; Łukasiak, J.; Scalise, G. In vitro activity and killing effect of citropin 1.1 against gram-positive pathogens causing skin and soft tissue infections. *Antimicrob. Agents Chemother.* **2005**, *49*, 2507–2509. [CrossRef] [PubMed]
46. Giacometti, A.; Cirioni, O.; Kamysz, W.; Silvestri, C.; Del Prete, M.S.; Licci, A.; D'Amato, G.; Lukasiak, J.; Scalise, G. In vitro activity of citropin 1.1 alone and in combination with clinically used antimicrobial agents against Rhodococcus equi. *J. Antimicrob. Chemother.* **2005**, *56*, 410–412. [CrossRef] [PubMed]
47. Cirioni, O.; Silvestri, C.; Pierpaoli, E.; Barucca, A.; Kamysz, W.; Ghiselli, R.; Scalise, A.; Brescini, L.; Castelli, P.; Orlando, F.; et al. IB-367 pre-treatment improves the in vivo efficacy of teicoplanin and daptomycin in an animal model of wounds infected with meticillin-resistant Staphylococcus aureus. *J. Med. Microbiol.* **2013**, *62*, 1552–1558. [CrossRef]
48. Bucki, R.; Leszczyńska, K.; Namiot, A.; Sokołowski, W. Cathelicidin LL-37: A multitask antimicrobial peptide. *Arch. Immunol. Ther. Exp. (Warsz.)* **2010**, *58*, 15–25. [CrossRef]

49. Nagaoka, I.; Tamura, H.; Reich, J. Therapeutic Potential of Cathelicidin Peptide LL-37, an Antimicrobial Agent, in a Murine Sepsis Model. *Int. J. Mol. Sci.* **2020**, *21*, 5973. [CrossRef]
50. Simonetti, O.; Cirioni, O.; Goteri, G.; Lucarini, G.; Kamysz, E.; Kamysz, W.; Orlando, F.; Rizzetto, G.; Molinelli, E.; Morroni, G.; et al. Efficacy of Cathelicidin LL-37 in an MRSA Wound Infection Mouse Model. *Antibiotics* **2021**, *10*, 1210. [CrossRef]
51. Simonetti, O.; Cirioni, O.; Ghiselli, R.; Goteri, G.; Orlando, F.; Monfregola, L.; De Luca, S.; Zizzi, A.; Silvestri, C.; Veglia, G.; et al. Antimicrobial properties of distinctin in an experimental model of MRSA-infected wounds. *Eur. J. Clin. Microbiol. Infect. Dis.* **2012**, *31*, 3047–3055. [CrossRef]
52. Simonetti, O.; Cirioni, O.; Goteri, G.; Ghiselli, R.; Kamysz, W.; Kamysz, E.; Silvestri, C.; Orlando, F.; Barucca, C.; Scalise, A.; et al. Temporin A is effective in MRSA-infected wounds through bactericidal activity and acceleration of wound repair in a murine model. *Peptides* **2008**, *29*, 520–528. [CrossRef]
53. Ghiselli, R.; Cirioni, O.; Giacometti, A.; Scalise, A.; Simonetti, O.; Mocchegiani, F.; Orlando, F.; Goteri, G.; Della Vittoria, A.; Filosa, A.; et al. Comparative efficacy of topical versus systemic teicoplanin in experimental model of wound infections. *J. Surg. Res.* **2008**, *144*, 74–81. [CrossRef] [PubMed]
54. Rizzetto, G.; Molinelli, E.; Radi, G.; Diotallevi, F.; Cirioni, O.; Brescini, L.; Giacometti, A.; Offidani, A.; Simonetti, O. Role of Daptomycin in Cutaneous Wound Healing: A Narrative Review. *Antibiotics* **2022**, *11*, 944. [CrossRef] [PubMed]
55. Simonetti, O.; Lucarini, G.; Orlando, F.; Pierpaoli, E.; Ghiselli, R.; Provinciali, M.; Castelli, P.; Guerrieri, M.; Di Primio, R.; Offidani, A.; et al. Role of Daptomycin on Burn Wound Healing in an Animal Methicillin-Resistant Staphylococcus aureus Infection Model. *Antimicrob. Agents Chemother.* **2017**, *61*, e00606-17. [CrossRef] [PubMed]
56. Silvestri, C.; Cirioni, O.; Arzeni, D.; Ghiselli, R.; Simonetti, O.; Orlando, F.; Ganzetti, G.; Staffolani, S.; Brescini, L.; Provinciali, M.; et al. In vitro activity and in vivo efficacy of tigecycline alone and in combination with daptomycin and rifampin against Gram-positive cocci isolated from surgical wound infection. *Eur J Clin Microbiol Infect Dis.* **2012**, *31*, 1759–1764. [CrossRef]
57. Simonetti, O.; Rizzetto, G.; Molinelli, E.; Cirioni, O.; Offidani, A. Review: A Safety Profile of Dalbavancin for On- and Off-Label Utilization. *Ther. Clin. Risk Manag.* **2021**, *17*, 223–232. [CrossRef]
58. Simonetti, O.; Lucarini, G.; Morroni, G.; Orlando, F.; Lazzarini, R.; Zizzi, A.; Brescini, L.; Provinciali, M.; Giacometti, A.; Offidani, A.; et al. New Evidence and Insights on Dalbavancin and Wound Healing in a Mouse Model of Skin Infection. *Antimicrob. Agents Chemother.* **2020**, *64*, e02062-19. [CrossRef]
59. Cormican, M.G.; Jones, R.N. Emerging resistance to antimicrobial agents in gram-positive bacteria. Enterococci, staphylococci and nonpneumococcal streptococci. *Drugs* **1996**, *51*, 6–12. [CrossRef]
60. Giacometti, A.; Cirioni, O.; Kamysz, W.; D'Amato, G.; Silvestri, C.; Del Prete, M.S.; Licci, A.; Lukasiak, J.; Scalise, G. In vitro activity and killing effect of temporin A on nosocomial isolates of Enterococcus faecalis and interactions with clinically used antibiotics. *J. Antimicrob. Chemother.* **2005**, *55*, 272–274. [CrossRef]
61. Cirioni, O.; Kamysz, E.; Ghiselli, R.; Kamysz, W.; Silvestri, C.; Orlando, F.; Rimini, M.; Brescini, L.; Gabrielli, E.; Marchionni, E.; et al. Lipopeptide Laur-CKK-NH2 dimer preserves daptomycin susceptibility and enhances its activity against Enterococcus faecalis. *J. Antimicrob. Chemother.* **2011**, *66*, 859–862. [CrossRef]
62. Van Duin, D.; Doi, Y. The global epidemiology of carbapenemase-producing Enterobacteriaceae. *Virulence* **2017**, *8*, 460–469. [CrossRef]
63. Simoni, S.; Morroni, G.; Brenciani, A.; Vincenzi, C.; Cirioni, O.; Castelletti, S.; Varaldo, P.E.; Giovanetti, E.; Mingoia, M. Spread of colistin resistance gene mcr-1 in Italy: Characterization of the mcr-1.2 allelic variant in a colistin-resistant blood isolate of Escherichia coli. *Diagn. Microbiol. Infect. Dis.* **2018**, *91*, 66–68. [CrossRef] [PubMed]
64. Cirioni, O.; Simonetti, O.; Pierpaoli, E.; Barucca, A.; Ghiselli, R.; Orlando, F.; Pelloni, M.; Trombettoni, M.M.; Guerrieri, M.; Offidani, A.; et al. Colistin enhances therapeutic efficacy of daptomycin or teicoplanin in a murine model of multiresistant Acinetobacter baumannii sepsis. *Diagn. Microbiol. Infect. Dis.* **2016**, *86*, 392–398. [CrossRef] [PubMed]
65. Morroni, G.; Simonetti, O.; Brenciani, A.; Brescini, L.; Kamysz, W.; Kamysz, E.; Neubauer, D.; Caffarini, M.; Orciani, M.; Giovanetti, E.; et al. In vitro activity of Protegrin-1, alone and in combination with clinically useful antibiotics, against Acinetobacter baumannii strains isolated from surgical wounds. *Med Microbiol Immunol.* **2019**, *208*, 877–883. [CrossRef] [PubMed]
66. Cirioni, O.; Simonetti, O.; Pierpaoli, E.; Barucca, A.; Ghiselli, R.; Orlando, F.; Pelloni, M.; Minardi, D.; Trombettoni, M.M.; Guerrieri, M.; et al. Enhanced Efficacy of Combinations of Pexiganan with Colistin Versus Acinetobacter Baumannii in Experimental Sepsis. *Shock* **2016**, *46*, 219–225. [CrossRef] [PubMed]
67. Cirioni, O.; Simonetti, O.; Morroni, G.; Brescini, L.; Kamysz, W.; Kamysz, E.; Orlando, F.; Pierpaoli, E.; Caffarini, M.; Orciani, M.; et al. Efficacy of Pexiganan Combination with Tigecycline in a Mouse Model of Pseudomonas aeruginosa Sepsis. *Curr. Top. Med. Chem.* **2018**, *18*, 2127–2132. [CrossRef] [PubMed]
68. Cirioni, O.; Silvestri, C.; Ghiselli, R.; Kamysz, W.; Minardi, D.; Castelli, P.; Orlando, F.; Kamysz, E.; Provinciali, M.; Muzzonigro, G.; et al. In vitro and in vivo effects of sub-MICs of pexiganan and imipenem on Pseudomonas aeruginosa adhesion and biofilm development. *Infez. Med.* **2013**, *21*, 287–295. [PubMed]
69. Cirioni, O.; Silvestri, C.; Ghiselli, R.; Orlando, F.; Riva, A.; Mocchegiani, F.; Chiodi, L.; Castelletti, S.; Gabrielli, E.; Saba, V.; et al. Protective effects of the combination of alpha-helical antimicrobial peptides and rifampicin in three rat models of Pseudomonas aeruginosa infection. *J. Antimicrob. Chemother.* **2008**, *62*, 1332–1338. [CrossRef] [PubMed]
70. Giacometti, A.; Cirioni, O.; Barchiesi, F.; Fortuna, M.; Scalise, G. In-vitro activity of cationic peptides alone and in combination with clinically used antimicrobial agents against Pseudomonas aeruginosa. *J. Antimicrob. Chemother.* **1999**, *44*, 641–645. [CrossRef]

71. Cirioni, O.; Giacometti, A.; Kamysz, W.; Silvestri, C.; Riva, A.; Della Vittoria, A.; Abbruzzetti, A.; Lukasiak, J.; Scalise, G. In vitro activities of tachyplesin III against Pseudomonas aeruginosa. *Peptides* **2007**, *28*, 747–751. [CrossRef]
72. Cirioni, O.; Ghiselli, R.; Silvestri, C.; Kamysz, W.; Orlando, F.; Mocchegiani, F.; Di Matteo, F.; Riva, A.; Lukasiak, J.; Scalise, G.; et al. Efficacy of tachyplesin III, colistin, and imipenem against a multiresistant Pseudomonas aeruginosa strain. *Antimicrob. Agents Chemother.* **2007**, *51*, 2005–2010. [CrossRef]
73. Minardi, D.; Ghiselli, R.; Cirioni, O.; Giacometti, A.; Kamysz, W.; Orlando, F.; Silvestri, C.; Parri, G.; Kamysz, E.; Scalise, G.; et al. The antimicrobial peptide tachyplesin III coated alone and in combination with intraperitoneal piperacillin-tazobactam prevents ureteral stent Pseudomonas infection in a rat subcutaneous pouch model. *Peptides* **2007**, *28*, 2293–2298. [CrossRef] [PubMed]
74. Cirioni, O.; Ghiselli, R.; Silvestri, C.; Kamysz, W.; Orlando, F.; Riva, A.; Kamysz, E.; Castelletti, S.; Rocchi, M.; Saba, V.; et al. Efficacy of the combination of tachyplesin III and clarithromycin in rat models of Escherichia coli sepsis. *Antimicrob. Agents Chemother.* **2008**, *52*, 4351–4355. [CrossRef] [PubMed]
75. Cirioni, O.; Ghiselli, R.; Tomasinsig, L.; Orlando, F.; Silvestri, C.; Skerlavaj, B.; Riva, A.; Rocchi, M.; Saba, V.; Zanetti, M.; et al. Efficacy of LL-37 and granulocyte colony-stimulating factor in a neutropenic murine sepsis due to Pseudomonas aeruginosa. *Shock* **2008**, *30*, 443–448. [CrossRef] [PubMed]
76. Cirioni, O.; Giacometti, A.; Ghiselli, R.; Bergnach, C.; Orlando, F.; Silvestri, C.; Mocchegiani, F.; Licci, A.; Skerlavaj, B.; Rocchi, M.; et al. LL-37 protects rats against lethal sepsis caused by gram-negative bacteria. *Antimicrob. Agents Chemother.* **2006**, *50*, 1672–1679. [CrossRef]
77. Morroni, G.; Sante, L.D.; Simonetti, O.; Brescini, L.; Kamysz, W.; Kamysz, E.; Mingoia, M.; Brenciani, A.; Giovanetti, E.; Bagnarelli, P.; et al. Synergistic effect of antimicrobial peptide LL-37 and colistin combination against multidrug-resistant *Escherichia coli* isolates. *Future Microbiol.* **2021**, *16*, 221–227. [CrossRef]
78. Cirioni, O.; Giacometti, A.; Silvestri, C.; Della Vittoria, A.; Licci, A.; Riva, A.; Scalise, G. In vitro activities of tritrpticin alone and in combination with other antimicrobial agents against Pseudomonas aeruginosa. *Antimicrob. Agents Chemother.* **2006**, *50*, 3923–3925. [CrossRef]
79. Simonetti, O.; Cirioni, O.; Ghiselli, R.; Orlando, F.; Silvestri, C.; Mazzocato, S.; Kamysz, W.; Kamysz, E.; Provinciali, M.; Giacometti, A.; et al. In vitro activity and in vivo animal model efficacy of IB-367 alone and in combination with imipenem and colistin against Gram-negative bacteria. *Peptides* **2014**, *55*, 17–22. [CrossRef]
80. Ghiselli, R.; Silvestri, C.; Cirioni, O.; Kamysz, W.; Orlando, F.; Calcinari, A.; Kamysz, E.; Casteletti, S.; Rimini, M.; Tocchini, M.; et al. Protective effect of citropin 1.1 and tazobactam-piperacillin against oxidative damage and lethality in mice models of gram-negative sepsis. *J. Surg. Res.* **2011**, *171*, 726–733.
81. Giacometti, A.; Cirioni, O.; Del Prete, M.S.; Paggi, A.M.; D'Errico, M.M.; Scalise, G. Combination studies between polycationic peptides and clinically used antibiotics against Gram-positive and Gram-negative bacteria. *Peptides* **2000**, *21*, 1155–1160. [CrossRef]
82. Giacometti, A.; Cirioni, O.; Del Prete, M.S.; Barchiesi, F.; Fortuna, M.; Drenaggi, D.; Scalise, G. In vitro activities of membrane-active peptides alone and in combination with clinically used antimicrobial agents against Stenotrophomonas maltophilia. *Antimicrob. Agents Chemother.* **2000**, *44*, 1716–1719. [CrossRef]
83. Giacometti, A.; Cirioni, O.; Greganti, G.; Quarta, M.; Scalise, G. In vitro activities of membrane-active peptides against gram-positive and gram-negative aerobic bacteria. *Antimicrob. Agents Chemother.* **1998**, *42*, 3320–3324. [CrossRef] [PubMed]
84. Burstein, V.L.; Beccacece, I.; Guasconi, L.; Mena, C.J.; Cervi, I.; Chiapello, L.S. Skin Immunity to Dermatophytes: From Experimental Infection Models to Human Disease. *Front. Immunol.* **2020**, *11*, 605644. [CrossRef] [PubMed]
85. Suleyman, G.; Alangaden, G.J. Nosocomial Fungal Infections: Epidemiology, Infection Control, and Prevention. *Infect. Dis. Clin. North Am.* **2021**, *35*, 1027–1053. [CrossRef] [PubMed]
86. Sacheli, R.; Hayette, M.P. Antifungal Resistance in Dermatophytes: Genetic Considerations, Clinical Presentations and Alternative Therapies. *J. Fungi* **2021**, *7*, 983. [CrossRef] [PubMed]
87. Kamysz, W.; Nadolski, P.; Kedzia, A.; Cirioni, O.; Barchiesi, F.; Giacometti, A.; Scalise, G.; Lukasiak, J.; Okrój, M. In vitro activity of synthetic antimicrobial peptides against Candida. *Pol. J. Microbiol.* **2006**, *55*, 303–307.
88. Simonetti, O.; Silvestri, C.; Arzeni, D.; Cirioni, O.; Kamysz, W.; Conte, I.; Staffolani, S.; Orsetti, E.; Morciano, A.; Castelli, P.; et al. In vitro activity of the protegrin IB-367 alone and in combination compared with conventional antifungal agents against dermatophytes. *Mycoses* **2014**, *57*, 233–239. [CrossRef]
89. Barchiesi, F.; Giacometti, A.; Cirioni, O.; Arzeni, D.; Kamysz, W.; Silvestri, C.; Licci, A.; Marigliano, A.; Della Vittoria, A.; Nadolski, P.; et al. In-vitro activity of the synthetic protegrin IB-367 alone and in combination with antifungal agents against clinical isolates of Candida spp. *J. Chemother.* **2007**, *19*, 514–518. [CrossRef]
90. Kamysz, E.; Simonetti, O.; Cirioni, O.; Arzeni, D.; Ganzetti, G.; Campanati, A.; Giacometti, A.; Gabrielli, E.; Silvestri, C.; Kamysz, W.; et al. In vitro activity of the lipopeptide PAL-Lys-Lys-NH2, alone and in combination with antifungal agents, against clinical isolates of Candida spp. *Peptides* **2011**, *32*, 99–103. [CrossRef]
91. Barchiesi, F.; Giacometti, A.; Cirioni, O.; Arzeni, D.; Silvestri, C.; Kamysz, W.; Abbruzzetti, A.; Riva, A.; Kamysz, E.; Scalise, G. In vitro activity of the synthetic lipopeptide PAL-Lys-Lys-NH(2) alone and in combination with antifungal agents against clinical isolates of Cryptococcus neoformans. *Peptides* **2007**, *28*, 1509–1513. [CrossRef]
92. Barchiesi, F.; Silvestri, C.; Arzeni, D.; Ganzetti, G.; Castelletti, S.; Simonetti, O.; Cirioni, O.; Kamysz, W.; Kamysz, E.; Spreghini, E.; et al. In vitro susceptibility of dermatophytes to conventional and alternative antifungal agents. *Med. Mycol.* **2009**, *47*, 321–326. [CrossRef]

93. Simonetti, O.; Arzeni, D.; Ganzetti, G.; Silvestri, C.; Cirioni, O.; Gabrielli, E.; Castelletti, S.; Kamysz, W.; Kamysz, E.; Scalise, G.; et al. In vitro activity of the lipopeptide derivative (Pal-Lys-Lys-NH), alone and in combination with antifungal agents, against clinical isolates of dermatophytes. *Br. J. Dermatol.* **2009**, *161*, 249–252. [CrossRef] [PubMed]
94. Simonetti, O.; Ganzetti, G.; Arzeni, D.; Campanati, A.; Marconi, B.; Silvestri, C.; Cirioni, O.; Gabrielli, E.; Lenci, I.; Kamysz, W.; et al. In vitro activity of Tachyplesin III alone and in combination with terbinafine against clinical isolates of dermatophytes. *Peptides* **2009**, *30*, 1794–1797. [CrossRef] [PubMed]
95. Fioriti, S.; Cirioni, O.; Simonetti, O.; Franca, L.; Candelaresi, B.; Pallotta, F.; Neubauer, D.; Kamysz, E.; Kamysz, W.; Canovari, B.; et al. In Vitro Activity of Novel Lipopeptides against Triazole-Resistant *Aspergillus fumigatus*. *J. Fungi* **2022**, *8*, 872. [CrossRef] [PubMed]
96. Cirioni, O.; Giacometti, A.; Barchiesi, F.; Scalise, G. In-vitro activity of lytic peptides alone and in combination with macrolides and inhibitors of dihydrofolate reductase against Pneumocystis carinii. *J. Antimicrob. Chemother.* **1998**, *42*, 445–451. [CrossRef] [PubMed]
97. Cirioni, O.; Giacometti, A.; Barchiesi, F.; Scalise, G. In-vitro activity of lytic peptides, inhibitors of ion transport systems and ionophorous antibiotics against Pneumocystis carinii. *J. Antimicrob. Chemother.* **1998**, *42*, 141–145. [CrossRef] [PubMed]

pharmaceutics

MDPI

Article

Structure of Lacticaseicin 30 and Its Engineered Variants Revealed an Interplay between the N-Terminal and C-Terminal Regions in the Activity against Gram-Negative Bacteria

Désiré Madi-Moussa [1], Barbara Deracinois [1], Radja Teiar [1], Yanyan Li [2], Marius Mihasan [3], Christophe Flahaut [1], Sylvie Rebuffat [2], Françoise Coucheney [1,*] and Djamel Drider [1,*]

[1] UMR Transfrontalière BioEcoAgro1158, Univ. Lille, INRAE, Univ. Liège, UPJV, JUNIA, Univ. Artois, Univ. Littoral Côte d'Opale, ICV—Institut Charles Viollette, 59000 Lille, France

[2] Laboratory Molecules of Communication and Adaptation of Microorganisms (MCAM, UMR 7245 CNRS-MNHN), National Museum of Natural History (MNHN), CNRS, CP 54, 57 Rue Cuvier, 75005 Paris, France

[3] Biochemistry and Molecular Biology Laboratory, Faculty of Biology, Alexandru Ioan Cuza University of Iasi, Carol I Blvd., No. 20A, 700506 Iasi, Romania

* Correspondence: francoise.coucheney@univ-lille.fr (F.C.); djamel.drider@univ-lille.fr (D.D.)

Abstract: Lacticaseicin 30 is one of the five bacteriocins produced by the Gram-positive *Lacticaseibacillus paracasei* CNCM I-5369. This 111 amino acid bacteriocin is noteworthy for being active against Gram-negative bacilli including *Escherichia coli* strains resistant to colistin. Prediction of the lacticaseicin 30 structure using the Alphafold2 pipeline revealed a largely helical structure including five helix segments, which was confirmed by circular dichroism. To identify the structural requirements of the lacticaseicin 30 activity directed against Gram-negative bacilli, a series of variants, either shortened or containing point mutations, was heterologously produced in *Escherichia coli* and assayed for their antibacterial activity against a panel of target strains including Gram-negative bacteria and the Gram-positive *Listeria innocua*. Lacticaseicin 30 variants comprising either the N-terminal region (amino acids 1 to 39) or the central and C-terminal regions (amino acids 40 to 111) were prepared. Furthermore, mutations were introduced by site-directed mutagenesis to obtain ten bacteriocin variants E6G, T7P, E32G, T33P, T52P, D57G, A74P, Y78S, Y93S and A97P. Compared to lacticaseicin 30, the anti-Gram-negative activity of the N-terminal peptide and variants E32G, T33P and D57G remained almost unchanged, while that of the C-terminal peptide and variants E6G, T7P, T52P, A74P, Y78S, Y93S and A97P was significantly altered. Finally, the N-terminal region was further shortened to keep only the first 20 amino acid part that was predicted to include the first helix. The anti-Gram-negative activity of this truncated peptide was completely abolished. Overall, this study shows that activity of lacticaseicin 30, one of the rare Gram-positive bacteriocins inhibiting Gram-negative bacteria, requires at least two helices in the N-terminal region and that the C-terminal region carries amino acids playing a role in modulation of the activity. Taken together, these data will help to design forthcoming variants of lacticaseicin 30 as promising therapeutic agents to treat infections caused by Gram-negative bacilli.

Keywords: *Escherichia coli*; antimicrobial activity; structure-activity relationship; helical conformation

Citation: Madi-Moussa, D.; Deracinois, B.; Teiar, R.; Li, Y.; Mihasan, M.; Flahaut, C.; Rebuffat, S.; Coucheney, F.; Drider, D. Structure of Lacticaseicin 30 and Its Engineered Variants Revealed an Interplay between the N-Terminal and C-Terminal Regions in the Activity against Gram-Negative Bacteria. *Pharmaceutics* **2022**, *14*, 1921. https://doi.org/10.3390/pharmaceutics14091921

Academic Editors: Scavello Francesco, Jean-Eric Ghia and Amiche Mohamed

Received: 3 July 2022
Accepted: 7 September 2022
Published: 12 September 2022

1. Introduction

The antibiotic crisis is now well acknowledged worldwide as a serious health problem. Antibiotics which enabled saving millions of lives in the world since their discovery are now facing a rapid decrease in efficiency due to the bacterial resistance crisis. The reasons usually reported to explain such a situation include the overuse and misuse of conventional antibiotics, and their inappropriate prescription. Moreover, possibilities to refill the antibiotic pipeline will be very limited in the near future due to reduced economic incentives expected by the major pharmaceutical companies [1,2]. In 2014, antimicrobial resistance (AMR) was estimated to cause 10 million deaths per year by 2050 [3]. To face this overwhelming situation, numerous alternatives to antibiotics have been explored, among which bacteriophage therapy [4,5], predatory bacteria [6], competitive exclusion of pathogens [7] and bacteriocins [8–10] are included. These approaches offer clear advantages, such as their specificity and low detrimental impact on beneficial microbial communities, unlike antibiotics which generally have collateral damage on commensal bacteria [11].

Bacteriocins are ribosomally synthesized antimicrobial peptides (AMPs) produced by Gram-positive and Gram-negative bacteria as well as Archaea [12,13]. They exhibit extensive variations in their molecular masses, inhibitory spectra, modes of action, and mechanisms of biosynthesis, export and self-protection of the producing strains [14]. These AMPs are considered to be significant actors of microbial competitions because of their role in colonizing niches, killing competing strains and their use of cross-talk or quorum-sensing networks within bacterial communities [15–17]. Bacteriocins from Gram-positive bacteria are predominantly produced by lactic acid bacteria (LAB). These bacteriocins, referred to here as LAB-bacteriocins, are safe for cells from the Eukarya domain [18,19]. They show most often a narrow spectra of activity, acting therefore selectively on members of species identical or closely related to the producer, and in rarer cases they exhibit broad spectra, thus targeting other species [20,21]. During the last few years, several classifications of LAB-bacteriocins have been proposed [17,22,23]. Among those, the classification proposed by Alvarez-Sieiro et al. [22], dividing LAB-bacteriocins into three main classes, is largely used. According to this classification, the class I of bacteriocins includes peptides that have undergone extensive post-translational modifications during their biosynthesis, resulting in the introduction of rare amino acids, such as lanthionines that are present in lanthipeptides. Class II includes unmodified bacteriocins having molecular masses below 10 kDa, while class III contains thermo-labile unmodified bacteriocins of more than 10 kDa with a bacteriolytic or non-lytic mechanism of action [23]. Depending on their structural and functional characteristics, many LAB-bacteriocins act on the cytoplasmic membrane of target bacteria by forming pores, leading to the leakage of ions and small essential molecules, or by degrading the cell walls. With continuing research, LAB-bacteriocins have been allocated with further functions such as antiviral activity or inhibition of proliferation of unscheduled and unregulated tumor cell lines [17,24]. Gram-negative bacteria are generally resistant to LAB-bacteriocins due to their outer membrane, which confers upon them supplementary protection against antimicrobial agents. Nonetheless, a limited number of LAB-bacteriocins possessing activity against Gram-negative bacteria, including *Escherichia coli,* have been reported in the literature [25–27]. However, if the mode of action of LAB-bacteriocins against Gram-positive bacteria is globally well documented [8,28], while their action against Gram-negative bacteria remains to be understood.

Here, we focused on *Lacticaseibacillus paracasei* CNCM I-5369, a strain isolated from an Algerian dairy product recently shown to produce five class II bacteriocins endowed with activity at pH 5 against Gram-negative bacteria including *E. coli* strains resistant to colistin [29]. Moreover, we have heterologously produced each of these bacteriocins, which are encoded by *orf010*, *orf012*, *orf023*, *orf030* and *orf038*. The bacteriocin encoded by *orf030*, here referred to as lacticaseicin 30, was obtained in large quantities in the soluble fraction, contrarily to the other produced peptides [30], and was shown to exhibit potent activity against *E. coli.*

In this study, the structure-activity relationship of lacticaseicin 30 was investigated. For this purpose, the predicted secondary structure of this bacteriocin, which includes five helices distributed over the 111 amino acid sequence, was used to design a series of variants, exhibiting either truncated sequences including one helix, two helices or three helices, or specific point mutations. All lacticaseicin 30 variants were assessed for their anti-Gram-negative activity.

2. Materials and Methods

2.1. Bacterial Strains, Plasmids and Culture Conditions

Bacterial strains and plasmids used in this work are given in the supplementary materials (Table S1). *E. coli* strains were grown in Luria–Bertani (LB) broth [31] at 37 °C, with shaking at 160 rpm and, when necessary, ampicillin at 100 µg/mL (Sigma Aldrich, St Louis, MO, USA) was added to the medium. Strains used as target organisms to assess antibacterial activity were cultivated, without shaking, in brain heart infusion (BHI, Sigma Aldrich, Saint Louis, MO, USA) medium at 37 °C for 12–18 h before use.

2.2. Construction of Lacticaseicin 30 Variant Peptides Carrying N-Terminal Part (N-Ter Lacticaseicin 30), or the Central and C-Terminal Parts (C-Ter Lacticaseicin 30) and Their Expression in E. coli Cells

All oligonucleotides used in this work are listed in the supplementary materials (Table S2). The molecular cloning and other standard techniques were used thereof to perform genetic construction of lacticaseicin 30 and variants. N-ter-lacticaseicin 30, C-ter lacticaseicin 30 and N-ter-H1 lacticaseicin 30 plasmids were described by Sambrook et al. [32]. The *orf030*, *orf030-nter*, *orf030-cter* and *orf030-nter-h1* were amplified by PCR using pT7-6his-030 plasmid as template and F-*Bam*HI-030 and R-030-*Hind*III (for *orf030*), F-*Bam*HI-030 and R-Nter_030-*Hind*III (for *orf030-nter*), F-*Bam*HI-Cter_030 and R-030-*Hind*III (for *orf030-cter*) and F-*Bam*HI-030 and R-Nter-H1_030-*Hind*III (for *orf030-nter-h1*) primers. Then, each PCR product was cloned between the *Bam*HI and *Hind*III sites of the pET-32b(+) plasmid.

Phusion High-Fidelity DNA Polymerase, restriction endonucleases and T4 ligase were obtained from ThermoFisher Scientific (Thermo Scientific, Waltham, MA, USA) and used in accordance with the manufacturer's instructions. Plasmids and PCR products were purified using NucleoSpin kits (Macherey-Nagel, Düren, Germany) and the final plasmid constructions were verified by PCR and sequencing (Eurofins, Ebersberg, Germany). The resulting sequences were analyzed using the SnapGenes tool (GSL Biotech LLC, San Diego, CA, USA).

2.3. Construction of Lacticaseicin 30 Variant Plasmids for Expression in E. coli Cells

Each lacticaseicin variant plasmid was generated by site-directed mutagenesis using the pT7-6his-030 plasmid as template and the appropriate primers (Table S2), the QuikChange II Site-Directed Mutagenesis Kit (Agilent Technologies, Santa Clara, CA, USA), and following the recommended instructions.

2.4. Expression and Purification of Lacticaseicin 30 and Its Variants

Each plasmid constructed in this work was expressed in *E. coli* Rosetta and grown at 37 °C in LB broth, supplemented with ampicillin, until reaching the mid-log phase. Expression was then induced by adding 0.5 mM isopropyl β-D-1-thiogalactopyranoside (IPTG, Sigma Aldrich, St Louis, MO, USA), and the cells were incubated for three additional hours at 37 °C with shaking at 160 rpm. Cells were then harvested by centrifugation and resuspended in the Tris–HCl buffer (20 mM Tris–HCl pH 8, 300 mM NaCl). Finally, cells were lysed by sonication (OmniRuptor 4000 Ultrasonic Homogenizer, Omni International, Kennessaw, GA, USA) on ice-cold water and centrifuged at 11,000× *g* for 1 h. The supernatant was loaded onto a nickel resin grafted on a nitrilo-tri-acetic matrix (Protino Ni-NTA Agarose, Machery-Nagel, Düren, Germany) column previously equilibrated with Tris–HCl buffer. The nickel resin was washed with 2 × 10 column volumes of the same buffer supplemented with 30 mM imidazole and the bacteriocins were eluted using 5 column volumes of the previous buffer supplemented with 200 mM imidazole. A desalting step was performed using PD miditrap^{TM} columns (GE Healthcare Life Science, Pollard, UK) to remove imidazole. The histidine-tag was removed with Tev-protease (Sigma Aldrich, St Louis, MO, USA), while the TRX-tag of lacticaseicin 30 peptide variants (Nter-lacticaseicin 30, Cter-lacticaseicin 30 and Nter-H1-lacticaseicin 30) was removed with enterokinase (New England Biolabs, Ipswich, MA, USA). To separate the tag and the peptide without the tag, additional Ni-NTA chromatography was performed. The purity was verified by Tricine-SDS-PAGE [33]. When necessary, desalted bacteriocin suspensions were lyophilized using the freeze dryer Lyolab 3000 (Thermo Scientific, Waltham, MA, USA) following the recommended instructions. The final concentration of each purified peptide was determined with the bicinchoninic acid assay protein kit (BCA, Sigma Aldrich, St Louis, MO, USA), as recommended by the supplier.

2.5. Matrix-Assisted Laser Desorption Ionization Time-of-Flight Mass Spectrometry (MALDI-TOF MS)

Prior to MALDI-TOF-MS analysis, a purified Ser-lacticaseicin 30 suspension was concentrated using Pierce™ C18 tips (Thermo Scientific, Waltham, MA, USA) according to the manufacturers' instructions. MALDI-TOF mass-to-charge (m/z) ratios of Ser-lacticaseicin 30 were obtained using α-cyano-4-hydroxycinnamic acid (HCCA, 10 mg/mL) as matrix on an Autoflex Speed™ (Bruker Daltonics, Bremen, Germany), running Flexcontrol 3.4 software (Bruker Daltonics). The Bruker bacterial test standard (Bruker Daltonics) was used to calibrate the mass spectrometer according to the manufacturer's recommendations. Mass spectra were acquired in the positive linear ion mode across a m/z ratio of 2000–20000 Da using the manufacturer's automatic method MBT_FC.par. The MALDI-TOF-MS spectra corresponded to an accumulation of 3000 laser shots, in 500 shot steps performed randomly on different areas of the spot. MALDI-TOF-MS spectra were managed using FlexAnalysis 3.4 software (Bruker Daltonics).

2.6. Reverse Phase-High Performance Liquid Chromatography (RP-HPLC) Coupled to Electrospray Ionization-Mass Spectrometry (ESI-MS)

Chloroform/methanol precipitation was performed, according to [34], from 200 µL of desalted Ser-lacticaseicin 30 suspension before RP-HPLC-ESI-MS analysis. Precipitated Ser-lacticaseicin 30 was redissolved in 40 µL of 25 mM ammonium bicarbonate buffer and 10 µL was chromatographically separated on an ACQUITY UPLC system (Waters Corporation) using a C18 column (150 × 3.0 mm, 2.6 µm, Uptisphere CS Evolution, Interchim, Montluçon, France). Eluent A was milliQ H_2O containing formic acid (0.1%, v/v) and eluent B was acetonitrile (ACN) containing formic acid (0.1%, v/v). The ACN gradient (flow rate 0.5 mL/min) was as follows: from 5% to 40% eluent B over 45 min, from 40% to 95% eluent B over 5 min, followed by washing and equilibrating procedures with 95% and 5% solvent B for 5 min each, respectively. The eluate was directed into the electrospray ionization source of the Q-TOF Synapt G2-Si™ (Waters Corporation, Manchester, UK). MS analysis was performed in sensitivity, positive ion and data-dependent analysis (DDA) modes. The source temperature was set at 150 °C and the capillary and cone voltages were set to 3000 and 60 V, respectively. MS data were collected for m/z values in the range of 50 to 2000 Da with a scan time of 0.2 s and a lock mass correction with 556.632 m/z, corresponding to simply charged leucine encephalin. The mass spectrum corresponding to the sum of 400 ms acquisition was deconvoluted using UniDec software [35]. The molecular mass range was fixed to an upper limit of 20 kDa while the charge range was fixed between 1 and 50.

2.7. Analysis of the Ser-Lacticaseicin 30 Amino Acid Sequence by Peptide Fingerprinting

In total, 10 µL of native Ser-lacticaseicin 30, redissolved in 25 mM ammonium bicarbonate buffer, were hydrolyzed by 1 µL of trypsin/Lys-C mixture at 0.1 µg/µL for 3 h at 37 °C and again by adding another 1 µL of the enzyme mixture for 16 h at 37 °C and centrifuged for 10 min at 8000× *g*. The pellet was removed and the peptides in the supernatant were subjected to fingerprinting using the Q-TOF Synapt G2-Si™ spectrometer and the same HPLC and MS acquisition conditions as described above. A maximum of 10 precursor ions was chosen for MS/MS analysis with an intensity threshold of 10,000. MS/MS data were collected using the collision-induced dissociation (CID) mode and a scan time of 0.1 s at an energy collision of 8 V to 9 V for low m/z and at a range of 40 V to 90 V for high m/z.

For peptide identification, database searches were performed via PEAKS Studio Xpro software (Bioinformatics Solutions Inc., Waterloo, Canada) using the UniProtKB/Swiss-Prot-TrEMBL databases restricted to *Lactobacillus* (https://www.uniprot.org/, accessed on April 2022). Mass tolerance of 35 ppm and MS/MS tolerance of 0.2 Da were allowed. Data searches were performed, assigning trypsin as protease and three missed cleavage sites were allowed. Variable methionine oxidations were also considered. The relevance of peptide identities was judged according to their identification score returned by PEAKS Studio Xpro using a p-value of 0.05 ($p < 0.05$) and a false discovery rate (FDR) < 0.1%.

2.8. Alphafold2 Structure Prediction of Lacticaseicin 30 and Its Truncated Variants

AlphaFold2 [36] running locally at an Ubuntu 20.04 workstation and the full_dbs preset were used to predict structures based on the corresponding FASTA sequences of lacticaseicin 30, N-ter-lacticaseicin 30, N-ter-H1-lacticaseicin 30 and C-ter-lacticaseicin 30. AMBER99SB [37] relaxation and OpenMM [38] energy minimization were performed and the five models that were generated were ranked based on their pLDDT confidence score. The highest-ranked conformation was selected for each peptide and the structures were visualized using UCSF ChimeraX [39]. Structural alignments were also performed in ChimeraX using the maker command and the following default parameters: chain pairing:bb; Needleman–Wunsch alignment algorithm; BLOSUM-62 similarity matrix (Figure S1).

2.9. Circular Dichroism Spectroscopy

Circular dichroism measurements were performed in a 1 mm quartz cuvette using the JASCO 810 spectrophotometer (Jasco, Tokyo, Japan) without or with 10 mM of sodium dodecyl sulfate (SDS) micelle. Acquisition of the spectra was performed at 25 °C, in the wavelength range 190–260 nm for all samples and under nitrogen atmosphere. Before the measurement, the lyophilized bacteriocin was dissolved in phosphate buffer at pH 7 or pH 5 [40] to obtain a final concentration of 12 µg/mL. All spectra were recorded with appropriate blank subtractions and averaging of three independent measurements. The amounts of structural motifs were calculated using the BeStSel server [41]. Prediction of the secondary and three-dimensional structures of lacticaseicin 30 was performed using the I-Tasser tool [42,43].

2.10. Antimicrobial Activity

The antimicrobial activity of lacticaseicin 30 and its variants (500 µg/mL) was tested by the critical dilution micromethod using different target strains [44]. Samples were acidified to pH 5 and sterilized by filtration. Serial double dilutions of the filtrate samples were made in 200 µL volumes of BHI in a sterile 96-well Falcon microtiter plate (Corning, Tewksbury, MA, USA). Each well was inoculated with 2 µL of overnight culture of the target strain. Then, microplates were incubated at 37 °C for 24 h without agitation. Bacterial growth was estimated by measuring absorbance at 620 nm using a microtiter plate scanner (Biotech Instruments Inc., Winooski, VT, USA). The antibacterial activity was determined in arbitrary units per milliliter (AU/mL) according to the following formula: $2^n \times$ (1000 µL/deposited volume), with n corresponding to the highest number of dilution at which growth inhibition of the sensitive strain is observed [45]. Therefore, the minimum inhibitory concentration (MIC) was directly determined from the bacteriocin activity and defined as the lowest concentration of bacteriocin resulting in no visible turbidity after 24 h of incubation.

3. Results

3.1. Heterologous Expression and Characterization of Lacticaseicin 30

The pT7-6his-030 plasmid [30] was transferred to *E. coli* Rosetta strain. Production of lacticaseicin 30 was induced by addition of isopropylthio-β-galactoside (IPTG). Lacticaseicin 30 was purified by Ni-NTA chromatography, and the 6his-tag was removed using the TEV-protease. The purified and his-tag-depleted bacteriocin, produced by heterologous expression, was characterized by mass spectrometry (MS) (Figure 1) to confirm the bacteriocin amino acid sequence. According to the plasmid construction and the Tev-protease manufacturer's instructions, the histidine-tag removal with Tev-protease leaves a serine residue at the N-terminal position (Ser-lacticaseicin 30), and therefore the calculated average theoretical molecular mass (MM) of the Ser-lacticaseicin is 12,339.2 Da (using ProtParam (https://web.expasy.org/protparam/, accessed on April 2022)). As shown in Figure 1A, the most intense mass signal (mass-to-charge ratio (m/z) of $[M + H]^+$ bacteriocin ions = 12,350.2) and the second main mass signal (m/z = 6178.3) corresponding to $[M + 2H]^{2+}$ bacteriocin ions matched well with the theoretical MM of the Ser-lacticaseicin 30. Consequently, the MALDI-TOF-MS measured MM was assessed as 12,349.2 Da for the heterologous expressed Ser-lacticaseicin 30. To accurately measure the molecular mass of Ser-lacticaseicin 30, this latter was concentrated by organic solvent precipitation, redissolved in ammonium bicarbonate buffer and subjected to reverse-phase high-performance liquid chromatography coupled to an electrospray ionization (ESI) high-resolution mass spectrometer. Figure 1B shows the ESI-MS spectrum and the corresponding deconvoluted mass spectrum, revealing an MM of 12,340 Da, confirming (i) that the purified and his-tag-depleted bacteriocin, produced by heterologous expression is the Ser-lacticaseicin 30, and (ii) that no further proteolytic maturation occurred. The entire amino acid sequence of the heterologously expressed lacticaseicin 30 was confirmed by peptide fingerprinting (Figure 1C). Indeed, after trypsin/Lys-C hydrolysis, the Ser-lacticaseicin-30-issued peptides were subjected to an RP-HPLC-MS/MS analysis, and the obtained LC-MS/MS data were submitted to PEAKS Studio Xpro for peptide identification against the UniProtKB/Swiss-Prot-TrEMBL databases restricted to *Lactobacillus* (accessed on April 2022). As shown in Figure 1C, PEAKS studio Xpro returned the identification of pre-bacteriocin from *Lactobacillus casei* with a protein identification score (−10logP) of 172.53, a sequence coverage of 88% for 20 identified peptides. Only three short-size amino acid sequences were not identified: (i) as expected due to the addition of a serine residue at the N-terminal extremity, the first five N-terminal amino acids; (ii) the six amino acid sequence AEPALR; and (iii) the dipeptide YR in the C-terminal region. Note that the methionine oxidations are due to the fingerprinting experimental procedure, and therefore the methionine residues of purified Ser-lacticaseicin are not oxidized. All together, these MS data evidenced that the amino acid sequence of the heterologously expressed, purified and His-tag-depleted bacteriocin corresponds to Ser-lacticaseicin 30.

The antimicrobial activity of the purified lacticaseicin 30 was tested at pH 5, with and without the 6×his-tag, and the MICs were determined using *E. coli* ATCC 8739 as the target strain. Lacticaseicin 30 displayed the same MIC value of 40 μg/mL, with and without the 6×his-tag.

3.2. Conformational Analysis of Lacticaseicin 30

The secondary and three-dimensional (3D) structures of lacticaseicin 30 were predicted using AlphaFold2 [36], a machine learning approach that was proven to deliver highly accurate and reliable protein structure prediction [46]. As seen on Figure 2, the predicted secondary structure of lacticaseicin 30 is characterized by the presence of five helices (73 amino acids out of 111, i.e., 66%), several connecting coiled regions (38 amino acids out of 111, i.e., 34%) and five potential H-bonds indicating interactions between these distinct helical segments. The predicted largely helical structure was further confirmed by circular dichroism (CD). CD spectra were obtained at pH 7 and 5, in the presence or absence of sodium dodecyl sulfate (SDS) micelles used as a model of the anionic membrane bilayer [47]. The CD spectra showed that lacticaseicin 30 adopts a helical structure at both pH 5 and 7, and in the absence or presence of SDS micelles (Figure 3A). The content of the different secondary structures, in terms of helix, antiparallel and parallel β-sheets and turns, was evaluated using the BestSel software (Figure 3B). The percentage of α-helix in the presence of SDS micelles was 29% and 63% at pH 7 and 5, respectively, while it was estimated at 41% and 54%, respectively, in the absence of SDS micelles. These results show that interaction with a membrane-like environment is not required to trigger the organization of lacticaseicin 30 in a helical structure, which is intrinsically acquired even in buffer. Moreover, they show that an acidic pH environment favors a higher proportion of helix, suggesting it helps in folding the active structure of lacticaseicin 30 (Figure 3B).

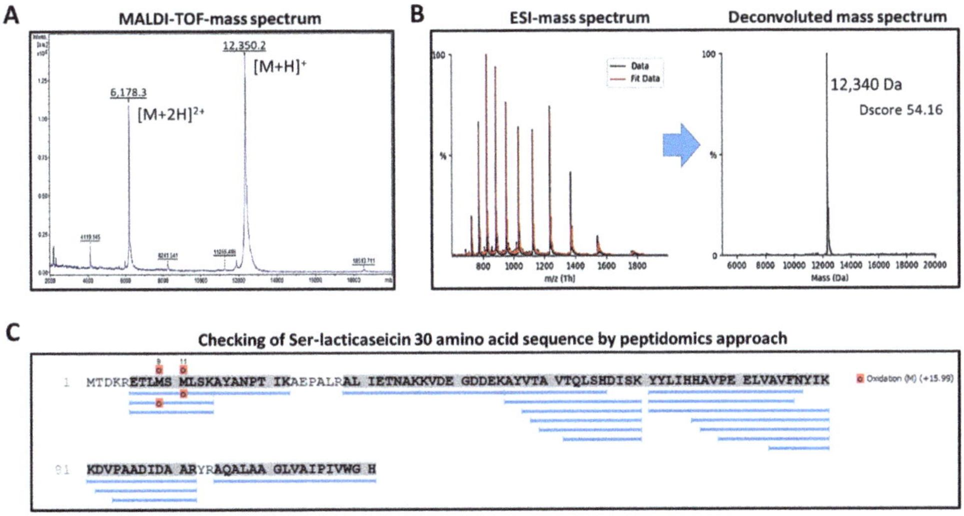

Figure 1. Mass spectrometry analyses of 6his-tag-depleted Ser-lacticaseicin 30 (average molecular mass 12,339.2 Da). (**A**) Linear positive MALDI-TOF mass spectrum illustrating the $[M + H]^+$ and $[M + 2H]^{2+}$ mass signals obtained. The measured MALDI-TOF-MS molecular mass was evaluated as 12,349.2 Da. (**B**) ESI mass spectrum and deconvoluted molecular mass (UniDec DScore: 54.16) revealing a molecular mass of 12,340 Da. (**C**) Peptide mapping of the 6his-tag-depleted Ser-lacticaseicin 30 after tryptic hydrolysis. Blue lines correspond to identified peptides along the lacticaseicin 30 sequence following the bioinformatic retreatment of RP-HPLC-MS/MS data using PEAKS Studio Xpro and the UniProtKB/Swiss-Prot-TrEMBL protein databases restricted to *Lactobacillus* (https://www.uniprot.org/, accessed on April 2022).

3.3. Design of Lacticaseicin 30 Variants

Lacticaseicin 30 variants were designed and constructed in order to locate the regions and amino acids specifically involved in the anti-Gram-negative activity, since such activity is rarely reported for LAB-bacteriocins. To this end, the DNA coding for the N-terminal region carrying the first two helices (from Met1 to Asp39), or the central and C-terminal regions carrying the last three helices (from Glu40 to His111) were cloned into the pET-32b plasmid, leading to the N-ter-lacticaseicin 30 and C-ter-lacticaseicin 30 variants, respectively (Figure 4). In addition, a shortened N-terminal region (from Met1 to Thr20) carrying only the first α-helix, designed as N-ter-H1-lacticaseicin 30 (Figure 4), was also expressed in the same system, considering nonetheless that thioredoxin (TRX) and 6his-tags located upstream of the multicloning site improve solubility of the variant peptides and their purification by the Ni-NTA chromatography.

In addition, several mutants were generated by site-directed mutagenesis, in order to identify the amino acids which, play a role in the anti-Gram-negative activity, and to understand the involvement of the predicted helical segments. To this purpose, charged (Glu, Asp) or aromatic (Tyr) amino acids located inside each α-helix were replaced by uncharged amino acids (Gly, Ser) (Figures 4 and S1). Furthermore, amino acids located in the center of each helix were substituted with proline, which is known for its role in breaking or introducing kinks in helices due to the lack of amide proton (Figures 4 and S1).

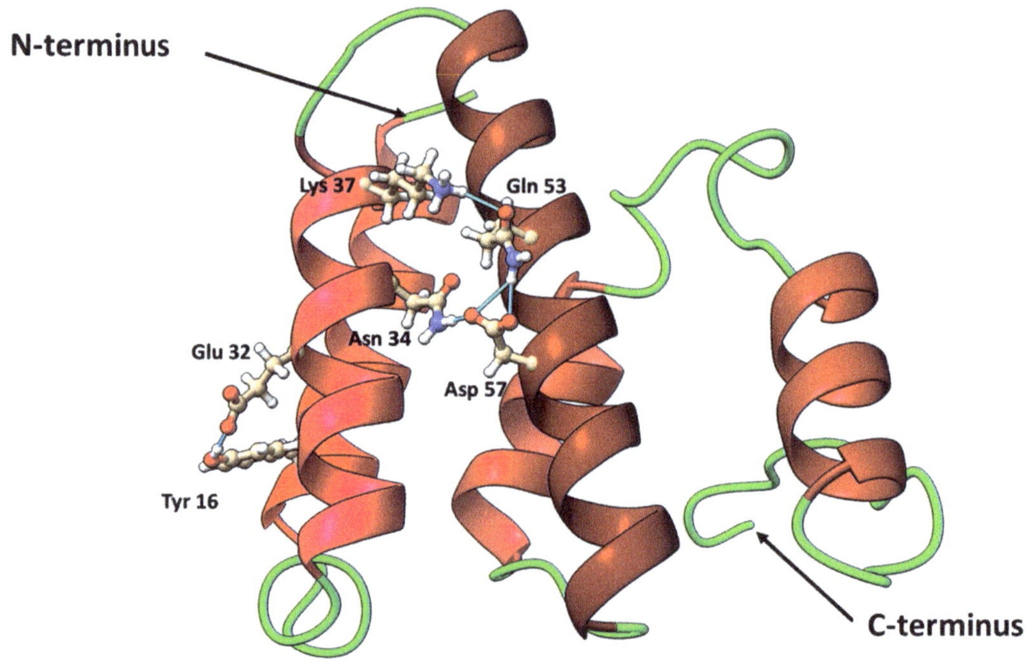

Figure 2. AlphaFold2 predicted structure of lacticaseicin 30 showing five helices (various shades of red) connected by coil regions (green). Five predicted H-bonds (blue) stabilize the position of the three helices.

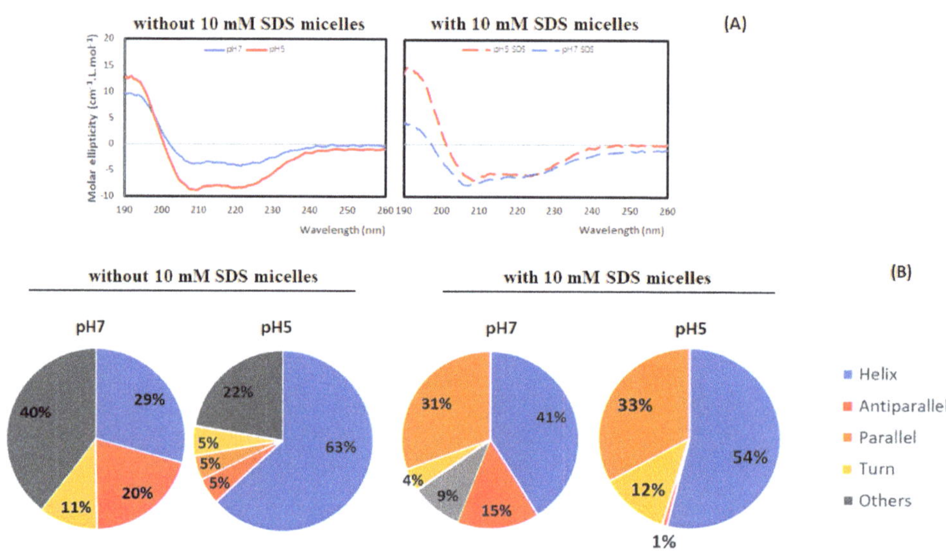

Figure 3. (**A**) Circular dichroism spectra of lacticaseicin 30 at pH 7 (blue) and pH 5 (red) recorded in the absence or presence of 10 mM SDS micelles. (**B**) The secondary structure content (%) of lacticaseicin 30 at pH 7 and 5 was predicted using the BestSel Software.

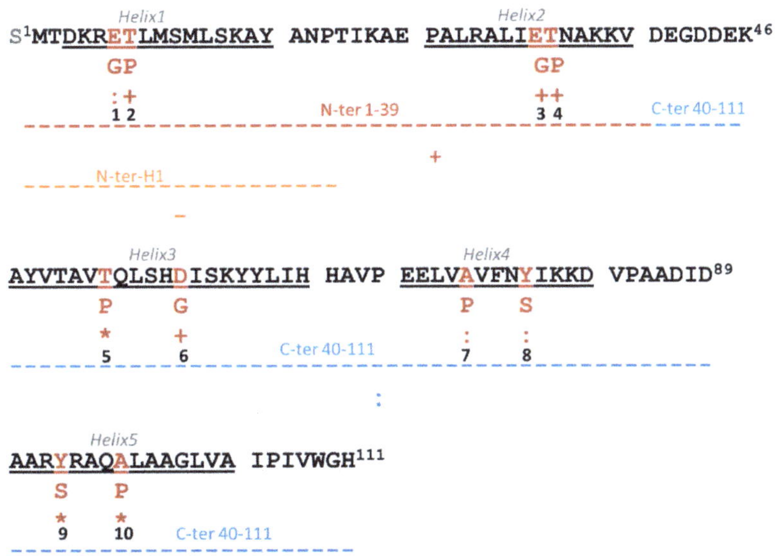

Figure 4. Amino acid sequence and predicted three-dimensional structure of lacticaseicin 30. The predicted helical regions (helix 1 to helix 5) are underlined. Truncated forms of lacticaseicin 30 are shown by colored dotted lines (red: N-ter-Lacticaseicin 30, orange: N-ter-H1-Lacticaseicin 30 and blue: C-ter-Lacticaseicin 30). Mutated amino acids of variants are in red. Lacticaseicin 30 variants generated by directed mutagenesis are designated by a number between 1 and 10 (1: E6G, 2: T7P, 3: E32G, 4: T33P, 5: T52P, 6: D57G, 7: A74P, 8: Y78S, 9: Y93S and 10: A97P). Activities in the same range as native lacticaseicin 30 and total absence of activity are shown by + and −, respectively; * and ":" indicate successively a gradual decrease of activity.

3.4. The N-Terminal Region Is Sufficient to Exert Anti-Gram-Negative Activity

After being produced, purified and quantified, lacticaseicin 30 and its variants, N-ter-lacticaseicin 30, C-ter-lacticaseicin 30 and N-ter-H1-lacticaseicin 30, were assessed for their antibacterial activity, which was determined at pH5 against four Gram-negative bacterial strains and *Listeria innocua* CIP 80.11 as Gram-positive bacterium (Table 1). Lacticaseicin 30 inhibited the growth of all target bacteria (Table 1), and particularly of *Escherichia coli* ATCC 8739 and *Proteus vulgaris* ATCC 33420, against which MIC values were the lowest (40 µg/mL), contrary to those obtained against *Pseudomonas aeruginosa* ATCC 27853 (160 µg/mL). The antibacterial activity against Gram-negative bacteria was clearly evidenced. Remarkably, the shortened variants, N-ter-lacticaseicin 30 and C-ter-lacticaseicin 30, both exhibited significant activity against the aforementioned Gram-negative target bacteria, as well as against the Gram-positive *Listeria innocua* CIP 80.11 (Table 1). The MIC values obtained with the N-ter-lacticaseicin 30 and C-ter-lacticaseicin 30 variants revealed that the N-terminal 1–39 region of lacticaseicin 30 was sufficient for the anti-Gram-negative activity against most of the targets (*P. aeruginosa* was no longer inhibited efficiently), whereas that of the truncated N-terminal peptide N-ter-H1-lacticasecin 30 was fully abolished, arguing that the first helix at the bacteriocin N-terminus is insufficient for the anti-Gram-negative activity.

Table 1. Minimum inhibitory concentrations (MIC) values (µg/mL) of lacticaseicin 30 and its truncated forms.

Peptides	Molecular Mas (Da)	*Escherichia coli* ATCC 8739	*Salmonella enterica* Serotype Newport 6962	*Proteus vulgaris* ATCC 3342	*Pseudomonas aeruginosa* ATCC 27853	*Listeria innocua* CIP 80.11
Lacticaseicin 30	12,252.09	40	80	40	160	100
N-ter Lacticaseicin 30	4236.03	40	80	40	320	100
C-ter Lacticaseicin 30	8034.08	160	160	160	320	100
N-ter-H1 Lacticaseicin 30	1905.27	-	-	-	-	100

3.5. Amino Acids Critical for the Anti-Gram-Negative Activity of Lacticaseicin 30

Different variants of lacticaseicin 30 were generated by site-directed mutagenesis using the pT7-6his-030 plasmid. These substitutions were performed in regions expected to play a key role in the folding of the bacteriocin. Importantly, these amino acid substitutions were introduced in the middle of each predicted α-helix in order to induce a conformational change. The antibacterial activity of lacticaseicin 30 and its variants was measured against the four selected Gram-negative bacteria (Table 2). Overall, amino acid substitutions led to a decrease in the antibacterial activity, except for E32G, T33P and D57G peptides; these MIC values remained unchanged and similar to those of lacticaseicin 30 (Table 2). Furthermore, when Glu6, Tyr78 and Tyr93 are mutated, they induce a net decrease of activity, suggesting a role in the anti-Gram-negative activity. Moreover, the MIC values obtained for lacticaseicin 30 and its truncated forms suggested that the Gram-negative activity (*E. coli*, *Salmonella*, *Proteus* and *Pseudomonas*) require the presence of helix 1. Helix 2 and its acidic residue E32 appeared to not be a determinant for the anti-Gram-negative activity evaluated in this study. Helix 3 and its aromatic residue T52 seems to be weakly required for the anti-Gram-negative activity. Helix 4 and its aromatic residue Y78 is, however, important for the anti-Gram-negative activity. Remarkably, helix 5 is important for activity against *Escherichia coli* and *Proteus* but is less important against *Salmonella* and *Pseudomonas* (Tables 1 and 2).

Table 2. Minimum inhibitory concentrations (µg/mL) of lacticaseicin 30 and its variant peptides.

Peptides	Molecular Mass (Da)	Putative Helix	*Escherichia coli* ATCC 8739	*Salmonella* Newport ATCC 6962	*Proteus vulgaris* ATCC 33420	*Pseudomonas aeruginosa* ATCC 27853
Lacticaseicin 30	12,252.09	-	40	80	40	100
E6G	12,180.03	H1	100	160	100	400
T7P	12,248.10	H1	60	160	60	200
E32G	12,180.03	H2	40	80	40	100
T33P	12,248.10	H2	40	80	40	100
T52P	12,248.10	H3	70	140	70	200
D57G	12,194.06	H3	40	80	40	100
A74P	12,278.13	H4	100	160	100	200
Y78S	12,176.00	H4	100	200	100	400
Y93S	12,176.00	H5	70	140	70	200
A97P	12,278.13	H5	60	100	60	200

4. Discussion

A notable number of AMPs including some bacteriocins exhibit potent and broad-spectrum antimicrobial activities. Some of them are cationic and perturb the permeability of the bacterial membrane bilayers. The presence of positive charges and hydrophobic residues constitutes a common trait of the AMPs which interact with the bacterial membrane, causing pores and membrane depolarization or altering the microbial metabolic pathways or inhibiting DNA/RNA/protein synthesis [40,48–50]. Bacteriocins can be bacteriostatic or bactericidal, induce a rapid killing effect and are thought to have a lower propensity to develop resistance than conventional antibiotics [10,20,49,51]. Overall, the modes of action of LAB-bacteriocins acting against Gram-positive bacteria have been well studied [8,28], but those used by LAB-bacteriocins acting against Gram-negative bacteria are much less documented [25,26,29], and remain to be fully explored.

Recently, we have isolated the *L. paracasei* CNCM I-5369 strain, and shown its potential to produce five distinct class II bacteriocins and inhibit Gram-negative bacteria, including strains of *E. coli* resistant to colistin [29]. Here, we focused on lacticaseicin 30, one of these bacteriocins, which was then used as a model to understand this original activity. Of note, its antibacterial activity may result from the combined action between the peptide itself and an acidic pH (pH 5) [29,30]. According to the classification proposed by Alvarez-Sieiro et al. [22], lacticaseicin 30 appeared to have the characteristics of class IIc bacteriocins. Indeed, this bacteriocin is synthesized without a leader sequence and does not undergo post-translational modifications, suggesting it could be a novel leaderless bacteriocin. Pérez-Ramos et al. [21] reported that leaderless bacteriocins disrupt the cell membrane of target bacteria and most of them do not require any docking molecule for their antimicrobial activity.

As with a few AMPs already reported [52,53], lacticaseicin 30 contains a higher number of negatively charged residues (Asp and Glu) than of positively charged ones (Arg and Lys), with an isoelectric point (p*I*) of 6.05, and many hydrophobic residues. At a neutral pH (pH > p*I*), the peptide is anionic while at an acidic pH (pH < p*I*), it is cationic. As previously mentioned, hydrophobic and cationic residues are one of the main characteristics of AMPs, and the presence of cationic residues can mediate interactions with negatively charged bacterial lipids, while the hydrophobic residues could contribute to the membrane perturbation [54].

The lacticaseicin 30 amino acid sequence is structurally organized into five distinct helices (H1 to H5) (Figure 2), based on the AlphaFold2 predictions and the circular dichroism spectroscopy data obtained in the absence or presence of SDS micelles at pH 5 or 7 (Figure 3).

As indicated in Table 1, the activity against Gram-negative bacteria appeared to be exerted in a strain-dependent manner. On the other hand, we observed an increase of the MIC values with the shortened peptide carrying the C-terminal region of lacticaseicin 30 (C-ter-lacticaseicin 30), in comparison with the native peptide (Table 1). Accordingly, the antibacterial susceptibility has decreased 4-fold against *E. coli* ATCC 8739 and 2-fold against *Salmonella enterica* serovar Newport ATCC 6962 and *Proteus vulgaris* ATCC 33420. Remarkably, these shortened analogs displayed similar activity against *Pseudomonas aeruginosa* ATCC 27853. It is worthy of note that antibacterial assays performed on different *E. coli* strains carrying or not carrying modifications on their LPS were conducted with novel bacteriocins including thereof the class II lacticaseicin 30 [29], and the results obtained did not suggest LPS as the main target of these novel class II bacteriocins. Nonetheless, the activity against *Listeria innocua* CIP 80.11, used as the Gram-positive target strain, remained unchanged. These results delineate the dual role of the N-terminal region composed of helix 1 and helix 2, while the C-terminal region and more particularly helices 4 and 5 could modulate the antibacterial activity according to the target bacteria, suggesting another mechanism of action that remains to be determined. This study is in line with that of Van Kraaij et al. [55], who showed the role of the C-terminal region of nisin across the membrane. Similarly, Johnsen et al. [56] and Rihakova et al. [57] revealed the involvement of the C-terminal region of pediocin-like bacteriocins (class IIa bacteriocins) in determining their antibacterial spectrum.

To gain further insights on the antibacterial activity of lacticaseicin 30, another analog named N-ter-H1-lacticaseicin 30, consisting of a shortened N-terminal region carrying the 1 to 20 amino acids and including only the first predicted α-helix, was designed, expressed and produced. Interestingly, the antibacterial activity remained unchanged against the Gram-positive target strain, whereas it was completely abolished against the Gram-negative target bacteria (Table 1). This result argues that at least two helices located in the N-terminal region are required for activity against Gram-negative bacteria. Furthermore, substitutions of selected amino acids have been performed by site-directed mutagenesis with the aim to understand their roles in the global antibacterial activity. Site-directed mutation refers to the redesign of natural antimicrobial peptides by adding, deleting or replacing one or several amino acid residues [58]. Therefore, two types of key mutations have been created. The first one consisted of replacing glutamic acid, aspartic acid or tyrosine by glycine or serine, and the second consisted of replacing threonine or alanine by proline. These mutations have been created inside each predicted α-helix. Then, the peptide variants E6G, T7P, D57G, A74P, Y78S, Y93S, A97P, E32G, T33P and T52P were tested for their activities against the target Gram-negative bacteria. Overall, they exhibited a loss in the antibacterial activity except for E32G, T33P and T52P, for which anti-Gram-negative activity remained unchanged (Table 2). Proline is a non-polar amino acid and proline-rich AMPs act differently from other AMPs. Indeed, some proline-rich AMPs have been shown to enter the bacterial cytoplasm through the inner membrane transporter SbmA instead of killing bacteria through membrane disruption [59]. Once in the cytoplasm, some proline-rich AMPs target ribosomes and block the binding of aminoacyl-tRNA to the peptidyltransferase center and interfere with protein synthesis [60], while others bind and inhibit DnaK [61]. In the case of lacticaseicin 30, which is devoid of proline residues, the substitution of T7, T52, A74 or A97 with prolines decreases the activity of the corresponding peptides, indicating that the mechanism implied in the anti-Gram-negative activity of lacticaseicin 30 is perturbed by the presence of proline. On the other hand, glycine is generally classified as a non-polar amino acid that induces flexibility of the peptide backbones [62,63].

5. Conclusions

To sum up, lacticaseicin 30 is predicted to adopt a secondary structure characterized by five helices. The generation of peptide variants carrying single point mutations, or truncated sequences, enabled some amino acids that play a major role in the structure of lacticaseicin 30 and its activity against Gram-negative bacteria, including strains of *E. coli* resistant to colistin. Moreover, this study permitted us to establish that at least two helices located in the N-terminal region are required for the anti-Gram-negative activity, while the C-terminal region would serve as a modulator of the activity, conferring its selectivity. These promising achievements open a new avenue in the characterization of LAB-bacteriocins endowed with activity against Gram-negative bacteria. Further experiments consisting of designing novel variants with enhanced antibacterial activity directed especially against resistant strains constitute our next goal.

Supplementary Materials: The following supporting information can be downloaded at: https://www.mdpi.com/article/10.3390/pharmaceutics14091921/s1, Table S1: Bacterial strains and plasmids used in this study; Table S2: Sequences of oligonucleotide primers used in this study; Figure S1: Structural alignment of the predicted structures of native lacticaseicin 30 (brown) and its truncated derivatives N-ter-lacticaseicin 30 (transparent red, RMSD between 37 pruned atom pairs 0.381 angstroms; across all 39 pairs: 1.535 angstroms), N-ter-H1- lacticaseicin 30 (transparent green, RMSD between 20 pruned atom pairs 0.797 angstroms; across all 20 pairs: 0.797 angstroms) and C-ter-lacticaseicin 30 (transparent blue, RMSD between 31 pruned atom pairs 0.888 angstroms; across all 73 pairs 8.364 angstroms). Substituted amino acids in the variant peptides are visible as balls and sticks in their position in the native peptide.

Author Contributions: All authors contributed to the study conception and design. Material preparation, data collection and analysis were performed by D.M.-M., B.D., R.T., Y.L. and M.M. The first draft of the manuscript was written by D.M.-M. and all authors commented on previous versions of the manuscript. All authors have read and agreed to the published version of the manuscript.

Funding: This research was funded by la Région des Hauts-de-France, through ALIBIOTECH CPER/FEDER 2016/2021.

Institutional Review Board Statement: Not applicable.

Informed Consent Statement: Not applicable.

Data Availability Statement: All data generated or analyzed during this study are included in this published article.

Acknowledgments: The authors are grateful to Yanath Belguesmia and Marc Maresca for the critical reading of the manuscript. The MALDI-TOF/MS experiments were performed on the REALCAT platform funded by the French government grant managed by the French National Research Agency (ANR) as part of the "Investments for the Future" program (ANR-11-EQPX-0037). We also acknowledge the Hauts-de-France region, the ERDF, the Ecole Centrale de Lille, and the Foundation Centrale Initiatives for financial support with the acquisition of the REALCAT platform equipment.

Conflicts of Interest: The authors declare no conflict of interest.

References

1. Michael, C.A.; Dominey-Howes, D.; Labbate, M. The antimicrobial resistance crisis: Causes, consequences, and management. *Front. Public Health* **2014**, *2*, 145. [CrossRef] [PubMed]
2. Viswanathan, V.K. Off-Label Abuse of Antibiotics by Bacteria. *Gut. Microbes* **2014**, *5*, 3–4. [CrossRef] [PubMed]
3. O'Neill, J. *Tackling Drug-Resistant Infections Globally: Final Report and Recommendations*; Review on Antimicrobial Resistance: London, UK, 2016.
4. Zalewska-Piątek, B.; Piątek, R. Bacteriophages as Potential Tools for Use in Antimicrobial Therapy and Vaccine Development. *Pharmaceuticals* **2021**, *14*, 331. [CrossRef] [PubMed]
5. Chan, B.K.; Abedon, S.T.; Loc-Carrillo, C. Phage Cocktails and the Future of Phage Therapy. *Future Microbiol.* **2013**, *8*, 769–783. [CrossRef]
6. Kadouri, D.E.; To, K.; Shanks, R.M.Q.; Doi, Y. Predatory Bacteria: A Potential Ally against Multidrug-Resistant Gram-Negative Pathogens. *PLoS ONE* **2013**, *8*, e63397. [CrossRef]
7. Schneitz, C. Competitive Exclusion in Poultry—30 Years of Research. *Food Control* **2005**, *16*, 657–667. [CrossRef]
8. Cotter, P.D.; Ross, R.P.; Hill, C. Bacteriocins—A Viable Alternative to Antibiotics? *Nat. Rev. Microbiol.* **2013**, *11*, 95–105. [CrossRef]
9. Ołdak, A.; Zielińska, D. Bacteriocins from Lactic Acid Bacteria as an Alternative to Antibiotics. *Postepy Hig. Med. Dosw. Online* **2017**, *71*, 328–338. [CrossRef]
10. Soltani, S.; Hammami, R.; Cotter, P.D.; Rebuffat, S.; Said, L.B.; Gaudreau, H.; Bédard, F.; Biron, E.; Drider, D.; Fliss, I. Bacteriocins as a New Generation of Antimicrobials: Toxicity Aspects and Regulations. *FEMS Microbiol. Rev.* **2021**, *45*, fuaa039. [CrossRef]
11. Allen, H.K. *Alternatives to Antibiotics: Why and How*; NAM Perspectives: Washington, DC, USA, 2017.
12. *Prokaryotic Antimicrobial Peptides: From Genes to Applications*; Drider, D.; Rebuffat, S. (Eds.) Springer: New York, NY, USA, 2011; ISBN 978-1-4419-7691-8.
13. Flaherty, R.A.; Freed, S.D.; Lee, S.W. The Wide World of Ribosomally Encoded Bacterial Peptides. *PLoS Pathog.* **2014**, *10*, e1004221. [CrossRef]
14. Salazar, F.; Ortiz, A.; Sansinenea, E. Characterisation of Two Novel Bacteriocin-like Substances Produced by *Bacillus amyloliquefaciens* ELI149 with Broad-Spectrum Antimicrobial Activity. *J. Glob Antimicrob. Resist.* **2017**, *11*, 177–182. [CrossRef] [PubMed]
15. Yang, S.-C.; Lin, C.-H.; Sung, C.T.; Fang, J.-Y. Antibacterial Activities of Bacteriocins: Application in Foods and Pharmaceuticals. *Front. Microbiol.* **2014**, *5*, 241. [CrossRef]
16. Dobson, A.; Cotter, P.D.; Ross, R.P.; Hill, C. Bacteriocin Production: A Probiotic Trait? *Appl. Environ. Microbiol.* **2012**, *78*, 1–6. [CrossRef]
17. Chikindas, M.L.; Weeks, R.; Drider, D.; Chistyakov, V.A.; Dicks, L.M. Functions and Emerging Applications of Bacteriocins. *Curr. Opin. Biotechnol.* **2018**, *49*, 23–28. [CrossRef] [PubMed]
18. Belguesmia, Y.; Madi, A.; Sperandio, D.; Merieau, A.; Feuilloley, M.; Prévost, H.; Drider, D.; Connil, N. Growing Insights into the Safety of Bacteriocins: The Case of Enterocin S37. *Res. Microbiol.* **2011**, *162*, 159–163. [CrossRef] [PubMed]
19. Eveno, M.; Savard, P.; Belguesmia, Y.; Bazinet, L.; Gancel, F.; Drider, D.; Fliss, I. Compatibility, Cytotoxicity, and Gastrointestinal Tenacity of Bacteriocin-Producing Bacteria Selected for a Consortium Probiotic Formulation to Be Used in Livestock Feed. *Probiotics Antimicrob. Proteins* **2021**, *13*, 208–217. [CrossRef] [PubMed]
20. Simons, A.; Alhanout, K.; Duval, R.E. Bacteriocins, Antimicrobial Peptides from Bacterial Origin: Overview of Their Biology and Their Impact against Multidrug-Resistant Bacteria. *Microorganisms* **2020**, *8*, 639. [CrossRef]
21. Pérez-Ramos, A.; Madi-Moussa, D.; Coucheney, F.; Drider, D. Current Knowledge of the Mode of Action and Immunity Mechanisms of LAB-Bacteriocins. *Microorganisms* **2021**, *9*, 2107. [CrossRef]
22. Alvarez-Sieiro, P.; Montalbán-López, M.; Mu, D.; Kuipers, O.P. Bacteriocins of Lactic Acid Bacteria: Extending the Family. *Appl. Microbiol. Biotechnol.* **2016**, *100*, 2939–2951. [CrossRef]
23. Cotter, P.D.; Hill, C.; Ross, R.P. Bacteriocins: Developing Innate Immunity for Food. *Nat. Rev. Microbiol.* **2005**, *3*, 777–788. [CrossRef]
24. Drider, D.; Bendali, F.; Naghmouchi, K.; Chikindas, M.L. Bacteriocins: Not Only Antibacterial Agents. *Probiotics Antimicrob. Proteins* **2016**, *8*, 177–182. [CrossRef] [PubMed]

25. Messaoudi, S.; Kergourlay, G.; Dalgalarrondo, M.; Choiset, Y.; Ferchichi, M.; Prévost, H.; Pilet, M.-F.; Chobert, J.-M.; Manai, M.; Dousset, X. Purification and Characterization of a New Bacteriocin Active against Campylobacter Produced by *Lactobacillus salivarius* SMXD51. *Food Microbiol.* **2012**, *32*, 129–134. [CrossRef] [PubMed]
26. Stern, N.J.; Svetoch, E.A.; Eruslanov, B.V.; Perelygin, V.V.; Mitsevich, E.V.; Mitsevich, I.P.; Seal, B.S. Isolation of a *Lactobacillus salivarius* Strain and Purification of Its Bacteriocin, Which Is Inhibitory to *Campylobacter jejuni* in the Chicken Gastrointestinal System. *Antimicrob. Agents Chemother* **2006**, *50*, 3111–3116. [CrossRef] [PubMed]
27. Todorov, S.D.; Dicks, L.M.T. *Lactobacillus plantarum* Isolated from Molasses Produces Bacteriocins Active against Gram-Negative Bacteria. *Enzym. Microb. Technol.* **2005**, *36*, 318–326. [CrossRef]
28. Naghmouchi, K.; Drider, D.; Fliss, I. Action of Divergicin M35, a Class IIa Bacteriocin, on Liposomes and *Listeria*. *J. Appl. Microbiol.* **2007**, *102*, 1508–1517. [CrossRef] [PubMed]
29. Belguesmia, Y.; Bendjeddou, K.; Kempf, I.; Boukherroub, R.; Drider, D. Heterologous Biosynthesis of Five New Class II Bacteriocins from *Lactobacillus paracasei* CNCM I-5369 with Antagonistic Activity against Pathogenic *Escherichia coli* Strains. *Front. Microbiol.* **2020**, *11*, 1198. [CrossRef] [PubMed]
30. Madi-Moussa, D.; Coucheney, F.; Drider, D. Expression of Five Class II Bacteriocins with Activity against *Escherichia coli* in *Lacticaseibacillus paracasei* CNCM I-5369, and in a Heterologous Host. *Biotechnol. Rep.* **2021**, *30*, e00632. [CrossRef] [PubMed]
31. Sezonov, G.; Joseleau-Petit, D.; D'Ari, R. *Escherichia coli* physiology in Luria-Bertani broth. *J. Bacteriol.* **2007**, *189*, 8746–8749. [CrossRef]
32. Sambrook, J.; Russell, D.W. *Molecular Cloning: A Laboratory*, 3rd ed.; Cold Spring Harbor Laboratory Press: New York, NY, USA, 2001.
33. Schägger, H. Tricine–SDS-PAGE. 1. *Nat. Protoc.* **2006**, *1*, 16–22. [CrossRef]
34. Wessel, D.; Flügge, U.I. A Method for the Quantitative Recovery of Protein in Dilute Solution in the Presence of Detergents and Lipids. *Anal. Biochem.* **1984**, *138*, 141–143. [CrossRef]
35. Marty, M.T.; Baldwin, A.J.; Marklund, E.G.; Hochberg, G.K.A.; Benesch, J.L.P.; Robinson, C.V. Bayesian Deconvolution of Mass and Ion Mobility Spectra: From Binary Interactions to Polydisperse Ensembles. *Anal. Chem.* **2015**, *87*, 4370–4376. [CrossRef] [PubMed]
36. Jumper, J.; Evans, R.; Pritzel, A.; Green, T.; Figurnov, M.; Ronneberger, O.; Tunyasuvunakool, K.; Bates, R.; Žídek, A.; Potapenko, A.; et al. Highly Accurate Protein Structure Prediction with AlphaFold. 7873. *Nature* **2021**, *596*, 583–589. [CrossRef] [PubMed]
37. Hornak, V.; Abel, R.; Okur, A.; Strockbine, B.; Roitberg, A.; Simmerling, C. Comparison of Multiple Amber Force Fields and Development of Improved Protein Backbone Parameters. *Proteins Struct. Funct. Bioinform.* **2006**, *65*, 712–725. [CrossRef]
38. Eastman, P.; Swails, J.; Chodera, J.D.; McGibbon, R.T.; Zhao, Y.; Beauchamp, K.A.; Wang, L.-P.; Simmonett, A.C.; Harrigan, M.P.; Stern, C.D.; et al. OpenMM 7: Rapid Development of High-Performance Algorithms for Molecular Dynamics. *PLoS Comput. Biol.* **2017**, *13*, e1005659. [CrossRef] [PubMed]
39. Pettersen, E.F.; Goddard, T.D.; Huang, C.C.; Meng, E.C.; Couch, G.S.; Croll, T.I.; Morris, J.H.; Ferrin, T.E. UCSF ChimeraX: Structure Visualization for Researchers, Educators, and Developers. *Protein Sci.* **2021**, *30*, 70–82. [CrossRef]
40. Green, A.A. The Preparation of Acetate and Phosphate Buffer Solutions of Known PH and Ionic Strength. *J. Am. Chem. Soc.* **1933**, *55*, 2331–2336. [CrossRef]
41. Micsonai, A.; Wien, F.; Bulyáki, É.; Kun, J.; Moussong, É.; Lee, Y.-H.; Goto, Y.; Réfrégiers, M.; Kardos, J. BeStSel: A Web Server for Accurate Protein Secondary Structure Prediction and Fold Recognition from the Circular Dichroism Spectra. *Nucleic Acids Res.* **2018**, *46*, W315–W322. [CrossRef]
42. Roy, A.; Kucukural, A.; Zhang, Y. I-TASSER: A Unified Platform for Automated Protein Structure and Function Prediction. *Nat. Protoc.* **2010**, *5*, 725–738. [CrossRef]
43. Yang, F.; Hou, C.; Zeng, X.; Qiao, S. The Use of Lactic Acid Bacteria as a Probiotic in Swine Diets. *Pathogens* **2015**, *4*, 34–45. [CrossRef]
44. Daba, H.; Pandian, S.; Gosselin, J.F.; Simard, R.E.; Huang, J.; Lacroix, C. Detection and activity of a bacteriocin produced by Leuconostoc mesenteroides. *Appl. Environ. Microbiol.* **1991**, *57*, 3450–3455. [CrossRef]
45. Batdorj, B.; Dalgalarrondo, M.; Choiset, Y.; Pedroche, J.; Métro, F.; Prévost, H.; Chobert, J.-M.; Haertlé, T. Purification and Characterization of Two Bacteriocins Produced by Lactic Acid Bacteria Isolated from Mongolian Airag. *J. Appl. Microbiol.* **2006**, *101*, 837–848. [CrossRef] [PubMed]
46. Goulet, A.; Cambillau, C. Present Impact of AlphaFold2 Revolution on Structural Biology, and an Illustration with the Structure Prediction of the Bacteriophage J-1 Host Adhesion Device. *Front. Mol. Biosci.* **2022**, *9*, 907452. [CrossRef] [PubMed]
47. Tulumello, D.V.; Deber, C.M. SDS Micelles as a Membrane-Mimetic Environment for Transmembrane Segments. *Biochemistry* **2009**, *48*, 12096–12103. [CrossRef] [PubMed]
48. Pommer, A.J.; Wallis, R.; Moore, G.R.; James, R.; Kleanthous, C. Enzymological Characterization of the Nuclease Domain from the Bacterial Toxin Colicin E9 from *Escherichia coli*. *Biochem. J.* **1998**, *334*, 387–392. [CrossRef]
49. Meade, E.; Slattery, M.A.; Garvey, M. Bacteriocins, Potent Antimicrobial Peptides and the Fight against Multi Drug Resistant Species: Resistance Is Futile? *Antibiotics* **2020**, *9*, 32. [CrossRef]
50. da Silva Sabo, S.; Vitolo, M.; González, J.M.D.; Oliveira, R.P. de S. Overview of *Lactobacillus plantarum* as a Promising Bacteriocin Producer among Lactic Acid Bacteria. *Food Res. Int.* **2014**, *64*, 527–536. [CrossRef]

51. Li, J.; Koh, J.-J.; Liu, S.; Lakshminarayanan, R.; Verma, C.S.; Beuerman, R.W. Membrane Active Antimicrobial Peptides: Translating Mechanistic Insights to Design. *Front. Neurosci.* **2017**, *11*, 73. [CrossRef]
52. Epand, R.M.; Vogel, H.J. Diversity of Antimicrobial Peptides and Their Mechanisms of Action. *Biochim. Biophys Acta* **1999**, *1462*, 11–28. [CrossRef]
53. Harris, F.; Dennison, S.; Phoenix, D.A. Anionic antimicrobial peptides from eukaryotic organisms. *Curr. Protein Pept. Sci.* **2009**, *10*, 585–606. [CrossRef]
54. Clark, S.; Jowitt, T.A.; Harris, L.K.; Knight, C.G.; Dobson, C.B. The Lexicon of Antimicrobial Peptides: A Complete Set of Arginine and Tryptophan Sequences. *Commun. Biol.* **2021**, *4*, 605. [CrossRef]
55. van Kraaij, C.; Breukink, E.; Noordermeer, M.A.; Demel, R.A.; Siezen, R.J.; Kuipers, O.P.; de Kruijff, B. Pore Formation by Nisin Involves Translocation of Its C-Terminal Part across the Membrane. *Biochemistry* **1998**, *37*, 16033–16040. [CrossRef] [PubMed]
56. Johnsen, L.; Fimland, G.; Nissen-Meyer, J. The C-Terminal Domain of Pediocin-like Antimicrobial Peptides (Class IIa Bacteriocins) Is Involved in Specific Recognition of the C-Terminal Part of Cognate Immunity Proteins and in Determining the Antimicrobial Spectrum. *J. Biol. Chem.* **2005**, *280*, 9243–9250. [CrossRef] [PubMed]
57. Rihakova, J.; Petit, V.W.; Demnerova, K.; Prévost, H.; Rebuffat, S.; Drider, D. Insights into Structure-Activity Relationships in the C-Terminal Region of Divercin V41, a Class IIa Bacteriocin with High-Level Antilisterial Activity. *Appl. Environ. Microbiol.* **2009**, *75*, 1811–1819. [CrossRef] [PubMed]
58. Torres, M.D.T.; Sothiselvam, S.; Lu, T.K.; Fuente-Nunez, C. Peptide Design Principles for Antimicrobial Applications. *J. Mol. Biol.* **2019**, *431*, 3547–3567. [CrossRef]
59. Mattiuzzo, M.; Bandiera, A.; Gennaro, R.; Benincasa, M.; Pacor, S.; Antcheva, N.; Scocchi, M. Role of the *Escherichia coli* SbmA in the Antimicrobial Activity of Proline-Rich Peptides. *Mol. Microbiol.* **2007**, *66*, 151–163. [CrossRef]
60. Seefeldt, A.C.; Nguyen, F.; Antunes, S.; Pérébaskine, N.; Graf, M.; Arenz, S.; Inampudi, K.K.; Douat, C.; Guichard, G.; Wilson, D.N.; et al. The Proline-Rich Antimicrobial Peptide Onc112 Inhibits Translation by Blocking and Destabilizing the Initiation Complex. *Nat. Struct Mol. Biol.* **2015**, *22*, 470–475. [CrossRef]
61. Welch, N.G.; Li, W.; Hossain, M.A.; Separovic, F.; O'Brien-Simpson, N.M.; Wade, J.D. (Re)Defining the Proline-Rich Antimicrobial Peptide Family and the Identification of Putative New Members. *Front. Chem.* **2020**, *8*, 607769. [CrossRef]
62. Kwon, Y.M.; Kim, H.J.; Kim, Y.I.; Kang, Y.J.; Lee, I.H.; Jin, B.R.; Han, Y.S.; Cheon, H.M.; Ha, N.G.; Seo, S.J. Comparative Analysis of Two Attacin Genes from Hyphantria Cunea. *Comp. Biochem. Physiol. B Biochem. Mol. Biol.* **2008**, *151*, 213–220. [CrossRef]
63. Lee, J.H.; Cho, K.S.; Lee, J.; Yoo, J.; Lee, J.; Chung, J. Diptericin-like Protein: An Immune Response Gene Regulated by the Anti-Bacterial Gene Induction Pathway in Drosophila. *Gene* **2001**, *271*, 233–238. [CrossRef]

pharmaceutics

MDPI

Article

Impacts of PEGylation and Glycosylation on the Biological Properties of Host Defense Peptide IDR1018

Hashem Etayash, Fione Yip and Robert E. W. Hancock *

Centre for Microbial Diseases and Immunity Research, Department of Microbiology and Immunology, University of British Columbia, 2259 Lower Mall Research Station, Vancouver, BC V6T 1Z4, Canada; etayashhashem@gmail.com (H.E.); mt3043mt@gmail.com (F.Y.)
* Correspondence: bob@hancocklab.com

Abstract: The multifunctional properties of host defense peptides (HDPs) make them promising drug candidates to tackle bacterial infections and tissue inflammation. However, these peptides tend to aggregate and can harm host cells at high doses, potentially limiting their clinical use and applications. In this study, we explored the influences of both pegylation and glycosylation on the biocompatibility and biological properties of HDPs, particularly the innate defense regulator IDR1018. Two peptide conjugates were designed by attaching either polyethylene glycol (PEG6) or a glucose moiety to the peptide towards the N-terminus. Significantly, both derivatives reduced the aggregation, hemolysis, and cytotoxicity of the parent peptide by orders of magnitude. In addition, while the pegylated conjugate, PEG6-IDR1018, retained an excellent immunomodulatory profile, similar to that observed for IDR1018 itself, the glycosylated conjugate, Glc-IDR1018, significantly outperformed the parent peptide in inducing anti-inflammatory mediators, MCP1 and IL-1RA and in suppressing the level of lipopolysaccharide-induced proinflammatory cytokine IL-1β. Conversely, the conjugates led to a partial reduction in antimicrobial and antibiofilm activity. These findings underline the impacts of both pegylation and glycosylation on the biological properties of the HDP IDR1018 and indicate the potential of glycosylation to enhance the design of highly effective immunomodulatory peptides.

Keywords: antimicrobial peptides; biofilm; anti-inflammatory; PEGylation; glycosylation

Citation: Etayash, H.; Yip, F.; Hancock, R.E.W. Impacts of PEGylation and Glycosylation on the Biological Properties of Host Defense Peptide IDR1018. *Pharmaceutics* **2023**, *15*, 1391. https://doi.org/10.3390/pharmaceutics15051391

Academic Editors: Scavello Francesco, Amiche Mohamed and Jean-Eric Ghia

Received: 17 March 2023
Revised: 18 April 2023
Accepted: 27 April 2023
Published: 1 May 2023

1. Introduction

Bacterial infections are the second leading cause of death worldwide with an estimated 7.7 million deaths annually, of which 5 million are associated with antimicrobial resistance (AMR) to existing antibiotics [1]. They are a constant threat to global health systems and a continuing threat to human lives. Certainly, highly effective therapeutic interventions are needed more than ever to target bacterial infections and offset the rise in AMR. While various innovative approaches to discovering and developing new antibacterial agents have been proposed or undertaken such as the use of antibodies [2], aptamers [3], nucleic acid materials [4], nanoparticles of metal oxides [5–7], small synthetic chemical compounds [8,9], and others [10], antimicrobial host defense peptides (HDPs), which are part of the innate immune system, have been proven as a promising potential therapeutic strategy to confront bacterial infections and tackle the clinical threats of biofilm-forming drug-resistant strains due to the multiple advantages they offer [11]. HDPs are relatively small polymers (usually, 4–50 amino acid residues) that can be easily synthesized and modified at minimal costs. Unlike antibodies and small chemical compounds of nucleic acid-based drugs, antimicrobial HDPs often exhibit broad-spectrum activity against a wide range of bacteria species, including Gram-positive and Gram-negative pathogens, and usually, they tend to have a low propensity to develop antimicrobial resistance due to their multifaceted mechanism of actions [12,13]. These HDPs are also biodegradable, and tend to degrade easily; thus, they do not persist in the body and do not pose any undesirable side effects [13]. Furthermore, the ability of these antimicrobial peptides to disperse and eradicate mature bacterial biofilms and high-density bacterial infections both in vitro and in vivo gives them advantages over current antibiotics and other molecules as excellent antimicrobial and antibiofilm drug candidates [11]. Also, the ability of HDPs to act as anti-infective immune modulators with promising anti-inflammatory activity makes them clinically attractive candidates for multipurpose applications [14,15]. However, despite their advantages, as pharmaceutical products, HDPs endure several constraints that limit their ease of accessibility and use, including their systemic aggregation and toxic side effects typically at higher doses, in addition to their poor stability against blood-borne proteases [13,16]. While various approaches including, for instance, formulation strategies, sequence modifications and designing HDP mimetics have been explored to circumvent some of these shortcomings of HDPs, only limited successes for different peptides were achieved [13,17–23]. In this study, we decided to explore the special effects of both pegylation and glycosylation on the physiochemical properties as well as the biological activity properties of the HDP, IDR1018, which is a synthetic multifaceted peptide with immunomodulatory and antibiofilm activities [24,25]. Pegylation and glycosylation are modifications utilized in peptide/protein drug design and are often used to improve the stability and pharmacokinetic assets of potential drug candidates while reducing toxicities and potentially harmful side effects [26–29]. Pegylation is a process by which peptides are chemically conjugated to polyethylene glycol (PEG) in order to change their physicochemical or biological properties [30]. As PEG and its derivatives are inert, water-soluble, non-toxic and non-immunogenic, they are widely used to overcome limitations associated with biopharmaceutical products including water solubility issues, aggregations, toxicity against mammalian cells and immunogenicity [30]. Pegylation offers significant advantages for biopharmaceutical products, as attaching PEG improves proteolytic stability, helps mitigate the immunogenicity, increases resistance to bacterial-secreted enzymes, boosts blood circulation half-lives and enhances biodistribution as well as drug bioavailability [28,30]. Nevertheless, despite these advantages, pegylation is often associated with a partial or complete reduction in the antimicrobial activity of HDPs [30]. Glycosylation is a process by which a sugar moiety is chemically attached to biopharmaceutical molecules such as peptides, proteins, antibodies, etc., in order to change their physicochemical properties or produce better bioactive compounds [31]. As with pegylation, glycosylation can have a significant influence on the properties of HDPs, for example, modifying toxicities, resistance to proteolytic degradation, pharmacokinetics and dynamic properties [27,31,32]. However, as with pegylation, glycosylation does not always improve the antibacterial activity of HDPs since the attaching surges change the chemical structure, hydrophobicity and overall charge of the peptides, which can impact the insertion and interaction of peptides with bacterial membranes [31].

Indeed, the overall influence of pegylation and glycosylation on drug design is remarkable with many clinically accepted drugs being pegylated and glycosylated. Since little is known about the influence of both conjugations on the multifaceted properties of HDPs, especially the immunomodulatory functionality, we aimed in this study to look at their overall impacts by covalently modifying the antimicrobial HDP IDR1018 with short-chain PEG (PEG6) and a glucose moiety (N-acetyl glucosamine (GlcNAc)).

2. Materials and Methods

2.1. Peptide Synthesis, Pegylation, and Glycosylation

IDR1018 peptide (sequence, VRLIVAVRIWRR-NH_2) was purchased from GenScript (Piscataway, NJ, USA) at >95% purity. The pegylated IDR1018 [PEG6; VRLIVAVRIWRR-NH_2] as well as the glycosylated IDR1018 [Glc-IDR1018; T(GlcNAc)VRLIVAVRIWRR-NH_2], were obtained at ≥95% purity from Biomatik LLC (Wilmington, DE, USA) (Table 1).

Table 1. IDR1018 and its conjugated derivatives, HPLC, and mass spectra data.

Peptide and Derivatives	Sequence	HPLC (Ret. Time (min))	MW (Calc.)	MW (Obs.)	Purity (%)
IDR1018	VRLIVAVRIWRR-NH_2	9.1	1535.8	1535.8	>95.0
PEG6-IDR1018	PEG6-VRLIVAVRIWRR-NH_2	17.9	1872.3	1872.6	>95.0
Glc-IDR1018	T(GlcNAc)VRLIVAVRIWRR-NH_2	14.3	1841.4	1841.3	>92.0

HPLC, high-performance liquid chromatography; Ret. Time, retention time in minutes; MW, molecular weight; Calc., calculated; Obs., observed.

2.2. Aggregation Assay

Solutions of IDR1018 peptide and the conjugates were tested for aggregation in saline (0.9% NaCl), 5% dextrose, and 10% RPMI tissue culture medium at 1 mg mL^{-1}. The solutions were placed in 96 well plates and sterile water was used as a negative control. The turbidity was determined for each sample as described in earlier reports [19,33]. Similarly, the conjugates were screened for aggregation in the presence of sodium salts of polyatomic anions (citrate or phosphate) at various strengths (0.1−1000 mM). The % of aggregates was assessed relative to the OD_{600} of sterile water.

Furthermore, isolated peripheral blood mononuclear cells (PBMCs) seeded in 96-well flat-bottom tissue culture plates (Corning Inc., Corning, NY, USA) at a concentration of 2×10^6 in RMPI media were treated with the two conjugates at a final concentration of 32 μg mL^{-1}, incubated for 4 h at 37 °C in 5% CO_2, and visualized on a Nikon Eclipse TS100 microscope. Experiments were performed in triplicate in three independent experiments. Representative images of the microscopy are presented in Figure 1.

2.3. Hemolysis Test

Red blood cells (RBCs) were isolated and used to test the hemolytic activity of the peptide conjugates according to documented procedures [27,34] with some modifications. The blood samples were collected from healthy donors according to the University of British Columbia ethics certification and guidelines. Informed consent was obtained from all donors involved in the study. A 0.5% (*v*/*v*) RBC suspension in PBS was transferred into 96 well plates, where pre-diluted peptides series were incubated. The samples were placed at 37 °C for 1 h prior to reading the hemoglobin release at 414 nm and 546 nm. Triton X-100 (1% final concentration) was used as a positive control (100% hemolysis). The % hemolysis was then determined relative to the negative control (untreated samples).

2.4. Cytotoxicity and Immunomodulatory Assays

The cytotoxicity of the conjugates was assessed in vitro against PBMCs as described earlier using the lactate dehydrogenase (LDH)-based assay [35]. Blood samples were collected from healthy donors with ethics approval and PBMCs were isolated following our previously reported protocol [35]. The cells were resuspended in RPMI medium supplemented with 10% FBS and seeded in tissue culture-treated plates at a final density of 1×10^5 cells $well^{-1}$. Subsequently, the cells were treated with the conjugates or IDR1018 and incubated for 24 h at 37 °C in 5% CO_2, to determine the LDH release. Triton X-100 (1% final concentration) was used as a toxic (positive) control.

To evaluate the capacity of the conjugates to modulate the immune system, suspended PBMCs (a final density of 1×10^5 cells $well^{-1}$) were treated with the conjugates at 8 μg mL^{-1} in the presence or absence of 20 ng mL^{-1} *P. aeruginosa* LPS, incubated for 24 h, centrifuged, and then the supernatants were collected to quantifying the levels of monocyte chemoattractant protein 1 (MCP-1), the tumor necrosis factor-alpha (TNFα), and interleukins (IL) IL-1β, IL-10, and the anti-inflammatory IL-1 receptor antagonist (IL-1RA) using specific ELISA assays (ELISA kits, eBioscience, Carlsbad, CA, USA). All experiments were conducted in triplicate in three independent experiments and average values were presented.

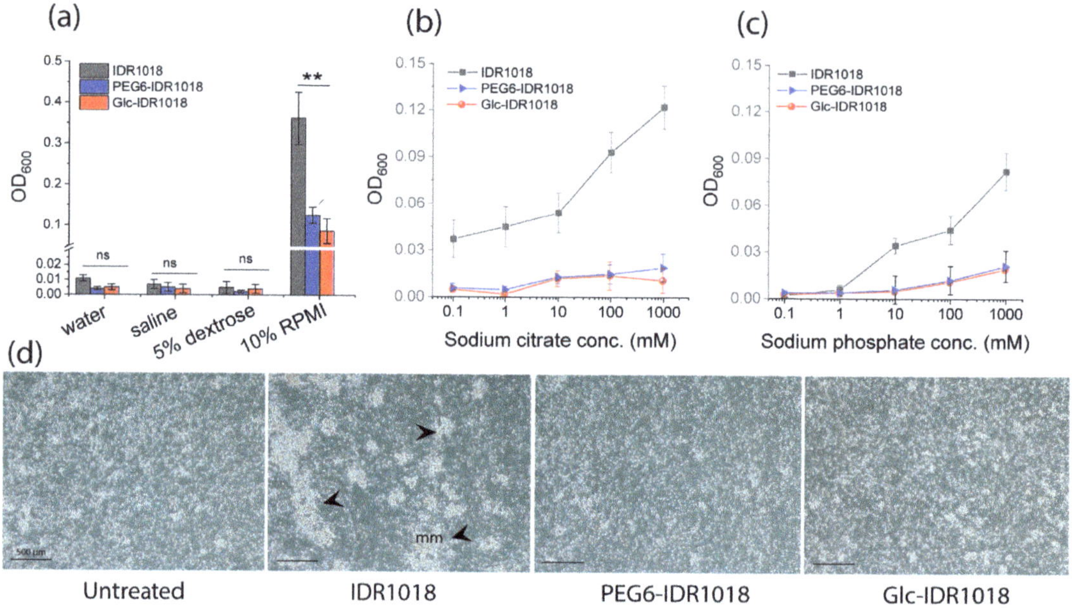

Figure 1. In vitro aggregation experiments of IDR1018 and its conjugates as tested in various solvents under different conditions. The aggregations of the peptides are indicated by the level of turbidity when dissolved in water, saline, 5% dextrose, or RPMI medium (**a**), or when dissolved in various concentrations of sodium citrate (**b**), and sodium phosphate (**c**). Microscopic images show PBMCs after treatment with IDR1018 or its conjugates, indicating aggregation by IDR1018 cf. the conjugates (**d**). Black arrows point to clusters that formed when PBMCs were treated with IDR1018 in a tissue culture medium. Data were described as means ± standard deviations, and significant values (**) were determined using the analysis of variance (ANOVA) test; ns in (**a**) indicates no significance. The scale bar is 500 μm.

2.5. Minimal Inhibitory Concentration (MIC)

The MIC was conducted using the liquid broth inhibition assay as previously described [36]. MHB was used as a medium for bacterial growth, and bacteria (either MRSA SAP0017, *P. aeruginosa* PA14, *E. coli* ATTC489 or *A. baumannii* 5015) were suspended at $\sim 1 \times 10^6$ CFU mL^{-1}. The peptide conjugates were incubated with the bacterial cultures at various concentrations for ~18 h prior determination of bacterial growth optical density at 600 nm (OD_{600}). The MIC was defined as the lowest concentration that led to no growth in the wells. All values are modal values of three independent experiments.

2.6. Biofilm Biomass and Biofilm Eradication Assays

The ability of the designed conjugates to decrease the biomass of bacterial biofilms was assessed in 96-well microtitre plates using the crystal violet staining assay as described previously [37]. A bacterial suspension of 1×10^7 CFU mL^{-1} was used to establish biofilms prior to treatment with the peptides. Appropriate nutrient media for optimal biofilm formation were used, including TSB supplemented with 1% glucose for MRSA SAP0017, BM2 glucose for *P. aeruginosa* PA14, and LB media supplemented with 1% glucose for *E. coli* ATTC489 and *A. baumannii* 5015. In the crystal violet-based assay, untreated bacterial culture and sterile media were used as positive and negative controls, where the % of biofilm biomass was considered 100 and 0%, respectively. The % of biomass in the treated samples was determined relative to the positive and negative controls. A biofilm eradication assay [17] was also performed, for which the residual viable biofilm cell count (CFU mL^{-1}) was determined. Experiments were conducted in triplicate, in three independent experiments, and average values were presented with ± standard deviations.

2.7. Statistical Analysis

GraphPad Prism 7.0 (GraphPad Software, La Jolla, CA, USA) and Origin Pro Software were used in the analysis. Whenever applicable, data were described as means ± standard deviations, and significant values were determined using the analysis of variance (ANOVA) test.

3. Results and Discussion

3.1. Peptide Conjugation Design

The antimicrobial HDP IDR1018 was chosen for this study due to its promise as a therapeutic agent [38]. The well-known properties of IDR1018, including its ease and cost of synthesis as a linear peptide with 12 amino acid residues, as well as easily accessible sites for modifications, make it an excellent candidate for testing new modification approaches [17,19,38]. The original IDR1018 peptide sequence (Table 1) was modified by the addition of either PEG6 or a glucose moiety (GlcNAc) at the N-terminus to generate new derivatives, namely PEG6-IDR1018 and Glc-IDR1018, respectively. In PEG6-IDR1018, the PEG was attached to the amine terminus of the amino acid, valine, while in the glycosylation strategy, the GlcNAc was introduced into the peptide through O-linked glycosylation in an additional threonine residue. The resulting peptide conjugates are described in Table 1 while the chemical structures are shown in Scheme 1.

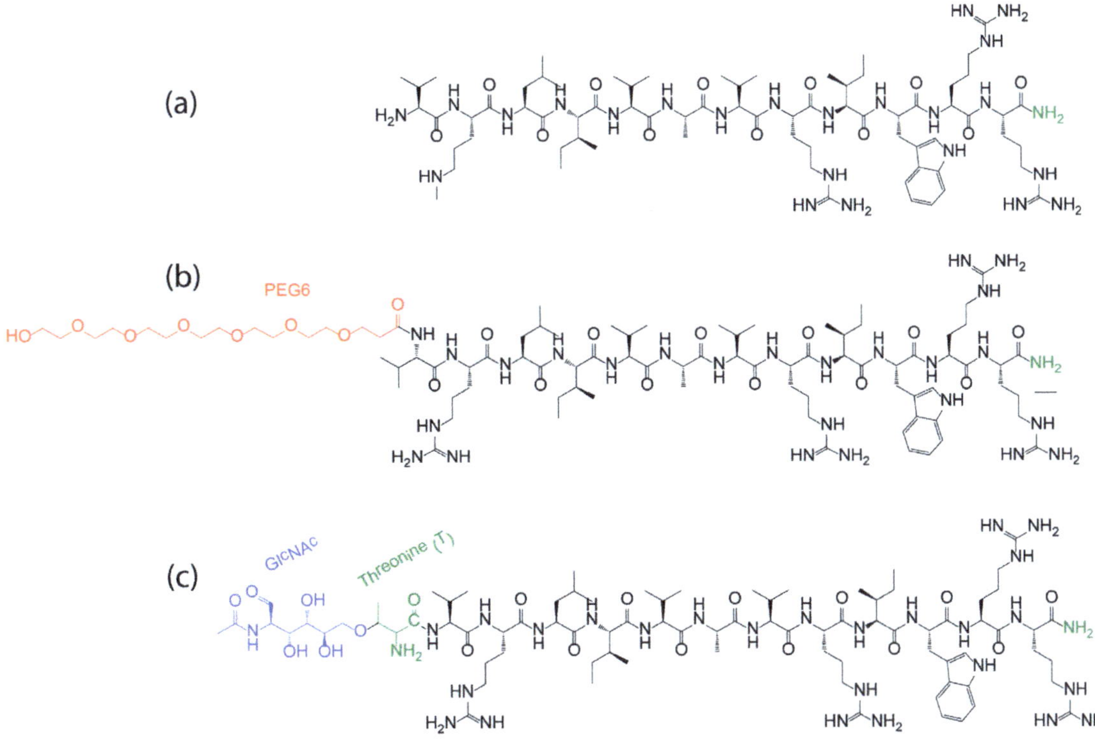

Scheme 1. IDR1018 conjugates, (**a**) the antimicrobial HDP IDR1018, (**b**) the pegylated derivative PEG6-IDR1018, where a carboxylated PEG chain was attached to the N-terminal of the peptide and (**c**) the glycosylated derivative, Glc-IDR1018, where N-acetyl glucosamine (GlcNAc) was attached to the hydroxyl (-OH) group of the amino acid threonine (T).

3.2. *In Vitro Peptide(s) Aggregation*

Aggregation is typically a limitation of HDPs since it not only impacts biological activity but also biocompatibility and immunogenicity [39,40]. In previous studies, the HDP IDR1018 was found to aggregate in various in vitro solvents as well as in vivo under certain conditions and as a consequence, a reduction in the immunomodulatory activity was observed under conditions that promote aggregation [19,33]. Here, we examined the aggregation tendency of peptide conjugates in multiple solutions, including water, saline (0.9% NaCl), 5% dextrose, and 10% RPMI tissue culture medium with 1% fetal bovine serum (FBS). While no aggregation (measured by turbidity) was observed in water, saline, or 5% dextrose for any sample, including IDR1018 (Figure 1a), a substantial increase in aggregation was detected in tissue culture medium for parent peptide IDR1018 but not for the pegylated and glycosylated derivatives.

When the peptides were prepared in sodium salts of abundant polyatomic anions, such as citrate and phosphate (Figure 1b,c), a substantial increase in turbidity was observed with IDR1018 at solute concentrations as low as 10 mM. In contrast, no turbidity increase was detected with PEG6-IDR1018 or Glc-IDR1018 at higher solute concentrations, suggesting a significant reduction in the salt-induced peptide aggregation. When incubated with PBMCs in tissue culture, neither peptide conjugate, PEG6-IDR1018, and Glc-IDR1018 showed any obvious sign of aggregation, in strong contrast to IDR1018, which showed amorphous aggregates (Figure 1d). These results, indeed, were consistent with earlier reports, that demonstrated the utility of pegylation and glycosylation in increasing solubility and diminishing aggregation for other antimicrobial peptides [31,41].

3.3. *Hemolysis, Cytotoxicity and Immunomodulatory Activity*

Compared to the parent IDR1018 peptide, which appeared to be hemolytic against RBCs as well as toxic to PBMCs at high doses (up to 20% at 128 μg mL^{-1}), neither of the conjugated derivatives exhibited any hemolysis or cytotoxicity at the highest concentration tested (Figure 2a,b). Consistent with other previous studies for other peptides [30,31,42,43], pegylation and glycosylation worked well in reducing HDP toxicity.

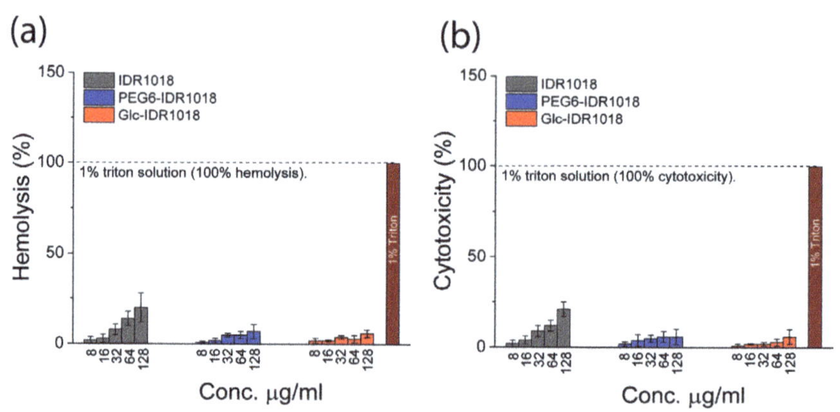

Figure 2. Hemolysis and Cytotoxicity of the peptide conjugates vs. IDR1018 itself. (**a**) displays the hemolysis analysis against RBCs while (**b**) depicts the cytotoxicity against PBMCs. Experiments were conducted in triplicate in three independent experiments and mean values ± standard deviation are presented.

The capacity of conjugated peptide derivatives to modulate the innate immune system was tested by assessing the induction and suppression of chemokines and cytokines from human PBMCs. These tested modulators including MCP1, TNFα, IL-1β, IL-1RA and IL-10, were selected based on previous studies where IDR1018 demonstrated an excellent ability to modulate them in vitro and in vivo [17,19,24,38,44,45]. In addition, these immunity modulators are involved in multiple systemic mechanisms essential to suppress inflammations and indirectly defend against bacterial infections. For instance, MCP-1 attracts macrophages and enhances the recruitment of monocytes at the sites of infections, injuries and inflammations [46], while the anti-inflammatory cytokines, IL-1RA and IL-10 act as natural inhibitors of the harmful effect of pro-inflammatory immune mediators such as IL-1α and IL-1β [47]. Interestingly, in the absence of an LPS stimulant, PEG6-IDR1018 maintained a similar ability to modulate chemokine and cytokine expression to that observed for unmodified IDR1018. However, Glc-IDR1018 exhibited a significant increase in the expression of the immune mediators relative to unmodified IDR1018. Thus Glc-IDR1018 boosted the level of MCP1 and IL-1RA by 2–5 fold,1.5 ng mL^{-1} and 0.8 ng mL^{-1}, respectively, compared to only 0.7 and 0.15 ng mL^{-1} induced by the unmodified IDR1018 (Figure 3a,b). These peptides did not induce TNFα (Figure 3c).

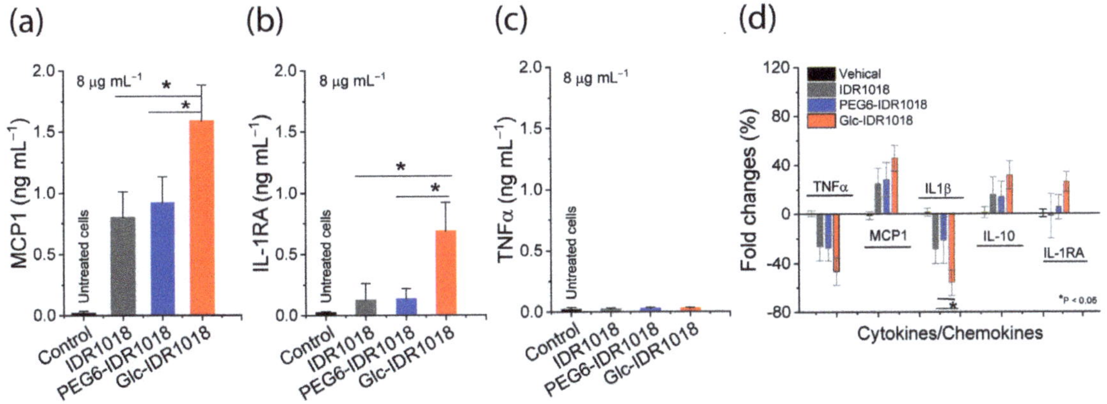

Figure 3. Enhanced effect of glycosylation on the immunomodulatory activities of IDR1018-derived peptides. (**a–c**) The effect of the peptides at 8 μg mL^{-1} in inducing chemokine MCP1 and cytokines IL-1RA and TNFα production by PBMCs in the absence of LPS stimulation. (**d**) the effect of peptides on the immune mediator expression when co-treated with 20 ng mL^{-1} LPS (% of fold changes relative to LPS-stimulated cells). Data are presented as mean ± standard deviation from three independent experiments (* represents $p < 0.05$, based on ANOVA analysis with Tukey's correction for multiple testing).

When stimulated with LPS which amplified the secretion of immune mediators, substantial reductions in the proinflammatory TNFα were detected due to peptide action, with no significant difference between all three peptides. However, the expression of IL-1β was significantly suppressed when cells were treated with Glc-IDR1018 cf. the other peptides (Figure 3d). Moreover, remarkable peptide-induced increases in MCP1, IL-1RA and IL-10 induction levels were also detected with substantially higher expression values associated with Glc-IDR1018 treatment. In agreement with previous reports, the results indicated that pegylation had no marked effect on the immunomodulatory properties of the HDP IDR1018 [43,48]. However, this was the first report we have observed indicating that glycosylation markedly improves the immunomodulatory properties of HDPs. While the mechanistic effects of glycosylation on the immunomodulatory properties of IDR1018 as well as the underlying biophysical processes remain to be elucidated, the overall results showed exciting data with excellent outcomes for this peptide conjugate in modulating the immune system with efficacy exceeding that for previously reported HDPs [15,19].

3.4. Antimicrobial Activity

The two peptide conjugates (Table 1) were intended to mimic and/or enhance the activity of the parent IDR1018 while improving biocompatibility and physicochemical properties. Despite an enhancement or retention of immunomodulatory activities, there was a 2 to 8-fold reduction in antimicrobial activity (i.e., higher MIC values) when compared to the parent IDR1018 (Table 2).

Table 2. Activities of the peptide conjugates vs. planktonic bacteria.

Bacteria	MIC (μg mL^{-1})		
	IDR1018	PEG6-IDR1018	Glc-IDR1018
S. aureus MRSA0017	16	32	64
P. aeruginosa PA14	16	64	128
E. coli ATTC489	16	64	64
A. baumannii 5015	16	64	64

The MICs of the PEG6-IDR1018 ranged between 32–64 μg mL^{-1}, while the MICs of the Glc-IDR1018 ranged between 64–128 μg mL^{-1} based on the strains tested. According to the literature, [26,29,43,49,50], such decreases in antimicrobial activity are not uncommon for PEGylated and glycosylated peptides. The activity of the antimicrobial peptide Nisin, for instance, was reduced by conjugation with polyethylene glycol [50]. Likewise, the activity of other antimicrobial HDPs, such as α defensine 1 [51], tachyplestin I [52], magainin 2 [49], aurein 2.2 [53] and others [26], all exhibited reduced antimicrobial activity when pegylated. It is worth noting that previous results for pegylation and glycosylation appear to depend on the type of ligation, the PEG or glycan composition and structure, the nature of the PEG or the sugar-peptide bond and their lengths, and the overall structural conformation of the generated molecules [54,55]. In a study by Falciani C et al., for instance, pegylation at the C-terminus of the antimicrobial peptide M33 was shown to improve the stability of the peptide against *Pseudomonas aeruginosa* elastase [56]. While on the contrary, conjugation at the N-terminus of HDPs, LL-37 and cecropin A, derivatives (PEG-CaLL), showed to improve the antimicrobial activity against various bacterial strains, including *B. anthracis* (including vegetative and spore forms), *Escherichia coli* and *Staphylococcus aureus* when compared to the LL-37 peptide [57]. By analogy, in a study by Talat et al., the activity of the glycosylated peptides was shown to be impacted by the stereochemistry of attached sugars [58]. The study also showed that β-linked sugars induce more flexible and conformationally unstrained conjugates in contrast to the α-linked counterparts which cause more rigid and highly stable conjugates [58]. Although the N and C termini of peptides tend to show conformational flexibility, further studies would be required to investigate if these conjugations impacted the overall HDP IDR1018 chemical structure including its secondary structure, in order to understand the potential reasons behind the reduced activity. It would also be worthwhile investigating the impact of these conjugations on other antimicrobial HDPs, including IDR1018 analogs.

3.5. Antibiofilm Activity

Effects on bacterial biomass were first tested using a crystal violet staining assay. While none of the peptides exhibited complete eradication of the biomass staining in the wells at the MIC values, the results showed a reduced ability of the designated conjugates to reduce the adhered biofilm mass by >90%, at much higher concentrations compared to the original IDR1018 (Figure 4). The best antibiofilm activity of the conjugates was observed against *S. aureus* MRSA0017, where ≥90% of the biomass reduction was observed at 64 μg mL^{-1} for both PEG6-IDR1018 and Glc-IDR1018 peptides. The results were supported by determining the viable biofilm cell counts (CFU) following biofilm treatments in the biofilm eradication assay (where biofilms were pre-formed), indicating comparable outcomes to the biomass dispersal results (Table 3). The peptide conjugates exhibited a 3 to 16-fold reduction in the antibiofilm activity (i.e., higher minimal biofilm eradication concentration values or MBEC = 64–256 μg mL^{-1}) compared to the unmodified IDR1018 (MBEC = 16 μg mL^{-1}) for all tested strains. Thus, in agreement with the MIC results, the pegylation and glycosylation led to maintenance but a weakening of the antibiofilm properties of the IDR1018. It is clear from these results that pegylation and glycosylation modifications did not generate more efficacious antimicrobial or antibiofilm peptides and that their influences on the HDPs' antimicrobial and antibiofilm activity seemed peptide-specific.

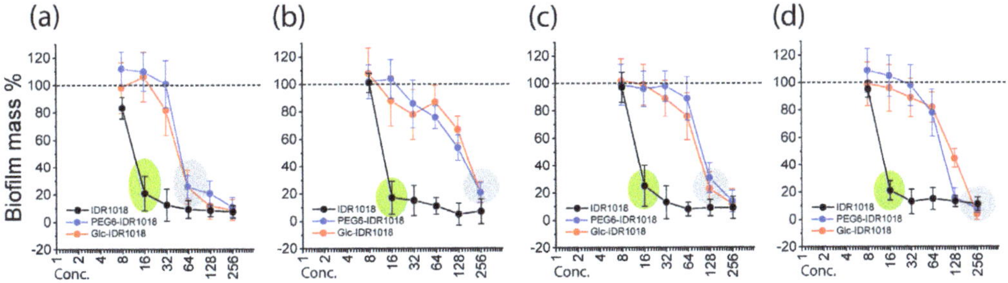

Figure 4. Effect of the peptide conjugates in reducing bacterial biofilm mass as assessed by the crystal violet staining assay. (**a**–**d**) show the antibiofilm activity against *S. aureus* MRSA0017, *P. aeruginosa* PA14, *E. coli* ATTC489, and *A. baumannii* 5015, respectively. The % of biofilm mass reduction was determined relative to untreated samples. The data represents the mean values ± standard deviation. All experiments were conducted three times. The green and grey shades indicate relative MBECs.

Table 3. Activities of the peptide conjugates vs. preformed bacterial biofilms (minimal biofilm eradication concentrations, MBECs).

Bacteria	MBEC (μg mL^{-1})		
	IDR1018	**PEG6-IDR1018**	**Glc-IDR1018**
S. aureus MRSA0017	16	64	64
P. aeruginosa PA14	16	256	256
E. coli ATTC489	16	128	128
A. baumannii 5015	16	128	256

4. Conclusions

Two conjugated derivatives of the antimicrobial HDP IDR1018 were designed based on N-terminal pegylation and glycosylation, in order to improve the biocompatibility and biological properties of the original peptide sequence. While additional studies remain to be explored, such as using different sugar moieties, different lengths of PEG chains, and different ligation approaches, the current conjugates significantly reduced the aggregation, hemolysis, and cytotoxicity of the parent peptide. Moreover, while pegylation conserved the immunomodulatory properties of the peptide, glycosylation generated a superior immunomodulating conjugate with a strong ability to stimulate the release of chemokine MCP-1 and the anti-inflammatory cytokine IL-1RA while suppressing the LPS-induced production of proinflammatory cytokines TNFα and IL-1β, suggesting a novel anti-inflammatory drug candidate. Consistent with several previously reported studies, both conjugations led to partial reductions in antimicrobial as well as antibiofilm activity cf. the original IDR1018. Overall, the study reported here clearly demonstrates the usefulness of these conjugation approaches in optimizing the physicochemical properties of therapeutic peptides and highlight a new potential opportunity for enhancing immunomodulatory HDPs through glycosylation.

Author Contributions: Methodology, H.E.; Formal analysis, F.Y.; Investigation, H.E. and F.Y.; Writing—original draft, H.E.; Writing—review & editing, H.E. and R.E.W.H.; Supervision, H.E. and R.E.W.H.; Project administration, R.E.W.H.; Funding acquisition, R.E.W.H. All authors have read and agreed to the published version of the manuscript.

Funding: This research was funded by the Canadian Institutes for Health Research (CIHR) Foundation grant (FDN-154287). H.E. is the recipient of the Michael Smith Health BC Research Fellowship. R.E.W.H. holds a UBC Killam Professorship.

Institutional Review Board Statement: All experiments performed on human blood adhered to the Declaration of Helsinki and approved by the Institutional Review Board and Ethics Committee of the University of British Columbia, Vancouver, BC, Canada (protocol code H-2100727 and date of approval 7 January 2022).

Informed Consent Statement: In all experiments involved human samples, informed consent was obtained from all subjects involved in the study.

Data Availability Statement: Not applicable.

Conflicts of Interest: The parent peptide IDR-1018 has been filed by the University of British Columbia for patent protection with REWH as an inventor and is the subject of several issued patents US Patent US8343475, New Zealand patent 574758, Australian Patent 2007288080, European patent EP2061886, Denmark patent #2061886, Spain patent 2061886, and Canadian patent 2,660,668. It has been licensed to ABT Innovations Inc., in which REWH has shares. The other authors declare no conflict of interest.

References

1. Ikuta, K.S.; Swetschinski, L.R.; Aguilar, G.R.; Sharara, F.; Mestrovic, T.; Gray, A.P.; Weaver, N.D.; Wool, E.E.; Han, C.; Hayoon, A.G. Global mortality associated with 33 bacterial pathogens in 2019: A systematic analysis for the global burden of disease study 2019. *Lancet* **2022**, *400*, 2221–2248. [CrossRef] [PubMed]
2. El-Kafrawy, S.A.; Abbas, A.T.; Oelkrug, C.; Tahoon, M.; Ezzat, S.; Zumla, A.; Azhar, E.I. IgY antibodies: The promising potential to overcome antibiotic resistance. *Front. Immunol.* **2023**, *14*, 1065353. [CrossRef] [PubMed]
3. Li, W. Prospective Application of aptamer-based assays and therapeutics in bloodstream infections. *Mini Rev. Med. Chem.* **2020**, *20*, 831–840. [CrossRef]
4. Sun, Y.; Meng, L.; Zhang, Y.; Zhao, D.; Lin, Y. The application of nucleic acids and nucleic acid materials in antimicrobial research. *Curr. Stem Cell Res. Ther.* **2021**, *16*, 66–73. [CrossRef] [PubMed]
5. Pino, P.; Bosco, F.; Mollea, C.; Onida, B. Antimicrobial nano-zinc oxide biocomposites for wound healing applications: A Review. *Pharmaceutics* **2023**, *15*, 970. [CrossRef]
6. Akshaya, S.; Rowlo, P.K.; Dukle, A.; Nathanael, A.J. Antibacterial coatings for titanium implants: Recent trends and future perspectives. *Antibiotics* **2022**, *11*, 1719. [CrossRef]
7. Sun, C.; Wang, X.; Dai, J.; Ju, Y. Metal and metal oxide nanomaterials for fighting planktonic bacteria and biofilms: A review emphasizing on mechanistic aspects. *Int. J. Mol. Sci.* **2022**, *23*, 11348. [CrossRef]
8. Zha, G.F.; Preetham, H.D.; Rangappa, S.; Sharath Kumar, K.S.; Girish, Y.R.; Rakesh, K.P.; Ashrafizadeh, M.; Zarrabi, A.; Rangappa, K.S. Benzimidazole analogues as efficient arsenals in war against methicillin-resistance *Staphylococcus aureus* (MRSA) and its SAR studies. *Bioorg. Chem.* **2021**, *115*, 105175. [CrossRef]
9. Mousavifar, L.; Roy, R. Recent development in the design of small 'drug-like' and nanoscale glycomimetics against *Escherichia coli* infections. *Drug Discov. Today* **2021**, *26*, 2124–2137. [CrossRef]
10. Hemmati, F.; Rezaee, M.A.; Ebrahimzadeh, S.; Yousefi, L.; Nouri, R.; Kafil, H.S.; Gholizadeh, P. Novel strategies to combat bacterial biofilms. *Mol. Biotechnol.* **2021**, *63*, 569–586. [CrossRef]
11. Hancock, R.E.W.; Alford, M.A.; Haney, E.F. Antibiofilm activity of host defence peptides: Complexity provides opportunities. *Nat. Rev. Microbiol.* **2021**, *19*, 786–797. [CrossRef] [PubMed]
12. Magana, M.; Pushpanathan, M.; Santos, A.L.; Leanse, L.; Fernandez, M.; Ioannidis, A.; Giulianotti, M.A.; Apidianakis, Y.; Bradfute, S.; Ferguson, A.L.; et al. The value of antimicrobial peptides in the age of resistance. *Lancet Infect. Dis.* **2020**, *20*, e216–e230. [CrossRef]
13. Etayash, H.; Hancock, R.E.W. Host defense peptide-mimicking polymers and polymeric-brush-tethered host defense peptides: Recent developments, limitations, and potential success. *Pharmaceutics* **2021**, *13*, 1820. [CrossRef] [PubMed]
14. Kim, J.; Cho, B.H.; Jang, Y.S. Understanding the roles of host defense peptides in immune modulation: From antimicrobial action to potential as adjuvants. *J. Microbiol. Biotechnol.* **2023**, *33*, 288–298. [CrossRef]
15. Hancock, R.E.W.; Haney, E.F.; Gill, E.E. The immunology of host defence peptides: Beyond antimicrobial activity. *Nat. Rev. Immunol.* **2016**, *16*, 321–334. [CrossRef]

16. Ting, D.S.J.; Beuerman, R.W.; Dua, H.S.; Lakshminarayanan, R.; Mohammed, I. Strategies in translating the therapeutic potentials of host defense peptides. *Front. Immunol.* **2020**, *11*, 983. [CrossRef]
17. Etayash, H.; Alford, M.; Akhoundsadegh, N.; Drayton, M.; Straus, S.K.; Hancock, R.E.W. Multifunctional antibiotic-host defense peptide conjugate kills bacteria, eradicates biofilms, and modulates the innate immune response. *J. Med. Chem.* **2021**, *64*, 16854–16863. [CrossRef] [PubMed]
18. Etayash, H.; Qian, Y.; Pletzer, D.; Zhang, Q.; Xie, J.; Cui, R.; Dai, C.; Ma, P.; Qi, F.; Liu, R.; et al. Host defense peptide-mimicking amphiphilic β-peptide polymer (Bu:DM) exhibiting anti-biofilm, immunomodulatory, and in vivo anti-infective activity. *J. Med. Chem.* **2020**, *63*, 12921–12928. [CrossRef]
19. Etayash, H.; Pletzer, D.; Kumar, P.; Straus, S.K.; Hancock, R.E.W. Cyclic derivative of host-defense peptide IDR-1018 improves proteolytic stability, suppresses inflammation, and enhances in vivo activity. *J. Med. Chem.* **2020**, *63*, 9228–9236. [CrossRef]
20. Gonçalves da Costa Sousa, M.; Conceição de Almeida, G.; Martins Mota, D.C.; Andrade da Costa, R.; Dias, S.C.; Limberger, S.N.; Ko, F.; Lin, L.T.; Haney, E.F.; Etayash, H.; et al. Antibiofilm and immunomodulatory resorbable nanofibrous filing for dental pulp regenerative procedures. *Bioact. Mater.* **2022**, *16*, 173–186. [CrossRef]
21. Thapa, R.K.; Diep, D.B.; Tønnesen, H.H. Topical antimicrobial peptide formulations for wound healing: Current developments and future prospects. *Acta Biomater.* **2020**, *103*, 52–67. [CrossRef] [PubMed]
22. Gaglio, S.C.; Jabalera, Y.; Montalbán-López, M.; Millán-Placer, A.C.; Lázaro-Callejón, M.; Maqueda, M.; Carrasco-Jimenez, M.P.; Laso, A.; Aínsa, J.A.; Iglesias, G.R.; et al. Embedding biomimetic magnetic nanoparticles coupled with peptide AS-48 into PLGA to treat intracellular pathogens. *Pharmaceutics* **2022**, *14*, 2744. [CrossRef] [PubMed]
23. van Gent, M.E.; Ali, M.; Nibbering, P.H.; Kłodzińska, S.N. Current advances in lipid and polymeric antimicrobial peptide delivery systems and coatings for the prevention and treatment of bacterial infections. *Pharmaceutics* **2021**, *13*, 1840. [CrossRef] [PubMed]
24. Wieczorek, M.; Jenssen, H.; Kindrachuk, J.; Scott, W.R.; Elliott, M.; Hilpert, K.; Cheng, J.T.; Hancock, R.E.W.; Straus, S.K. Structural studies of a peptide with immune modulating and direct antimicrobial activity. *Chem. Biol.* **2010**, *17*, 970–980. [CrossRef]
25. Hilchie, A.L.; Wuerth, K.; Hancock, R.E.W. Immune modulation by multifaceted cationic host defense (antimicrobial) peptides. *Nat. Chem. Biol.* **2013**, *9*, 761–768. [CrossRef] [PubMed]
26. Rezende, S.B.; Oshiro, K.G.N.; Júnior, N.G.O.; Franco, O.L.; Cardoso, M.H. Advances on chemically modified antimicrobial peptides for generating peptide antibiotics. *Chem. Commun.* **2021**, *57*, 11578–11590. [CrossRef] [PubMed]
27. Grimsey, E.; Collis, D.W.P.; Mikut, R.; Hilpert, K. The effect of lipidation and glycosylation on short cationic antimicrobial peptides. *Biochim. Biophys. Acta Biomembr.* **2020**, *1862*, 183195. [CrossRef]
28. Manteghi, R.; Pallagi, E.; Olajos, G.; Csóka, I. Pegylation and formulation strategy of anti-microbial peptide (AMP) according to the quality by design approach. *Eur. J. Pharm. Sci.* **2020**, *144*, 105197. [CrossRef]
29. Bednarska, N.G.; Wren, B.W.; Willcocks, S.J. The importance of the glycosylation of antimicrobial peptides: Natural and synthetic approaches. *Drug Discov. Today* **2017**, *22*, 919–926. [CrossRef]
30. Harris, J.M.; Chess, R.B. Effect of pegylation on pharmaceuticals. *Nat. Rev. Drug Discov.* **2003**, *2*, 214–221. [CrossRef]
31. Bellavita, R.; Braccia, S.; Galdiero, S.; Falanga, A. Glycosylation and lipidation strategies: Approaches for improving antimicrobial peptide efficacy. *Pharmaceuticals* **2023**, *16*, 439. [CrossRef] [PubMed]
32. Moradi, S.V.; Hussein, W.M.; Varamini, P.; Simerska, P.; Toth, I. Glycosylation, an effective synthetic strategy to improve the bioavailability of therapeutic peptides. *Chem. Sci.* **2016**, *7*, 2492–2500. [CrossRef] [PubMed]
33. Haney, E.F.; Wu, B.C.; Lee, K.; Hilchie, A.L.; Hancock, R.E.W. Aggregation and its influence on the immunomodulatory activity of synthetic innate defense regulator peptides. *Cell Chem. Biol.* **2017**, *24*, 969–980.e964. [CrossRef]
34. Rodríguez, A.; Villegas, E.; Montoya-Rosales, A.; Rivas-Santiago, B.; Corzo, G. Characterization of antibacterial and hemolytic activity of synthetic pandinin 2 variants and their inhibition against *Mycobacterium tuberculosis*. *PLoS ONE* **2014**, *9*, e101742. [CrossRef] [PubMed]
35. Etayash, H.; Haney, E.F.; Hancock, R.E.W. Assessing biofilm inhibition and immunomodulatory activity of small amounts of synthetic host defense peptides synthesized using SPOT-array technology. *Nat. Protoc.* **2021**, *16*, 1850–1870. [CrossRef] [PubMed]
36. Wiegand, I.; Hilpert, K.; Hancock, R.E.W. Agar and broth dilution methods to determine the minimal inhibitory concentration (MIC) of antimicrobial substances. *Nat. Protoc.* **2008**, *3*, 163–175. [CrossRef]
37. Haney, E.F.; Trimble, M.J.; Hancock, R.E.W. Microtiter plate assays to assess antibiofilm activity against bacteria. *Nat. Protoc.* **2021**, *16*, 2615–2632. [CrossRef]
38. Mansour, S.C.; de la Fuente-Núñez, C.; Hancock, R.E.W. Peptide IDR-1018: Modulating the immune system and targeting bacterial biofilms to treat antibiotic-resistant bacterial infections. *J. Pept. Sci.* **2015**, *21*, 323–329. [CrossRef]
39. Zou, R.; Zhu, X.; Tu, Y.; Wu, J.; Landry, M.P. Activity of antimicrobial peptide aggregates decreases with increased cell membrane embedding free energy cost. *Biochemistry* **2018**, *57*, 2606–2610. [CrossRef]
40. Zai, Y.; Xi, X.; Ye, Z.; Ma, C.; Zhou, M.; Chen, X.; Siu, S.W.I.; Chen, T.; Wang, L.; Kwok, H.F. Aggregation and its influence on the bioactivities of a novel antimicrobial peptide, Temporin-PF, and its analogues. *Int. J. Mol. Sci.* **2021**, *22*, 4509. [CrossRef]
41. Kumar, P.; Pletzer, D.; Haney, E.F.; Rahanjam, N.; Cheng, J.T.J.; Yue, M.; Aljehani, W.; Hancock, R.E.W.; Kizhakkedathu, J.N.; Straus, S.K. Aurein-derived antimicrobial peptides formulated with pegylated phospholipid micelles to target methicillin-resistant *Staphylococcus aureus* skin infections. *ACS Infect. Dis.* **2019**, *5*, 443–453. [CrossRef] [PubMed]

42. Moreira Brito, J.C.; Carvalho, L.R.; Neves de Souza, A.; Carneiro, G.; Magalhães, P.P.; Farias, L.M.; Guimarães, N.R.; Verly, R.M.; Resende, J.M.; Elena de Lima, M. PEGylation of the antimicrobial peptide LyeTx I-b maintains structure-related biological properties and improves selectivity. *Front. Mol. Biosci.* **2022**, *9*, 1001508. [CrossRef]
43. Singh, S.; Papareddy, P.; Mörgelin, M.; Schmidtchen, A.; Malmsten, M. Effects of PEGylation on membrane and lipopolysaccharide interactions of host defense peptides. *Biomacromolecules* **2014**, *15*, 1337–1345. [CrossRef] [PubMed]
44. Steinstraesser, L.; Hirsch, T.; Schulte, M.; Kueckelhaus, M.; Jacobsen, F.; Mersch, E.A.; Stricker, I.; Afacan, N.; Jenssen, H.; Hancock, R.E.W.; et al. Innate defense regulator peptide 1018 in wound healing and wound infection. *PLoS ONE* **2012**, *7*, e39373. [CrossRef] [PubMed]
45. Rivas-Santiago, B.; Castañeda-Delgado, J.E.; Rivas Santiago, C.E.; Waldbrook, M.; González-Curiel, I.; León-Contreras, J.C.; Enciso-Moreno, J.A.; del Villar, V.; Mendez-Ramos, J.; Hancock, R.E.W.; et al. Ability of innate defence regulator peptides IDR-1002, IDR-HH2 and IDR-1018 to protect against *Mycobacterium tuberculosis* infections in animal models. *PLoS ONE* **2013**, *8*, e59119. [CrossRef]
46. Deshmane, S.L.; Kremlev, S.; Amini, S.; Sawaya, B.E. Monocyte chemoattractant protein-1 (MCP-1): An overview. *J. Interf. Cytokine Res.* **2009**, *29*, 313–326. [CrossRef]
47. Yazdi, A.S.; Ghoreschi, K. The interleukin-1 family. *Adv. Exp. Med. Biol.* **2016**, *941*, 21–29.
48. Haney, E.F.; Wuerth, K.C.; Rahanjam, N.; Safaei Nikouei, N.; Ghassemi, A.; Alizadeh Noghani, M.; Boey, A.; Hancock, R.E.W. Identification of an IDR peptide formulation candidate that prevents peptide aggregation and retains immunomodulatory activity. *Pept. Sci.* **2019**, *111*, e24077. [CrossRef]
49. Imura, Y.; Nishida, M.; Matsuzaki, K. Action mechanism of PEGylated magainin 2 analogue peptide. *Biochim. Biophys. Acta* **2007**, *1768*, 2578–2585. [CrossRef]
50. Guiotto, A.; Pozzobon, M.; Canevari, M.; Manganelli, R.; Scarin, M.; Veronese, F.M. PEGylation of the antimicrobial peptide nisin A: Problems and perspectives. *Farmaco* **2003**, *58*, 45–50. [CrossRef]
51. Wu, Z.; Li, X.; Ericksen, B.; de Leeuw, E.; Zou, G.; Zeng, P.; Xie, C.; Li, C.; Lubkowski, J.; Lu, W.Y.; et al. Impact of pro segments on the folding and function of human neutrophil alpha-defensins. *J. Mol. Biol.* **2007**, *368*, 537–549. [CrossRef] [PubMed]
52. Imura, Y.; Nishida, M.; Ogawa, Y.; Takakura, Y.; Matsuzaki, K. Action mechanism of tachyplesin I and effects of PEGylation. *Biochim. Biophys. Acta* **2007**, *1768*, 1160–1169. [CrossRef]
53. Kumar, P.; Takayesu, A.; Abbasi, U.; Kalathottukaren, M.T.; Abbina, S.; Kizhakkedathu, J.N.; Straus, S.K. Antimicrobial peptide-polymer conjugates with high activity: Influence of polymer molecular weight and peptide sequence on antimicrobial activity, proteolysis, and biocompatibility. *ACS Appl. Mater. Interfaces* **2017**, *9*, 37575–37586. [CrossRef] [PubMed]
54. Talat, S.; Thiruvikraman, M.; Kumari, S.; Kaur, K.J. Glycosylated analogs of formaecin I and drosocin exhibit differential pattern of antibacterial activity. *Glycoconj. J.* **2011**, *28*, 537–555. [CrossRef] [PubMed]
55. Li, T.; Yang, N.; Teng, D.; Mao, R.; Hao, Y.; Wang, X.; Wang, J. C-terminal mini-PEGylation of a marine peptide N6 had potent antibacterial and anti-inflammatory properties against *Escherichia coli* and *Salmonella* strains in vitro and in vivo. *BMC Microbiol.* **2022**, *22*, 128. [CrossRef]
56. Falciani, C.; Lozzi, L.; Scali, S.; Brunetti, J.; Bracci, L.; Pini, A. Site-specific pegylation of an antimicrobial peptide increases resistance to Pseudomonas aeruginosa elastase. *Amino Acids* **2014**, *46*, 1403–1407. [CrossRef]
57. Morris, C.J.; Beck, K.; Fox, M.A.; Ulaeto, D.; Clark, G.C.; Gumbleton, M. Pegylation of antimicrobial peptides maintains the active peptide conformation, model membrane interactions, and antimicrobial activity while improving lung tissue biocompatibility following airway delivery. *Antimicrob. Agents Chemother.* **2012**, *56*, 3298–3308. [CrossRef]
58. Coltart, D.M.; Royyuru, A.K.; Williams, L.J.; Glunz, P.W.; Sames, D.; Kuduk, S.D.; Schwarz, J.B.; Chen, X.T.; Danishefsky, S.J.; Live, D.H. Principles of mucin architecture: Structural studies on synthetic glycopeptides bearing clustered mono-, di-, tri-, and hexasaccharide glycodomains. *J. Am. Chem. Soc.* **2002**, *124*, 9833–9844. [CrossRef]

pharmaceutics

MDPI

Article

Pulmonary Safety Profile of Esc Peptides and Esc-Peptide-Loaded Poly(lactide-co-glycolide) Nanoparticles: A Promising Therapeutic Approach for Local Treatment of Lung Infectious Diseases

Floriana Cappiello [1,†], Bruno Casciaro [1,†], Maria Rosa Loffredo [1], Elena Puglisi [1], Qiao Lin [2], Dandan Yang [2], Gemma Conte [3], Ivana d'Angelo [4], Francesca Ungaro [3], Loretta Ferrera [5], Raffaella Barbieri [6], Laura Cresti [7], Alessandro Pini [7], Yuanpu Peter Di [2,*] and Maria Luisa Mangoni [1,*]

1 Department of Biochemical Sciences, Laboratory Affiliated to Istituto Pasteur Italia-Fondazione Cenci Bolognetti, Sapienza University of Rome, 00185 Rome, Italy
2 Department of Environmental and Occupational Health, University of Pittsburgh, Pittsburgh, PA 15261, USA
3 Department of Pharmacy, University of Napoli Federico II, 80131 Napoli, Italy
4 Department of Environmental, Biological and Pharmaceutical Sciences and Technologies (DiSTABiF), University of Campania Luigi Vanvitelli, 81100 Caserta, Italy
5 U.O.C. Genetica Medica, IRCCS, Istituto Giannina Gaslini, 16147 Genoa, Italy
6 Biophysic Institute, Consiglio Nazionale delle Ricerche (CNR), 16149 Genoa, Italy
7 Department of Medical Biotechnologies, University of Siena, 53100 Siena, Italy
* Correspondence: peterdi@pitt.edu (Y.P.D.); marialuisa.mangoni@uniroma1.it (M.L.M.)
† These authors contributed equally to the work.

Citation: Cappiello, F.; Casciaro, B.; Loffredo, M.R.; Puglisi, E.; Lin, Q.; Yang, D.; Conte, G.; d'Angelo, I.; Ungaro, F.; Ferrera, L.; et al. Pulmonary Safety Profile of Esc Peptides and Esc-Peptide-Loaded Poly(lactide-co-glycolide) Nanoparticles: A Promising Therapeutic Approach for Local Treatment of Lung Infectious Diseases. *Pharmaceutics* **2022**, *14*, 2297. https://doi.org/10.3390/pharmaceutics14112297

Academic Editors: Scavello Francesco, Jean-Eric Ghia and Amiche Mohamed

Received: 16 September 2022
Accepted: 21 October 2022
Published: 26 October 2022

Abstract: In recent years, we have discovered Esc(1-21) and its diastereomer (Esc peptides) as valuable candidates for the treatment of Pseudomonas lung infection, especially in patients with cystic fibrosis (CF). Furthermore, engineered poly(lactide-co-glycolide) (PLGA) nanoparticles (NPs) were revealed to be a promising pulmonary delivery system of antimicrobial peptides. However, the "ad hoc" development of novel therapeutics requires consideration of their stability, tolerability, and safety. Hence, by means of electrophysiology experiments and preclinical studies on healthy mice, we demonstrated that neither Esc peptides or Esc-peptide-loaded PLGA NPs significantly affect the integrity of the lung epithelium, nor change the global gene expression profile of lungs of treated animals compared to those of vehicle-treated animals. Noteworthy, the Esc diastereomer endowed with the highest antimicrobial activity did not provoke any pulmonary pro-inflammatory response, even at a concentration 15-fold higher than the efficacy dosage 24 h after administration in the free or encapsulated form. The therapeutic index was ≥70, and the peptide was found to remain available in the bronchoalveolar lavage of mice, after two days of incubation. Overall, these studies should open an avenue for a new up-and-coming pharmacological approach, likely based on inhalable peptide-loaded NPs, to address CF lung disease.

Keywords: antimicrobial peptides; cystic fibrosis; lung infection; biodegradable nanocarrier; therapeutic index; transepithelial resistance; lung gene expression; mouse bronchoalveolar lavage

1. Introduction

The rising appearance of drug-resistant microorganisms has undermined the success achieved in the last century in the field of medicine via the utilization of antibiotics [1–4]. Antimicrobial peptides (AMPs) with a rapid-membrane-perturbing activity make bacteria less prone to develop resistance and constitute a promising source of new effective therapeutics [5–7]. Because of their biocompatibility, seldom accumulation in body tissues, and the harmlessness of their degradation products, peptides have become a unique class of therapeutic agents [8]; to date, more than 80 peptides have reached the global market [9].

In parallel, extensive research has been carried out in terms of the discovery, production, and optimization of AMPs, and an average of 70 AMPs are in preclinical/clinical evaluation [10,11]. AMPs are generally produced by all living species as key effectors of the innate immune system [12]; amphibian skin represents an invaluable wealthy storehouse of such molecules [13]. Studies conducted in our laboratory have led to the discovery of a frog-skin-derived membrane-active AMP, i.e., Esc(1-21), with bactericidal activity against both the planktonic and biofilm forms of the human pathogen *Pseudomonas aeruginosa* [14]. Nowadays, lung colonization by the sessile form of this bacterium holds a challenging threat, mostly in patients affected by cystic fibrosis (CF) [15,16], a genetic disease caused by mutations in the gene encoding the CF transmembrane conductance regulator (CFTR) protein, which controls chloride and bicarbonate transport mainly at the apical membrane of secretory epithelia, including those at the airways [17,18]. The most common mutation is the deletion of phenylalanine 508 (F508-del CFTR), which causes the production of an incorrectly folded protein that is rapidly degraded [19]. Furthermore, the small fraction of F508del-CFTR that reaches the plasma membrane [20] exhibits a defect in channel gating. As a result, the outflow of anions is inhibited, leading to increased water absorption by epithelial cells and the formation of a sticky airway mucus where inhaled microbes accumulate, giving rise to the onset of chronic pulmonary infection with serious respiratory dysfunctions.

During the last several years, by changing the stereochemical configuration of only two L-amino acids of Esc(1-21), i.e., Leu^{14} and Ser^{17}, to the corresponding D-enantiomers, we discovered that the resulting diastereomer Esc(1-21)-1c is: (i) more resistant to enzymatic degradation [21] and (ii) more efficient in restoring bronchial epithelium integrity [22,23]. This last feature is not shown by any traditional antibiotic and is expected to accelerate the healing of a damaged lung epithelium, especially in CF lungs, where wound-healing processes are highly compromised [24]. Interestingly, we also discovered an unprecedented property of AMPs, which is the ability of both Esc(1-21) and Esc(1-21)-1c (Esc peptides) to act as potentiators of the CFTR with ion conductance defects [25].

However, the "ad hoc" development of novel therapeutics based on peptide antibiotics and their translation from basic research to the clinic requires consideration of several underexplored aspects associated with the effective application of AMPs. These include studies on: (i) peptides' stability under conditions that reflect the physiology of the target site (e.g., the lung); (ii) tolerability for the assessment of a therapeutic window; (iii) identification and validation of an efficient delivery system of AMPs without eliciting undesirable local and/or systemic side effects. Previous studies highlighted Esc(1-21)-1c as the most efficient Esc peptide in reducing the lung bacterial burden in a mouse model of acute Pseudomonas lung infection, upon a single intratracheal (i.t.) instillation at a very low dosage (0.1 mg/kg, corresponding to 20 μM) with comparable efficacy to colistin, the last-resort antibiotic for the treatment of infections [26]. Nevertheless, studies aimed at evaluating Esc peptides' local toxicity or tolerability at higher therapeutic dosages are missing. Recently, we also demonstrated how polyvinyl-alcohol (PVA)-engineered poly(lactide-co-glycolide) (PLGA) nanoparticles (PVA-PLGA NPs) represent an enticing nanoformulation for pulmonary delivery of AMPs, able to (i) assist AMP diffusion through biological barriers, such as the mucus (which becomes a thick layer in CF) and (ii) prolong AMP antibacterial efficacy against Pseudomonas-induced lung infection [27]. Conceiving PVA-PLGA NPs as a promising inhalable formulation for the treatment of lung infections, and investigating their pulmonary safety, in terms of gene expression and tissue integrity compared to the free peptides, would be highly recommended [28].

Hence, in line with the above, we initially performed studies aimed at identifying the effect(s) of peptide-based PLGA formulation in comparison to the soluble free form of Esc peptides on both (i) the integrity of CFTR-expressing epithelium and (ii) the pulmonary host response after i.t. administration in healthy mice. Afterward, we assessed the maximum tolerated dosage and stability in the bronchoalveolar lavage for the Esc isoform endowed with the highest in vivo antimicrobial effectiveness, Esc(1-21)-1c.

2. Materials and Methods

2.1. Peptides

Synthetic Esc(1-21) (GIFSKLAGKKIKNLLISGLKG-NH_2) and its diastereomer, Esc(1-21)-1c, GIFSKLAGKKIKN(d-Leu)LI(d-Ser)GLKG-NH_2, were purchased from Biomatik (Wilmington, NC, USA). Briefly, the peptides were assembled by stepwise solid-phase synthesis using a standard F-moc protocol and purified via reverse-phase high-performance liquid chromatography (RP-HPLC) to a purity of 95%. The molecular mass was verified by mass spectrometry (Figure S1).

2.2. PVA-PLGA Nanoparticle Production and Characterization

NPs containing either synthetic Esc(1-21) or its diastereomer, Esc(1-21)-1c, at a theoretical loading of 2% (2 mg of peptide *per* 100 mg of NPs) were prepared by emulsion/solvent diffusion, as previously reported [27]. Briefly, 100 µL of an aqueous solution of each Esc peptide was added to a solution of uncapped PLGA 50:50 (Resomer® RG 502H, Mw 7000–17,000 Da, inherent viscosity 0.16–0.24 dL/g, Evonik, Germany) in methylene chloride (1 mL, 10 mg/mL) under vortex mixing (Reax top, Heidolph, Germany). The resulting water-in-oil emulsion was poured into 12.5 mL of ethanol to induce the production of NPs. Then, the NP dispersion was diluted with 12.5 mL of aqueous 0.1% (*w/v*) PVA (Mowiol® 40–88, average Mw ~205 000 Da, 87–89% hydrolyzed, Merck, Italy) and kept under magnetic stirring for 10 min at room temperature. The residual organic solvent was evaporated under vacuum at 30 °C (Rotavapor®, Heidolph VV 2000, Germany). NP colloidal dispersion was collected, adjusted to a final volume of 5 mL, and centrifuged at 7000× *g* for 20 min at 4 °C (Hettich Zentrifugen, Universal 16R, Hettich, Germany) to isolate NPs. The resulting NP pellet was diluted in ultrapure water up to the desired final concentration. When needed, Esc-peptide-loaded NPs were freeze-dried, adding trehalose as a cryoprotectant, with a NP/trehalose ratio of 1:25 *w*/*w*, as previously reported [27]. Control bare PVA-PLGA NPs were prepared in the absence of Esc peptides.

2.3. Cells

The following cell cultures were employed: CFBE41o- and Fischer rat thyroid (FRT) cells with stable expression of F508del-CFTR (F508del-CBFE41o- and F508del-FRT, respectively) as well as CFBE41o- cells expressing a wild-type CFTR (wt-CFBE41o-) [29]. CFBE41o- cells were cultured in minimal essential medium (MEM), and FRT cells were cultured in Coon's modified Ham's F-12 medium (Sigma-Aldrich, St. Luis, MO, USA), both supplemented with 10% fetal calf serum, 2 mM L-glutamine, and antibiotics (0.1 mg/mL of penicillin and streptomycin), at 37 °C and 5% CO_2 in 75 cm^2 flasks. Stable expression is maintained by adding puromycin (0.5 µg/mL or 2 µg/mL for wt-CBFE41o- or F508del-CBFE41o-, respectively) and 0.6 mg/mL zeocin for FRT cell line to the complete culture medium. CF and non-CF primary bronchial cells were from the University of Pittsburgh CF center cell culture core facility. They were cultured in flasks and maintained at air–liquid interface (ALI) until being fully differentiated as polarized cells for experiments.

2.4. Transepithelial Electrical Resistance (TEER) Experiments

Epithelial integrity was evaluated by measuring the transepithelial electrical resistance in CBFE41o- and FRT cells using the 24-transwell plates (24 Millicell plates PSHT010R1) and an electronic resistance system (Millicell ERS-2, EMD Millipore, Burlington, MA, USA). The medium was replaced every 3 days, and cells were used for TEER measurements after 7 days. Two days before the experiment, epithelia were incubated in their standard culture medium supplemented with 1 µM Vx-809 in the case of F508del- CBFE41o- and F508del-FRT. After 24 h, the medium was discarded, and epithelia were treated with Esc peptides, Esc-peptide-loaded PVA-PLGA NPs, or control bare PVA-PLGA NPs (in both the apical and basolateral compartments at different concentrations). After 24 h, the medium was replaced with a saline solution containing (in mM) 130 NaCl, 2.7 KCl, 1.5 KH_2PO_4, 1 $CaCl_2$, 0.5 $MgCl_2$, 10 glucose, and 10 Na-Hepes (pH, 7.4), which was added to both apical and basolateral compartments of the permeable supports. A sterile electrode was applied onto the apical side of the transwell insert containing the cells with/without treatments, and TEER was measured in basal conditions by the epithelial voltmeter and then converted to transepithelial conductance (TEEC) using the formula TEEC = 1/TEER [30].

2.5. Scanning Electron Microscopy

Primary cultured epithelial cells were seeded in 6.5 mm Transwell® with 0.4 μm-pore polyester membrane sterile insert (2×10^5 cells/insert/100 μL) while 350 μL of bronchial epithelial cell growth medium (Lonza, Basel, Switzerland) was added to the basolateral side of the transwell system (12 inserts in a 24 well plate). The apical medium was removed 3 days after seeding the cells on top of the inserts. The newly established air–liquid interface (ALI) cell culture system was cultured for an additional 8–10 days to achieve full confluency. Afterward, epithelia were washed apically with 100 μL of phosphate-buffered saline (PBS). A total of 100 μL of PBS supplemented or not with each Esc peptide at 20 μM was added on the apical compartment. Cells were then incubated for 5 h at 37 °C. Afterward, apical supernatant was removed, and samples were fixed with 2.5% glutaraldehyde in 0.01 M PBS for 60 min. The samples were then fixed in 1% osmium tetroxide (OsO_4) for 60 min and extensively washed with the same buffer and dehydrated with a graded ethanol series. After dehydration in hexamethyldisilazane and sputter coating, the samples were examined using a scanning electron microscope (Philips XL 30 CP instrument).

2.6. Gene Expression

All animal experiments were carried out based on a protocol (n. 20087639) approved by the Institutional Animal Care and Use Committee of the University of Pittsburgh according to the National Institutes of Health (NIH) guide for the care and use of laboratory animals. Seven-week-old female wild-type CD1 mice were anesthetized by isoflurane inhalation and instilled intratracheally with 50 μL of peptide solution (0.1 mg/kg, i.e., 20 μM) or PVA-PLGA NP suspension (either bare NPs or NPs loaded with each peptide at 0.1 mg/kg), in PBS. Control mice were given 50 μL of PBS without peptide. After 24 h, mice were euthanized for lung tissue isolation to extract RNA for differential expression analysis (sequencing work was performed by Novogene US Marketing).

2.7. Mouse Toxicity and Maximum Tolerated Dosage (MTD)

Five-week-old female wild-type CD1 mice were anesthetized by isoflurane inhalation and instilled intratracheally with 1.5 mg/kg (corresponding to 300 μM) of Esc(1-21)-1c in the free or encapsulated form in 50 μL of PBS. Mice instilled with 50 μL of PBS or bare PVA-PLGA NPs were included for control. After 1 day and 14 days, mice were euthanized for tissue isolation and examined for histopathology. Tissues were fixed in situ with 4% paraformaldehyde for 10 min with open chest cavity. They were then embedded in paraffin, and 5 μm-thick tissue slices were prepared by staining in hematoxylin and eosin and analyzed for pathological severity. In parallel, another group of seven animals was anesthetized by isoflurane inhalation and instilled intratracheally with escalating concentrations up to 7 mg/kg Esc(1-21)-1c in 50 μL of PBS and monitored for survival to determine the MTDs.

2.8. Stability Measurements in Bronchoalveolar Lavage Fluid

Bronchoalveolar lavage fluid (BAL) was collected from male and female C57BL/6 mice (4–5 months old) according to the procedure described in [31,32]. Mice were euthanized by means of CO_2 narcosis; a small incision at the level of the neck was made and a blunt needle connected to a syringe was inserted into the trachea. Then, lungs were washed with 1 mL of sterile PBS. BAL samples from each mouse were pooled, centrifuged at 1000 rpm (corresponding to $0.1\times g$) for 5 min, and stored at −80 °C. The supernatant was used for stability tests.

Peptides were dissolved in 400 μL of PBS at a concentration of 300 μM, and added to an equal volume of BAL at a final concentration of 150 μM. At different time points, 150 μL aliquots were withdrawn and added to 500 μL of acetonitrile, and then centrifuged. The supernatants were diluted with 160 μL of 0.1% trifluoroacetic acid (TFA)—water and analyzed by RP-HPLC and mass spectrometry. Liquid chromatography was performed on a Phenomenex Jupiter C_{18} analytical column (300 Å, 5 μm, 250 × 4.6 mm) in a 30 min gradient, using 0.1% TFA in water as solvent A and acetonitrile as solvent B. Mass spectrometry analysis of diluted samples was performed with a Bruker Daltonic-ultraflex-matrix-assisted laser desorption ionization tandem time-of-flight (MALDI-TOF/TOF) mass spectrometer.

2.9. *Statistical Analyses*

Quantitative data, collected from independent experiments, were expressed as the means ± standard errors of the means (S.E.M.). Statistical analysis was performed using one-way analysis of variance (ANOVA) with PRISM software version 8.0.1 (GraphPad, San Diego, CA, USA). Differences were considered statistically significant at a p value of <0.05. The levels of statistical significance are indicated in the legends of the figures.

3. Results

3.1. *In Vitro Effect of Esc Peptides and Esc-Peptide-Loaded PVA-PLGA NPs on F508del-CFTR-Expressing Epithelium*

Esc-peptide-loaded PVA-PLGA NPs, comprising a PLGA core to efficiently entrap and slowly release the peptide cargo and a PVA shell providing for mucus-/biofilm-penetrating properties, were produced as previously reported [27]. In the optimized formulation conditions, NPs display a hydrodynamic diameter lower than 300 nm, a low polydispersity index, a slight negative ζ-potential, a complete peptide entrapment (entrapment efficiency always around 100%), and a typical biphasic in vitro release profile of the entrapped Esc peptide, lasting for 3 days [27]. Conceiving PVA-PLGA NPs for inhalation [27], their effect on the lung epithelium integrity was initially tested, either in the free or loaded form, by measuring the transepithelial electrical conductance after 24 h of treatment. Bronchial epithelial cells expressing a functional copy of CFTR or its mutated F508del form (wt-CBFE41o- and F508del-CBFE41o-, respectively) were employed. Peptides in the soluble free form were also included for comparison. They were used at two different concentrations, i.e., 10 and 20 μM. Note that 10 μM corresponds to the best (minimal) concentration able to display a CFTR potentiator activity in bronchial epithelial cells [25], while 20 μM was the optimal peptide concentration showing pulmonary antimicrobial efficacy [26]. As reported in Figure 1, negligible changes in the transepithelial electrical conductance were attained when the epithelium was treated with the Esc-peptide-loaded PVA-PLGA NPs in comparison to untreated samples or samples treated with the free-peptide counterparts. This indicates that neither PVA-PLGA NPs or Esc peptides are harmful to epithelial cells expressing either wild-type or mutated CFTR nor cause paracellular leakage of ions, meaning that cell junctions remain well-tightened.

Remarkably, similar results were obtained for the FRT expressing F508del-CFTR, which have been extensively used for studies on CFTR protein (Figure S2).

Note that epithelial cells may be able to repair the damage induced by the administration of exogenous compounds in the long term (24 h). However, the harmless effect of Esc peptides on both normal and CF lung epithelia was also confirmed at a shorter time (5 h) by scanning electron microscopy (SEM). As shown in Figure S3, treatment of both normal and CF primary bronchial epithelial cells (grown in ALI to better mimic the human airway conditions and to drive differentiation towards a mucociliary phenotype) with 20 μM Esc peptides did not provoke any significant morphological change in the epithelial surface.

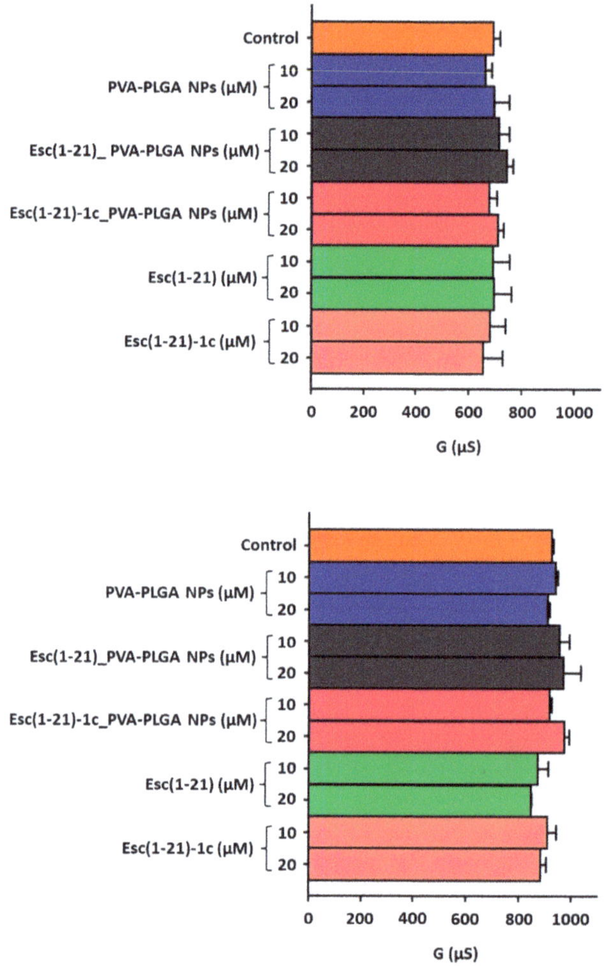

Figure 1. Effect of bare or Esc-peptide-loaded PVA-PLGA NPs on transepithelial conductance measured in F508del (upper panel) and wt (lower panel) CFBE41o- after 24 h of incubation compared to the free Esc peptides at two different concentrations. The controls were untreated cells. Data are expressed as mean ± S.E.M. from three independent experiments. No statistical difference was found.

3.2. In Vivo Studies: Effect of Free or Encapsulated Esc Peptides on Global Pulmonary Genetic Changes

To go in-depth into the potential clinical application of inhalable PVA-PLGA NPs as delivery systems of Esc peptides, we determined their pulmonary safety profile in healthy mice, when used either in their bare form or loaded with the Esc peptides. For a complete overview of the global genetic changes in the lungs of animals after 24 h exposure to peptide-loaded PVA-PLGA NPs compared to the free peptides or the vehicle PBS, we performed transcriptome studies by RNA-seq. To this aim, the standard immunocompetent outbred mouse strain CD-1 was used. Due to its ability to reflect the natural response of a non-immunocompromised human being, this is a common animal model for toxicity analysis. Esc-peptide-loaded PVA-PLGA NPs as well as free Esc peptides at a dosage of 0.1 mg/kg (~20 μM), which have already found to be active in vivo by provoking a 2-log_{10} reduction in lung bacterial burden [26], were intratracheally administered.

Only a minimal number of lung genes (up to 6 out of 25,000) were up- or downregulated by more than two-fold in animals treated with Esc peptides or with Esc-peptide-loaded PVA-PLGA NPs versus PBS-treated animals. The identified genes are shown in Table 1; the corresponding volcano plots of the differential gene expression are reported in Figure S4. Among them were a member of the non-protein-coding small nucleolar RNA gene family (*Snhg11*), three pseudogenes (i.e., *Eif4a-ps4*, *Gm10320*, and *Gm14150*), and two genes predicted to be structural constituents of ribosomes (i.e., *Rpl26* and *Rpl30*) or encoding a component of the cytoskeletal motor protein dynein (*Dynlt1f*) or alpha-synuclein (*Snca*), while the other genes were found to be involved in the positive regulation of protein phosphorylation (*Chil1*) or the signaling receptor-binding activity (*H2-T22*). Remarkably, none of these genes appear to be involved in toxicity-related processes.

Table 1. List of up- or downregulated genes in lungs of animals treated with Esc(1-21)-1c, Esc-peptide-loaded PVA-PLGA NPs, compared to those of PBS-treated animals.

	Effect	Gene Name	Gene Description	Gene Type	Log_2FC
Esc(1-21)	None				
Esc(1-21)-1c	Down	*Snhg11*	Small nucleolar RNA host gene 11	Protein coding	−1.949
	Up	*Rpl30*	Ribosomal protein L30	Protein coding	14.876
	Up	*Rpl26*	Ribosomal protein L26	Protein coding	20.049
	Up	*Chil1*	chitinase-like 1	Protein coding	34.248
	Up	*Eif4a-ps4*	-	Pseudogene	17.867
	Up	*Dynlt1f*	Dynein light-chain Tctex-type 1F	Protein coding	19.702
Esc(1-21)_PVA-PLGA NPs	Down	*H2-T22*	Histocompatibility 2 T region locus 22	Protein coding	−15.437
	Up	*Gm10320*	Predicted pseudogene 10320	Protein coding	24.607
Esc(1-21)-1c_PVA-PLGA NPs	Down	*Snca*	Synuclein alpha	Protein coding	−14.029
	Down	*Gm14150*	Predicted gene 14150	Pseudogene	−25.684

Up: upregulated; down: downregulated. Log_2 fold change is also indicated.

3.3. Esc(1-21)-1c Safety Profile in Lungs and Other Organs

We previously demonstrated that i.t. instillation of 0.1 mg/kg of Esc peptides either in the free form or upon encapsulation into PVA-PLGA NPs did not alter the lung mucociliary clearance in healthy mice [27]. However, the in vivo toxicity of Esc-peptide-loaded PVA-PLGA NPs upon pulmonary administration is an unexplored area. Therefore, we investigated the lung safety of the most promising Esc peptide, Esc(1-21)-1c, for therapeutic development. To this purpose, Esc(1-21)-1c was i.t. instilled in CD1 mice either in the free or encapsulated form at 1.5 mg/kg (~300 μM), corresponding to a 15-fold-higher concentration than that used in previous in vivo efficacy studies (0.1 mg/kg) [27]. As highlighted by the histological analysis in Figure 2, we did not detect any inflammatory response (there were no infiltrates of inflammatory cells) or lung tissue damage either after 1 day or 14 days from peptide administration in the free or encapsulated form. In parallel, no toxicity was detected for the corresponding amount of bare PVA-PLGA NPs, in line with our previous work showing an invariant expression of inflammation-associated genes (including IL-6, IL-10, or the tumor necrosis factor-α and NF-κB) in the lungs of mice after instillation of PVA-PLGA NPs encapsulated or not with Esc peptides [27].

In addition, no visible tissue injury, necrosis, or alteration in cell density was observed for other organs, such as the liver and kidneys, in comparison to control samples (Figure 3).

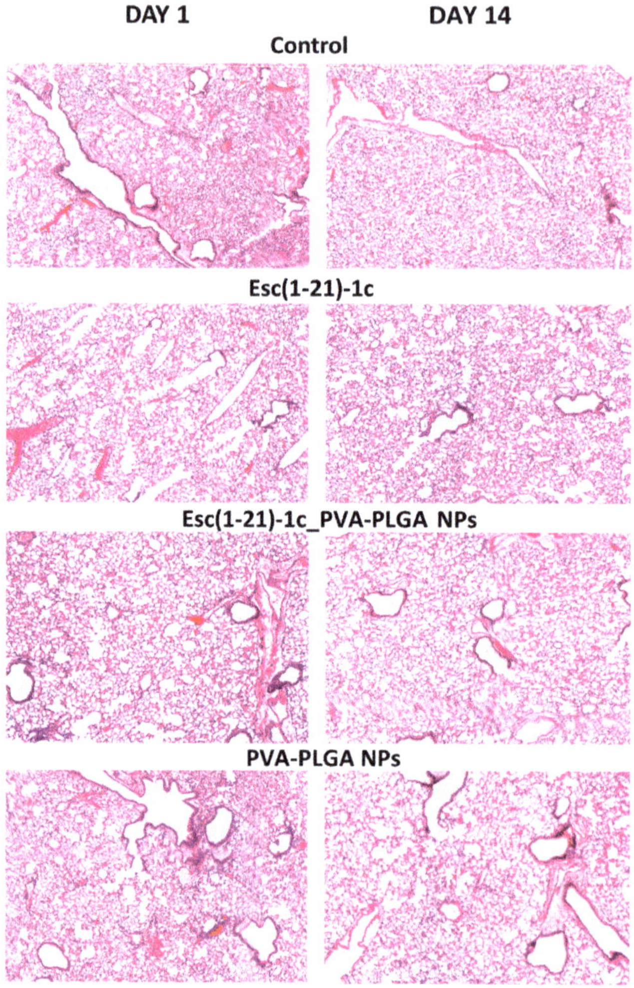

Figure 2. Histological analysis of lungs in CD1 mice, after 1 day and 14 days from i.t. administration of Esc(1-21)-1c either in the free or encapsulated form at a dosage of 1.5 mg/kg. The amount of unloaded PVA-PLGA NPs was the same as that present in Esc(1-21)-1c-loaded NPs. The results were compared to those obtained with PBS-treated mice (control), (magnification 4×).

To ensure that the lung safety of Esc(1-21)-1c 24 h after its administration was not due to its complete degradation in the lung environment, we studied its biostability at 300 μM in the presence of mouse BAL. As pointed out by the mass spectra in Figure 4, when Esc(1-21)-1c was incubated with BAL, a peak of molecular mass of 2185 Da corresponding to the full-length peptide [22] was detected even after 24 h. On the contrary, the all-L isomer was significantly less stable, as highlighted by the appearance of multiple degradation products and the lack of the peak corresponding to the entire peptide sequence after 6 h. However, both spectra showed the appearance of a peak (2129 Da) at 6 h, which became prevalent at 24 h. This peak indicates the loss of a Gly residue from the peptide sequences.

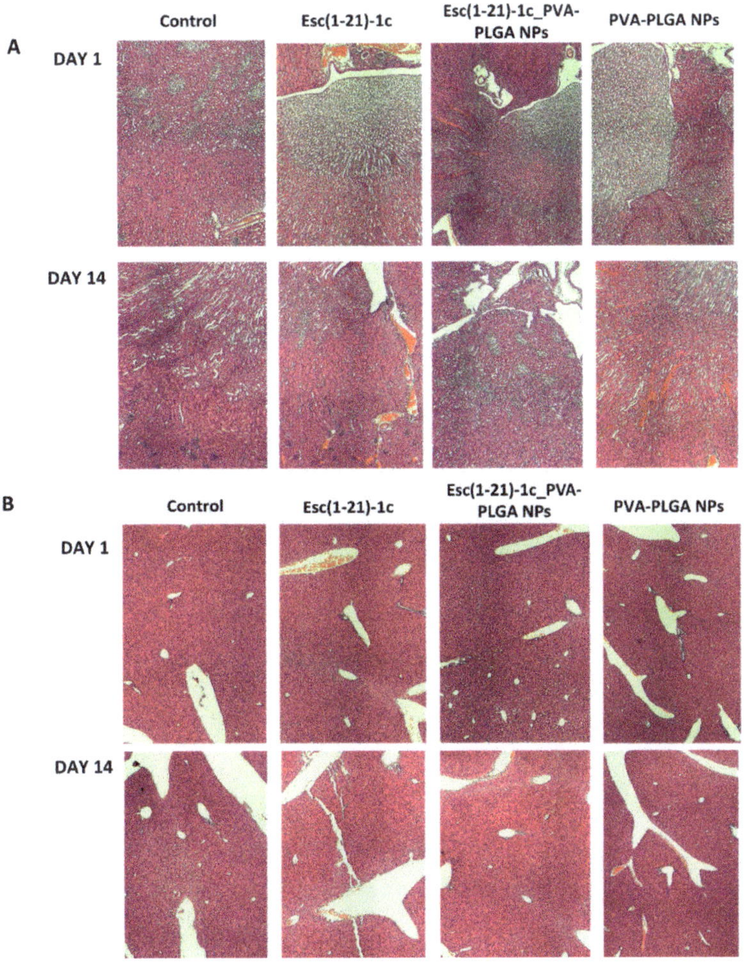

Figure 3. Representative images of kidney (**A**) and liver (**B**) tissues in CD1 mice, after 1 day and 14 days from i.t. administration of Esc(1-21)-1c either in the free or encapsulated form at a dosage of 1.5 mg/kg compared to the corresponding bare PVA-PLGA NPs or PBS-treated animals (control). Organs were harvested, fixed, and stained for histological evaluation (magnification 4×).

3.4. Determination of Maximum Tolerated Dosage

Development of peptides as anti-infective agents requires knowledge of their therapeutic index (TI), which is defined as the ratio between the maximum tolerated dosage, MTD (i.e., the highest concentration causing no obvious adverse effects and no mortality), and the therapeutic dosage mTd (i.e., the minimal dosage reducing bacterial burden by 2-$\log_{10}$ in the number of bacterial cells) [33]. Therefore, to identify the MTD of the most efficacious Esc(1-21)-1c, CD1 mice were i.t. instilled with increasing concentrations of the peptide, from 1.5 mg/kg to 7 mg/kg, and their survival was monitored for 14 days. Remarkably, the peptide was well-tolerated by the animals, which remained viable for the entire duration of the experiment and maintained the same range of motion both after a short time (1 h) and longer time (24 h) from its i.t. instillation at the highest dosage of 7 mg/kg. Notably, this concentration was 70-fold greater than the efficacious dose, indicating a TI of at least 70. When such a high concentration of Esc(1-21)-1c was used, the peptide was detectable in the mouse BAL after 48 h of incubation with BAL (Figure S5). This finding indicates that the harmlessness of the peptide at the long term cannot be attributed to its degradation.

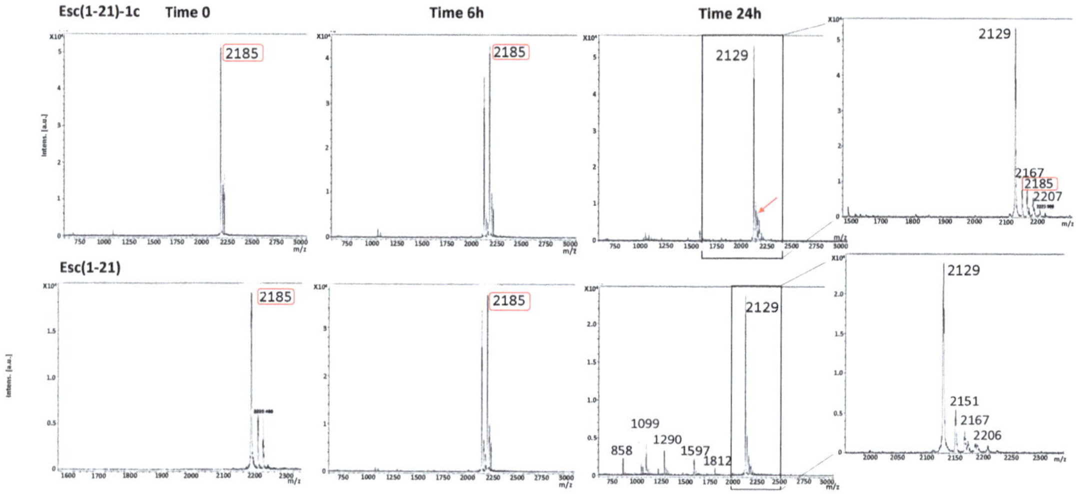

Figure 4. Mass spectra of Esc(1-21)-1c and the all-L Esc(1-21) after 6 and 24 h in BAL. Marked in red is the peak of molecular mass at 2185 corresponding to the full-length peptide. The arrow indicates the peak of molecular mass at 2185 found for Esc(1-21)-1c after 24 h of incubation with BAL.

Furthermore, histopathological examination of lung tissue did not reveal significant recruitment of inflammatory cells compared to PBS-treated mice within 48 h from i.t. instillation of Esc(1-21)-1c, in agreement with the absence of macroscopic damage to the lungs (Figure 5).

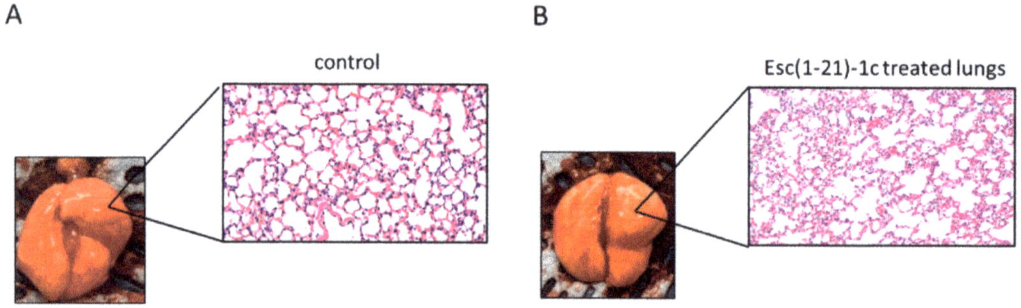

Figure 5. Histologic analysis of mouse lung tissues at 48 h after i.t. instillation of Esc(1-21)-1c at 7mg/kg. Lung tissues were harvested, fixed, and stained for histological evaluation without peptide treatment (**A**) or after peptide administration (**B**) (magnification 20×).

Similarly to what was found for the peptide dosage of 1.5 mg/kg, no alteration in the tissue structure was visible at the level of the liver, spleen, and kidneys after treatment with 7 mg/kg of Esc(1-21)-1c, compared to the control (Figure 6).

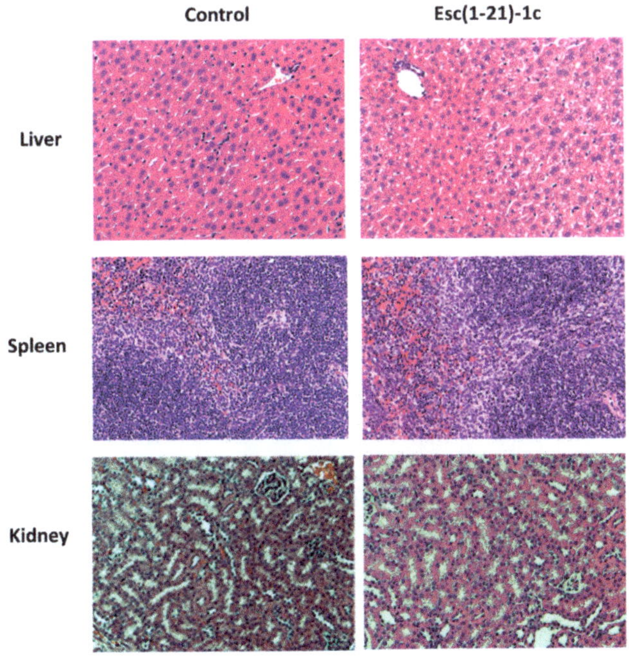

Figure 6. Representative images of liver, spleen, and kidneys from healthy mouse at 48 h after i.t. instillation of Esc(1-21)-1c at 7mg/kg. Tissues were harvested, fixed, and stained for histological evaluation without peptide treatment (control, left side) or after peptide administration (right side) (magnification 20×).

4. Discussion

The urgent health concern related to antibiotic resistance has driven a renewed interest in the clinical development of peptides compared to the previous two decades. Interestingly, besides having direct antimicrobial activity, some AMPs also display additional biological functions involved in the modulation of host immunity, including enhanced chemotaxis of immune cells, activation of immune cell differentiation, stimulation of wound healing, and angiogenesis, scavenging of bacterial endotoxins [34]. The broad spectrum and rapid bactericidal activity of AMPs combined with the low risk of inducing resistance makes them a valid option for alternative antimicrobial compounds. However, to the best of our knowledge, there are no AMPs in clinical trials for the development of new inhalable drugs against lung infections.

Here, to emphasize the attractive properties of Esc peptides for the treatment of lung diseases, especially in CF, likely via pulmonary delivery upon incorporation into PVA-PLGA NPs, we investigated their effect on the lung epithelial integrity and the host immune response in terms of gene expression and tissue damage at the target district. Remarkably, we discovered that both Esc peptides and PVA-PLGA (drug carrier) are harmless to epithelial cells. This was evidenced by the invariant transepithelial conductance in bronchial epithelia upon treatment, as measured by TEER experiments, which appear to be more sensitive than typical cytotoxicity tests by the 3-(4,5-dimethylthiazol-2-yl)-2,5-diphenyltetrazolium bromide (MTT) assay. Nonetheless, one of the most critical limitations of AMPs, particularly when embedded into a delivery system, is the lack of potential toxicity data in animal models. There are currently 15 FDA-approved PLA-/PLGA-based drug products available on the US market [35], and several studies on the usage of PLGA-based nanocarriers for pulmonary drug delivery are in progress [36,37]. Together with chitosan, PLGA is the most investigated polymer for the development of inhaled formulations aimed at extending the pulmonary exposure and pharmacological effect of encapsulated drugs [38]. This is because of its biodegradability [39], which can be modulated by varying the lactide/glycolide ratio, molecular weight, chemical structure, and biocompatibility [28]. We formerly highlighted how incorporating AMPs into PVA-PLGA NPs and their pulmonary administration represent a suitable approach to assist peptide diffusion through an artificial lung mucus, eradicate biofilm, and potentiate and extend the in vivo antimicrobial efficacy of Esc peptides in the lung [27]. The work from Haque and colleagues [40] reported that PLGA NPs do not appear to be absorbed into the lungs after pulmonary administration but rather are degraded into lower-molecular-weight constituents that are subsequently absorbed. PLGA NPs have also been used (i) to release loaded AMPs to the wounds to accelerate healing processes [41] and (ii) to deliver drugs in tumor combination therapy [42].

Nevertheless, a limited number of studies have been performed to confirm the safety of PLGA NPs in the short and long term, especially in the lung compartment [43–46].

Our previous work highlighted that PVA-PLGA NPs do not induce any recruitment of inflammatory cells into the lung alveoli of healthy mice nor changes in the expression of inflammation-associated genes 36 h after pulmonary administration [27]. Here, the transcriptomic analysis provided us with the first demonstration that Esc peptides, either in the free or encapsulated form in PVA-PLGA NPs, promote a response closer to the gene expression pattern of the PBS-treated host.

To the best of our knowledge, this is the first case showing the effect of peptide-loaded PVA-PLGA NPs on the global genetic profile of targeted tissues (i.e., lungs) upon administration in the conductive airways of mice, in comparison to the vehicle-treated animals. Interestingly, only 2 genes out of 25,000 were found to be up-/downregulated by more than 2-fold by the Esc-peptide-loaded PVA-PLGA versus the PBS-treated mice (i.e., *H2-T22* and the pseudogene *Gm10320* for Esc(1-21)-loaded NPs and *Snca* and the pseudogene *Gm14150* for Esc(1-21)-1c-loaded NPs). Moreover, as pointed out by histological analysis of lungs tissue, no sign of inflammation/damage was detected by NPs when used in the free or loaded form with Esc(1-21)-1c at a concentration 15-fold higher than the therapeutic dosage of 0.1 mg/kg [47–49], either after 1 day or 14 days from i.t. administration. Together, these data have contributed to further supporting the harmlessness of PVA-PLGA NPs in the respiratory tract.

The other important finding of this work is the discovery of the safety profile of Esc(1-21)-1c and its tolerability by animals, at a concentration 70-fold higher than the efficacy dose, without eliciting any detectable damage to the lung, spleen, liver, and kidney, unlike what has been described for the AMP D-BMAP18, which was found to cause lung edema when used at doses of 1 and 2 mg/kg [50]. In addition, we proved that the absence of toxicity was not due to the complete degradation of the peptide, which instead remained available in the BAL within the first 48 h. This is in contrast with the low biostability of other investigated AMPs for treatment of lung infections, such as (i) all-L-BMAP18 AMPs, which were degraded by pulmonary proteases in murine BAL fluids within the first 20 min of exposure [51] and did not display any in vivo antibacterial activity, and (ii) P-113, which showed a half-life of a few minutes in undiluted sputum [52].

5. Conclusions

In conclusion, we have filled the gap of some underexplored but relevant aspects for the translation of AMPs into new inhalable medicines by providing evidence of (i) the pulmonary safety profile of Esc peptides and PVA-PLGA NPs as a valuable nanoparticulate system for AMP delivery to the lungs; (ii) a prolonged biostability of Esc(1-21)-1c in the mouse BAL together, and (iii) its promising therapeutic index.

Overall, in addition to expanding our knowledge on the safety profile of Esc peptides for the development of new drugs to treat *P. aeruginosa* lung infection, our studies should open the avenue for a new up-and-coming pharmacological approach, likely based on inhalable peptide-loaded NPs, to address CF lung disease.

Supplementary Materials: The following supporting information can be downloaded at: https://www.mdpi.com/article/10.3390/pharmaceutics14112297/s1. Figure S1: HPLC and mass spectra of Esc peptides; Figure S2: Effect of bare or Esc-peptide-loaded PVA-PLGA NPs on F508del-FRT epithelium integrity after 24 h of incubation; Figure S3: SEM images of bronchial epithelia; Figure S4: Effect of free or encapsulated peptides on lung gene expression; Figure S5: Mass spectra of Esc(1-21)-1c after 48 h of incubation with mouse BAL.

Author Contributions: Conceptualization, B.C., Y.P.D., and M.L.M.; formal analysis, F.C., B.C., and L.F.; investigation, F.C., B.C., M.R.L., E.P., Q.L., D.Y., G.C., L.C., and R.B.; funding acquisition M.L.M.; resources, I.d., F.U., A.P., Y.P.D., and M.L.M.; supervision, Y.P.D. and M.L.M.; writing—original draft, F.C., B.C., and M.L.M.; writing—review and editing, Y.P.D. and M.L.M. All authors have read and agreed to the published version of the manuscript.

Funding: Fondazione Italiana per la Ricerca sulla Fibrosi Cistica (Project FFC 8/2019) Delegazione FFC di Imola e Romagna con Gruppo di sostegno FFC di Faenza.

Institutional Review Board Statement: Animal experiments were carried out based on a protocol (n. 20087639) approved by the Institutional Animal Care and Use Committee of the University of Pittsburgh according to the National Institutes of Health (NIH) guide for the care and use of laboratory animals.

Informed Consent Statement: Not applicable.

Data Availability Statement: Data are contained within the article or Supplementary Material.

Acknowledgments: M.L.M. thanks Luis Galietta (Telethon Institute of Genetics and Medicine, Pozzuoli, Italy) for providing cell lines.

Conflicts of Interest: The authors declare no conflict of interest.

References

1. Antimicrobial Resistance, C. Global burden of bacterial antimicrobial resistance in 2019: A systematic analysis. *Lancet* **2022**, *399*, 629–655. [CrossRef]
2. Lepape, A.; Jean, A.; De Waele, J.; Friggeri, A.; Savey, A.; Vanhems, P.; Gustin, M.P.; Monnet, D.L.; Garnacho-Montero, J.; Kohlenberg, A. European intensive care physicians' experience of infections due to antibiotic-resistant bacteria. *Antimicrob. Resist. Infect. Control* **2020**, *9*, 1. [CrossRef] [PubMed]
3. Boucher, H.W.; Talbot, G.H.; Bradley, J.S.; Edwards, J.E.; Gilbert, D.; Rice, L.B.; Scheld, M.; Spellberg, B.; Bartlett, J. Bad bugs, no drugs: No ESKAPE! An update from the Infectious Diseases Society of America. *Clin. Infect. Dis.* **2009**, *48*, 1–12. [CrossRef]
4. Barra, A.L.C.; Dantas, L.O.C.; Morao, L.G.; Gutierrez, R.F.; Polikarpov, I.; Wrenger, C.; Nascimento, A.S. Essential Metabolic Routes as a Way to ESKAPE From Antibiotic Resistance. *Front. Public Health* **2020**, *8*, 26. [CrossRef] [PubMed]
5. Browne, K.; Chakraborty, S.; Chen, R.; Willcox, M.D.; Black, D.S.; Walsh, W.R.; Kumar, N. A New Era of Antibiotics: The Clinical Potential of Antimicrobial Peptides. *Int. J. Mol. Sci.* **2020**, *21*, 7047. [CrossRef] [PubMed]
6. Mahlapuu, M.; Hakansson, J.; Ringstad, L.; Bjorn, C. Antimicrobial Peptides: An Emerging Category of Therapeutic Agents. *Front. Cell. Infect. Microbiol.* **2016**, *6*, 194. [CrossRef]
7. Mookherjee, N.; Anderson, M.A.; Haagsman, H.P.; Davidson, D.J. Antimicrobial host defence peptides: Functions and clinical potential. *Nat. Rev. Drug Discov.* **2020**, *19*, 311–332. [CrossRef] [PubMed]
8. Muttenthaler, M.; King, G.F.; Adams, D.J.; Alewood, P.F. Trends in peptide drug discovery. *Nat. Rev. Drug Discov.* **2021**, *20*, 309–325. [CrossRef]
9. Mahlapuu, M.; Bjorn, C.; Ekblom, J. Antimicrobial peptides as therapeutic agents: Opportunities and challenges. *Crit. Rev. Biotechnol.* **2020**, *40*, 978–992. [CrossRef]
10. Koo, H.B.; Seo, J. Antimicrobial peptides under clinical investigation. *Pept. Sci.* **2019**, *111*, e24122. [CrossRef]
11. Dijksteel, G.S.; Ulrich, M.M.; Middelkoop, E.; Boekema, B.K. Review: Lessons Learned From Clinical Trials Using Antimicrobial Peptides (AMPs). *Front. Microbiol.* **2021**, *12*, 616979. [CrossRef] [PubMed]
12. Boman, H.G. Peptide antibiotics and their role in innate immunity. *Annu. Rev. Immunol.* **1995**, *13*, 61–92. [CrossRef] [PubMed]
13. Mangoni, M.L. Temporins, anti-infective peptides with expanding properties. *Cell. Mol. Life Sci. CMLS* **2006**, *63*, 1060–1069. [CrossRef] [PubMed]
14. Luca, V.; Stringaro, A.; Colone, M.; Pini, A.; Mangoni, M.L. Esculentin(1-21), an amphibian skin membrane-active peptide with potent activity on both planktonic and biofilm cells of the bacterial pathogen Pseudomonas aeruginosa. *Cell. Mol. Life Sci. CMLS* **2013**, *70*, 2773–2786. [CrossRef]
15. Flume, P.A. Pulmonary complications of cystic fibrosis. *Respir. Care* **2009**, *54*, 618–627. [CrossRef]
16. Ciofu, O.; Tolker-Nielsen, T.; Jensen, P.O.; Wang, H.; Hoiby, N. Antimicrobial resistance, respiratory tract infections and role of biofilms in lung infections in cystic fibrosis patients. *Adv. Drug Deliv. Rev.* **2015**, *85*, 7–23. [CrossRef]

17. Sheppard, D.N.; Welsh, M.J. Structure and function of the CFTR chloride channel. *Physiol. Rev.* **1999**, *79*, S23–S45. [CrossRef]
18. Cheng, S.H.; Rich, D.P.; Marshall, J.; Gregory, R.J.; Welsh, M.J.; Smith, A.E. Phosphorylation of the R domain by cAMP-dependent protein kinase regulates the CFTR chloride channel. *Cell* **1991**, *66*, 1027–1036. [CrossRef]
19. Jensen, T.J.; Loo, M.A.; Pind, S.; Williams, D.B.; Goldberg, A.L.; Riordan, J.R. Multiple proteolytic systems, including the proteasome, contribute to CFTR processing. *Cell* **1995**, *83*, 129–135. [CrossRef]
20. Pranke, I.M.; Sermet-Gaudelus, I. Biosynthesis of cystic fibrosis transmembrane conductance regulator. *Int. J. Biochem. Cell Biol.* **2014**, *52*, 26–38. [CrossRef]
21. Di Grazia, A.; Cappiello, F.; Cohen, H.; Casciaro, B.; Luca, V.; Pini, A.; Di, Y.P.; Shai, Y.; Mangoni, M.L. D-Amino acids incorporation in the frog skin-derived peptide esculentin-1a(1-21)NH2 is beneficial for its multiple functions. *Amino Acids* **2015**, *47*, 2505–2519. [CrossRef] [PubMed]
22. Cappiello, F.; Di Grazia, A.; Segev-Zarko, L.A.; Scali, S.; Ferrera, L.; Galietta, L.; Pini, A.; Shai, Y.; Di, Y.P.; Mangoni, M.L. Esculentin-1a-Derived Peptides Promote Clearance of Pseudomonas aeruginosa Internalized in Bronchial Cells of Cystic Fibrosis Patients and Lung Cell Migration: Biochemical Properties and a Plausible Mode of Action. *Antimicrob. Agents Chemother.* **2016**, *60*, 7252–7262. [CrossRef] [PubMed]
23. Cappiello, F.; Ranieri, D.; Carnicelli, V.; Casciaro, B.; Chen, H.T.; Ferrera, L.; Di, Y.P.; Mangoni, M.L. Bronchial epithelium repair by Esculentin-1a-derived antimicrobial peptides: Involvement of metalloproteinase-9 and interleukin-8, and evaluation of peptides' immunogenicity. *Sci. Rep.* **2019**, *9*, 18988. [CrossRef] [PubMed]
24. Adam, D.; Bilodeau, C.; Sognigbe, L.; Maille, E.; Ruffin, M.; Brochiero, E. CFTR rescue with VX-809 and VX-770 favors the repair of primary airway epithelial cell cultures from patients with class II mutations in the presence of Pseudomonas aeruginosa exoproducts. *J. Cyst. Fibros.* **2018**, *17*, 705–714. [CrossRef]
25. Ferrera, L.; Cappiello, F.; Loffredo, M.R.; Puglisi, E.; Casciaro, B.; Botta, B.; Galietta, L.J.V.; Mori, M.; Mangoni, M.L. Esc peptides as novel potentiators of defective cystic fibrosis transmembrane conductance regulator: An unprecedented property of antimicrobial peptides. *Cell. Mol. Life Sci. CMLS* **2021**, *79*, 67. [CrossRef]
26. Chen, C.; Mangoni, M.L.; Di, Y.P. In vivo therapeutic efficacy of frog skin-derived peptides against Pseudomonas aeruginosa-induced pulmonary infection. *Sci. Rep.* **2017**, *7*, 8548. [CrossRef]
27. Casciaro, B.; d'Angelo, I.; Zhang, X.; Loffredo, M.R.; Conte, G.; Cappiello, F.; Quaglia, F.; Di, Y.P.; Ungaro, F.; Mangoni, M.L. Poly(lactide- co-glycolide) Nanoparticles for Prolonged Therapeutic Efficacy of Esculentin-1a-Derived Antimicrobial Peptides against Pseudomonas aeruginosa Lung Infection: In Vitro and in Vivo Studies. *Biomacromolecules* **2019**, *20*, 1876–1888. [CrossRef]
28. Mundargi, R.C.; Babu, V.R.; Rangaswamy, V.; Patel, P.; Aminabhavi, T.M. Nano/micro technologies for delivering macromolecular therapeutics using poly(D,L-lactide-co-glycolide) and its derivatives. *J. Control. Release* **2008**, *125*, 193–209. [CrossRef]
29. Sondo, E.; Tomati, V.; Caci, E.; Esposito, A.I.; Pfeffer, U.; Pedemonte, N.; Galietta, L.J. Rescue of the mutant CFTR chloride channel by pharmacological correctors and low temperature analyzed by gene expression profiling. *Am. J. Physiol.-Cell Physiol.* **2011**, *301*, C872–C885. [CrossRef]
30. Sondo, E.; Falchi, F.; Caci, E.; Ferrera, L.; Giacomini, E.; Pesce, E.; Tomati, V.; Mandrup Bertozzi, S.; Goldoni, L.; Armirotti, A.; et al. Pharmacological Inhibition of the Ubiquitin Ligase RNF5 Rescues F508del-CFTR in Cystic Fibrosis Airway Epithelia. *Cell Chem. Biol.* **2018**, *25*, 891–905.e898. [CrossRef]
31. Di, M.E.; Yang, D.; Di, Y.P. Using Bronchoalveolar Lavage to Evaluate Changes in Pulmonary Diseases. *Methods Mol. Biol.* **2020**, *2102*, 117–128. [CrossRef] [PubMed]
32. Van Hoecke, L.; Job, E.R.; Saelens, X.; Roose, K. Bronchoalveolar Lavage of Murine Lungs to Analyze Inflammatory Cell Infiltration. *J. Vis. Exp. JoVE* **2017**, *123*, e55398. [CrossRef] [PubMed]
33. Di, Y.P.; Lin, Q.; Chen, C.; Montelaro, R.C.; Doi, Y.; Deslouches, B. Enhanced therapeutic index of an antimicrobial peptide in mice by increasing safety and activity against multidrug-resistant bacteria. *Sci. Adv.* **2020**, *6*, eaay6817. [CrossRef] [PubMed]
34. Hancock, R.E.; Haney, E.F.; Gill, E.E. The immunology of host defence peptides: Beyond antimicrobial activity. *Nat. Rev. Immunol.* **2016**, *16*, 321–334. [CrossRef] [PubMed]
35. Wang, Y.; Qin, B.; Xia, G.; Choi, S.H. FDA's Poly (Lactic-Co-Glycolic Acid) Research Program and Regulatory Outcomes. *AAPS J.* **2021**, *23*, 92. [CrossRef]
36. Ernst, J.; Klinger-Strobel, M.; Arnold, K.; Thamm, J.; Hartung, A.; Pletz, M.W.; Makarewicz, O.; Fischer, D. Polyester-based particles to overcome the obstacles of mucus and biofilms in the lung for tobramycin application under static and dynamic fluidic conditions. *Eur. J. Pharm. Biopharm.* **2018**, *131*, 120–129. [CrossRef]
37. Emami, F.; Mostafavi Yazdi, S.J.; Na, D.H. Poly(lactic acid)/poly(lactic-co-glycolic acid) particulate carriers for pulmonary drug delivery. *J. Pharm. Investig.* **2019**, *49*, 667. [CrossRef]
38. Guo, Y.; Bera, H.; Shi, C.; Zhang, L.; Cun, D.; Yang, M. Pharmaceutical strategies to extend pulmonary exposure of inhaled medicines. *Acta Pharm. Sin. B* **2021**, *11*, 2565–2584. [CrossRef]
39. Rytting, E.; Nguyen, J.; Wang, X.; Kissel, T. Biodegradable polymeric nanocarriers for pulmonary drug delivery. *Expert Opin. Drug Deliv.* **2008**, *5*, 629–639. [CrossRef]
40. Haque, S.; Pouton, C.W.; McIntosh, M.P.; Ascher, D.B.; Keizer, D.W.; Whittaker, M.R.; Kaminskas, L.M. The impact of size and charge on the pulmonary pharmacokinetics and immunological response of the lungs to PLGA nanoparticles after intratracheal administration to rats. *Nanomed. Nanotechnol. Biol. Med.* **2020**, *30*, 102291. [CrossRef]

41. Chereddy, K.K.; Her, C.H.; Comune, M.; Moia, C.; Lopes, A.; Porporato, P.E.; Vanacker, J.; Lam, M.C.; Steinstraesser, L.; Sonveaux, P.; et al. PLGA nanoparticles loaded with host defense peptide LL37 promote wound healing. *J. Control. Release* **2014**, *194*, 138–147. [CrossRef] [PubMed]
42. Zhang, R.; He, J.; Xu, X.; Li, S.; Peng, H.; Deng, Z.; Huang, Y. PLGA-based drug delivery system for combined therapy of cancer: Research progress. *Mater. Res. Express* **2021**, *8*, 122002.
43. Scolari, I.R.; Volpini, X.; Fanani, M.L.; La Cruz-Thea, B.; Natali, L.; Musri, M.M.; Granero, G.E. Exploring the Toxicity, Lung Distribution, and Cellular Uptake of Rifampicin and Ascorbic Acid-Loaded Alginate Nanoparticles as Therapeutic Treatment of Lung Intracellular Infections. *Mol. Pharm.* **2021**, *18*, 807–821. [CrossRef] [PubMed]
44. Motawea, A.; Ahmed, D.A.M.; El-Mansy, A.A.; Saleh, N.M. Crucial Role of PLGA Nanoparticles in Mitigating the Amiodarone-Induced Pulmonary Toxicity. *Int. J. Nanomed.* **2021**, *16*, 4713–4737. [CrossRef]
45. Shen, Y.; TanTai, J. Co-Delivery Anticancer Drug Nanoparticles for Synergistic Therapy Against Lung Cancer Cells. *Drug Des. Dev. Ther.* **2020**, *14*, 4503–4510. [CrossRef]
46. Mura, S.; Hillaireau, H.; Nicolas, J.; Kerdine-Romer, S.; Le Droumaguet, B.; Delomenie, C.; Nicolas, V.; Pallardy, M.; Tsapis, N.; Fattal, E. Biodegradable nanoparticles meet the bronchial airway barrier: How surface properties affect their interaction with mucus and epithelial cells. *Biomacromolecules* **2011**, *12*, 4136–4143. [CrossRef]
47. Sung, J.C.; Pulliam, B.L.; Edwards, D.A. Nanoparticles for drug delivery to the lungs. *Trends Biotechnol.* **2007**, *25*, 563–570. [CrossRef]
48. Haque, S.; Whittaker, M.R.; McIntosh, M.P.; Pouton, C.W.; Kaminskas, L.M. Disposition and safety of inhaled biodegradable nanomedicines: Opportunities and challenges. *Nanomed. Nanotechnol. Biol. Med.* **2016**, *12*, 1703–1724. [CrossRef]
49. Beck-Broichsitter, M.; Merkel, O.M.; Kissel, T. Controlled pulmonary drug and gene delivery using polymeric nano-carriers. *J. Control. Release* **2012**, *161*, 214–224. [CrossRef]
50. Mardirossian, M.; Pompilio, A.; Degasperi, M.; Runti, G.; Pacor, S.; Di Bonaventura, G.; Scocchi, M. D-BMAP18 Antimicrobial Peptide Is Active In vitro, Resists to Pulmonary Proteases but Loses Its Activity in a Murine Model of Pseudomonas aeruginosa Lung Infection. *Front. Chem.* **2017**, *5*, 40. [CrossRef]
51. Mardirossian, M.; Pompilio, A.; Crocetta, V.; De Nicola, S.; Guida, F.; Degasperi, M.; Gennaro, R.; Di Bonaventura, G.; Scocchi, M. In vitro and in vivo evaluation of BMAP-derived peptides for the treatment of cystic fibrosis-related pulmonary infections. *Amino Acids* **2016**, *48*, 2253–2260. [CrossRef] [PubMed]
52. Sajjan, U.S.; Tran, L.T.; Sole, N.; Rovaldi, C.; Akiyama, A.; Friden, P.M.; Forstner, J.F.; Rothstein, D.M. P-113D, an antimicrobial peptide active against Pseudomonas aeruginosa, retains activity in the presence of sputum from cystic fibrosis patients. *Antimicrob. Agents Chemother.* **2001**, *45*, 3437–3444. [CrossRef] [PubMed]

pharmaceutics

MDPI

Review

Recent Advances in Multifunctional Antimicrobial Peptides as Immunomodulatory and Anticancer Therapy: Chromogranin A-Derived Peptides and Dermaseptins as Endogenous versus Exogenous Actors

Francesco Scavello [1,*], Mohamed Amiche [2] and Jean-Eric Ghia [3,4]

[1] IRCCS Humanitas Research Hospital, 20089 Rozzano, MI, Italy
[2] Laboratoire de Biogenèse des Signaux Peptidiques (BioSiPe), Institut de Biologie Paris-Seine, Sorbonne Université-CNRS, 75252 Paris, France
[3] Department of Immunology, Rady Faculty of Health Sciences, University of Manitoba, Winnipeg, MB R3E 0T5, Canada
[4] Section of Gastroenterology, Department of Internal Medicine, Rady Faculty of Health Sciences, University of Manitoba, Winnipeg, MB R3E 0T5, Canada
* Correspondence: francesco.scavello@humanitasresearch.it

Abstract: Antimicrobial peptides (AMPs) are produced by all living organisms exhibiting antimicrobial activities and representing the first line of innate defense against pathogens. In this context, AMPs are suggested as an alternative to classical antibiotics. However, several researchers reported their involvement in different processes defining them as Multifunctional AMPs (MF-AMPs). Interestingly, these agents act as the endogenous responses of the human organism against several dangerous stimuli. Still, they are identified in other organisms and evaluated for their anticancer therapy. Chromogranin A (CgA) is a glyco-phosphoprotein discovered for the first time in the adrenal medulla but also produced in several cells. CgA can generate different derived AMPs influencing numerous physiological processes. Dermaseptins (DRSs) are a family of α-helical-shaped polycationic peptides isolated from the skin secretions of several leaf frogs from the *Phyllomedusidae* family. Several DRSs were identified as AMPs and, until now, more than 65 DRSs have been classified. Recently, these exogenous molecules were characterized for their anticancer activity. In this review, we summarize the role of these two classes of MF-AMPs as an example of endogenous molecules for CgA-derived peptides, able to modulate inflammation but also as exogenous molecules for DRSs, exerting anticancer activities.

Keywords: antimicrobial peptides; multifunctional antimicrobial peptides; chromogranin A-derived peptides; immunomodulators; dermaseptins; anticancer activities

Citation: Scavello, F.; Amiche, M.; Ghia, J.-E. Recent Advances in Multifunctional Antimicrobial Peptides as Immunomodulatory and Anticancer Therapy: Chromogranin A-Derived Peptides and Dermaseptins as Endogenous versus Exogenous Actors. *Pharmaceutics* **2022**, *14*, 2014. https://doi.org/10.3390/pharmaceutics14102014

Academic Editor: Rakesh Tiwari

Received: 24 August 2022
Accepted: 20 September 2022
Published: 22 September 2022

1. Introduction

The first line of response of mammalians against pathogenic invasion is innate immunity [1]. It consists of different molecules produced and released by various cell types belonging to the organs, immune or other systems [1]. Among these molecules, we find proteins with direct antimicrobial activity or those which activate the complement proteins [1]. Furthermore, small peptides play an essential role thanks to the presence of specific cationic sequences that interact with the membrane of the pathogens [1,2]. These molecules are called host defense peptides and are more widely known as antimicrobial peptides (AMPs) [2]. They can not only act through different antimicrobial mechanisms of action but [1,2] also with different intensities against an extensive collection of pathogens [1–4]. However, in light of the potential human clinical application, they can be classified as endogenous AMPs produced by the human organism, such as Defensins, Cathelicidins and Dermcidins [3,4]. On the other hand, exogenous AMPs are produced by microorganisms themselves, plants, insects, amphibians and fishes or mammals but not identified in humans, such as Thionins, Piscidins, Cecropins and Dermaseptins [5–9]. However, increasing data show that these AMPs are multifunctional peptides (MF-AMPs) with different roles in cardiovascular, nervous and renal systems [10–12]. They are also reported as chemokines, vaccine adjuvants, regulators of the innate defense, immunomodulators and anticancer agents [13,14]. All these properties of MF-AMPs confer to these molecules a broad spectrum of potential applications in the biomedical field. The 20th century will be remembered as the time when antibiotic resistance became a world health problem [15]. Different studies reported this issue and recommended limiting the use of antibiotics to contain the evolution of resistant bacteria [15,16]. However, to the present day, the problem is not solved. The World Health Assembly recognized antibiotic resistance as an alarming issue in human medicine and a leading cause of worldwide death [15,17]. In this context, the MF-AMPs appear to be a possible alternative to antibiotics and have acquired an increasingly clinical interest [18]. At the same time, prevention of nosocomial infection and inflammatory activation of the intervention site after the implantation of medical devices is a significant hospital challenge [19]. Additionally, in this case, MF-AMPs are promising, and several studies reported their functionalization of biomaterials, conferring antimicrobial and anti-inflammatory properties [20]. Finally, some of these agents can modulate systemic and tissue inflammation and immune cell activation [21], supporting future clinical applications as immunomodulators. At the same time, several studies showed the activities of these agents against different models of cancer cells, demonstrating a profile of MF-AMPs with potential clinical application also in cancer therapy [14]. Therefore, the principal aim of this review is to report AMPs as multifunctional agents for future clinical applications. More specifically, we will focus on endogenous Chromogranin A-derived peptides for immunomodulation and Dermaseptins as exogenous agents for cancer therapy.

2. General Features, Mechanism of Action and Possible Clinical Application of MF-AMPs

AMPs are effective agents for killing or blocking the growth of free microorganisms due to the self-endogenous origins and natural factors produced by the organisms [4,22]. In addition, the endogenous AMPs are not toxic at high concentrations, and some of these are reported as multifunctional agents with immunomodulatory activity without generating pathogen resistance [22]. Among these, cationic AMPs target bacterial cell membranes in a non-specific manner for pathogens. Still, they are not toxic for host cells due to the specific electrostatic interaction with microbial membrane compounds, such as negatively charged lipids or specific microbial ligands for antimicrobial peptides [23,24]. These agents belong to the host defense response of vertebrates' innate immune system [23,24]. More than 2000 AMPs are known to be derived from several organisms and catalogued in the AMPs Database [25–27]. The peptides are classified based on four different biochemical characteristics: 1. net charge (anionic, cationic and neutral); 2. length of the primary structure (more or less than 100 amino acids); 3. typology of secondary or tertiary structure (linear with β-sheet, α-helix, extended loop and very complex structure, such as cyclic peptides); and 4. hydrophobicity (amphipathic, hydrophobic and hydrophilic) [23,28,29] (Figure 1).

These molecules' general mechanism of action consists of an initial phase with membrane targeting based on electrostatic interactions with lipids and membrane ligands [23,28]. After achieving a threshold concentration, the molecular interactions between the peptides and membrane induce a conformational phase transition of these molecules. This effect is characterized by a modification from the disordered structure of peptides, typical in aqueous environments, to α-helical or β-sheet conformation upon interaction with phospholipids in the pathogen membrane [23,28]. Then, the peptides can destabilize the membrane inducing electrostatic tension. Finally, membrane disintegration is obtained by pore formation with micellization or lipid segregation (Figure 1). This phenomenon is caused by different molecular mechanisms, such as self-association and multimerization, barrel-stave mechanism, toroid pore or wormhole mechanism and carpet mechanism [23,24,28]. In addition, due to their lipophilic profile, some AMPs do not degrade the membrane but may translocate across it. These peptides exert a cell-penetrating peptides (CPPs)-like function. They induce inhibition of intracellular functions by blocking the cytoskeleton growth, causing membrane dysfunction inside the cell or inhibiting extracellular biopolymers and DNA/RNA/protein synthesis [22,23,28] (Figure 1). In recent years, these agents have acquired an increasing interest in clinical applications. They may represent a novel and promising therapeutic alternative to conventional antibiotics for preventing and eradicating resistant pathogens. AMPs also prevent pathogens' biofilm formation due to antimicrobial activity [20,29]. In addition, AMPs are elective to destabilize existing biofilms targeting matrix proteins or signaling pathways for growth and essential metabolic processes or compounds [20,29]. In these aspects, another therapeutic utilization of AMPs is preventing infection in the surgical sites and microbial biofilms on biomaterials often associated with the onset of nosocomial infections. Different biomaterials functionalized with peptides and peptidomimetics agents are available for clinical applications. As a few examples, in addition to medically implanted devices, they can be used orally and for wound sites in the surgical area [30,31]. Other vital interests have been reported for endogenous MF-AMPs based on their properties to modulate systemic and tissue inflammation and immune cell activation [21], supporting future clinical applications as immunomodulators. On the other hand, numerously exogenous MF-AMPs are reported as anti-proliferative against several cancer cells with potential clinical application in cancer therapy [14].

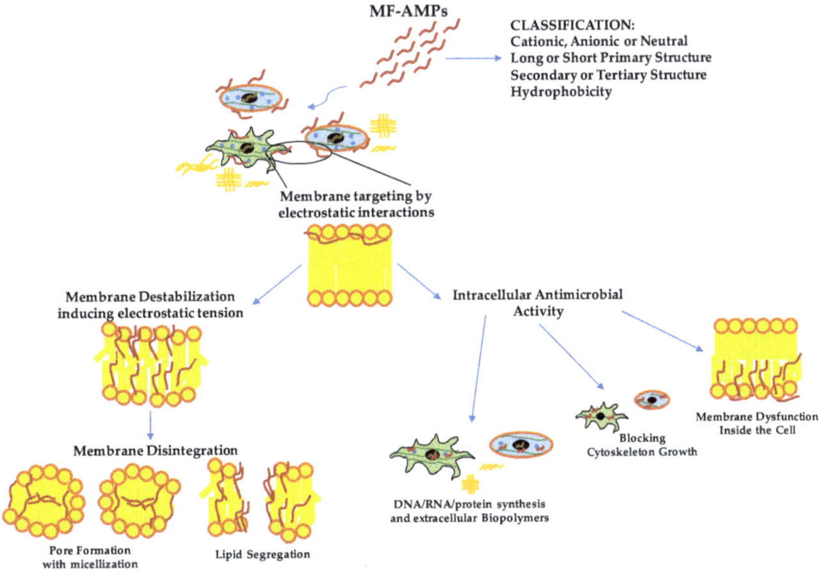

Figure 1. Antimicrobial Mechanisms of Action of Multifunctional Antimicrobial Peptides (MF-AMPs). Based on their biochemical characteristics, MF-AMPs can interact with the membrane lipids, inducing the pathogens' instability and rigidity of the cell membrane. Upon reaching the minimum inhibitory concentration (MIC) values, they cause pore formation and cell lysis. Some MF-AMPs exert a cell-penetrating peptides-like function. They can penetrate the membrane, blocking pathogens' growth by internal membrane dysfunction, cytoskeleton interaction and interference of biomolecules production. Deoxyribonucleic acid (DNA); ribonucleic acid (RNA).

3. CgA-Derived Peptides as Inflammatory Modulator Molecules

3.1. CgA-Derived AMPs

Chromogranin A (CgA) is a glycoprotein with 431 residues and a molecular weight of 49 kDa belonging to the Granin family, discovered for the first time at the end of the 1960s in the granules of adrenomedullary chromaffin cells [32]. In the last few decades, CgA has been identified in immune cells [33–35], neurons [36], cardiomyocytes [37], keratinocytes and fibroblasts [38]. This protein may influence different physiological processes. Its role was reported in cardiac function and cardio-protection [39], catecholamine storage and feedback release [40] and the modulation of vascular function [41] but also in cellular recruitment and the modulation of immune response [32,35,42]. However, this prohormone produces, by proteolytic processing, active biological peptides, such as Vasostatins (Vs), Prochromacin, Chromacin, Pancreastatin, WE 14, Catestatin (Cts), Parastatin and Serpinin [43,44]. At the end of the 1990s and early 2000s, several CgA-derived peptides were discovered as AMPs acting against several bacteria, fungi and yeast. The CgA-derived peptides have been found in biological fluids involved in host-defense responses, such as serum, saliva and neutrophils secretions, or against pathogens in the first barrier of the human body, such as the skin [33,34,38,45,46]. The CgA-derived peptides act as antimicrobial agents in the micro-molar range [33,46–48]. These concentrations are also reported in the biological fluid after stimulation with pathogen toxins or during infection [33,34,42,46]. Among the CgA-derived peptides, the Vs-I and Cts were first identified as antimicrobial agents; however, their antimicrobial domains were rapidly reported. Vs-I was initially identified as a vasoinhibitory agent [49]. For Vs-I, Lugardon characterized this peptide's antimicrobial activities against many pathogens [33,47,50]. However, after the incubation of Vs-I with endoproteinase Glu-C, a digested sequence CgA47-66, called Chromofungin (Chr), was identified and found highly active against several fungi and yeasts [47]. It has a global hydrophobicity and amphipathic character, allowing a strong interaction with the membrane. Specifically, Chr possesses a positive charge of +3.5, showing an amphipathic helix in the C-terminal part in the sequence CgA53-66 and at the N-terminal domain, a hydrophobic sequence corresponding to CgA48-51 and a hydrophilic structure CgA53-46, respectively [47,51]. The Vs-I and Chr antimicrobial mechanism of action is explained through the specific interaction of peptides with ergosterol, one of the main components of yeast and fungal membranes, inducing increased pressure and penetration into the membrane [47,51] (Figure 1). Other data demonstrated that Chr could inhibit Calcineurin activity by interacting with Calmodulin [47] (Figure 1). Within microbial cells, Vs-I and Chr may interfere with the Calcium/Calmodulin/Calcineurin signaling pathway by blocking the pathway implicated in virulence and skeleton development of cell walls [52]. Cts was identified as a catecholamine release-inhibitory peptide [53]. Cts is a small 21-amino-acid cationic peptide with a positive net charge of +5 within the bovine sequence (bCgA344-364) possessing a C-terminal hydrophobic sequence. Taylor and colleagues identified a smaller peptide (CgA344-358) derived from Cts with a more substantial inhibitory effect on catecholamine release [54]. This peptide was called Cateslytin (Ctl) by Briolat et al. and is also characterized by its antimicrobial activities with potent effects compared to Cts [46]. Ctl is also a positively charged (+5) arginine-rich antimicrobial peptide and, in an aqueous solution, is a linear peptide with a disordered structure. However, when interacting with the membrane, Ctl acquires an α-helical form [55]. Other studies with a system mimicking bacterial membrane demonstrated that Ctl could convert its structure into antiparallel β-sheets precipitating against the negatively charged part of the membranes [56]. Then, Ctl induced an increased rigidity, permeability gradient and membrane pore formation in the domains containing ergosterol [56–58] (Figure 1).

3.2. CgA-Derived Peptides and Immune Cells Activities and Inflammation

The involvement of CgA and its derived peptides in innate immunity is well known for its antimicrobial activity. In addition, the role of these MF-AMPs is also reported in immune cells, conferring them a complex profile of immunity modulators. This section analyzes this profile studied using in vitro and in vivo models. The role of CgA-derived peptides on immune cells was studied for the first time in 2009; in particular, the effects of Chr and Cts were evaluated on polymorphonuclear neutrophils. After the treatments with the peptides, Chr and Cts were observed inside the cells, demonstrating their ability to penetrate the mammalian membrane and the profile of CPPs [34]. Then, in the presence of extracellular calcium, the two peptides induced a transient calcium influx in the cells, binding Calmodulin-binding factors (W7 and CMZ) and activating iPLA2 [34]. In addition, the pharmacological block of these channels inhibited the calcium flux induced by Chr and Cts [34]. On the other hand, when extracellular calcium is absent, the peptides cannot induce calcium secretion [34]. Notably, the secretion of polymorphonuclear neutrophils treated with the two CgA-derived peptides induced the secretion of several important factors for innate immunity and inflammation, such as Lactotransferrin, Lysozyme, Neutrophil Gelatinase Associated Lipocalin and S100 calcium-binding protein A8/A9 [34]. Several studies reported the role of Vs-1 as an anti-atherogenesis and anti-inflammatory factor suppressing the adhesion of monocytes to endothelial cells by adhesion molecule down-regulation [59,60]. Xiong and coworkers reported the anti-inflammatory role of Vs-II (N-terminal fragment of CgA containing Vs-I; CgA1-113) in an apolipoprotein E-deficient ($ApoE^{-/-}$) mice model fed with a high-fat diet developing atherosclerosis. In this study, Vs-II treatment reduced the occurrence of atherosclerotic plaque and attenuated lesions [59]. Furthermore, Vs-II significantly reduced the production of pro-inflammatory cytokines in aortic tissue, such as Tumor Necrosis Factor-α (TNF-α), Monocyte Chemoattractant Protein-1 (MCP-1) and Vascular Cell Adhesion Molecule-1 (VCAM-1) [59]. The same authors demonstrated, by several in vivo analyses, that these anti-inflammatory properties are based on the ability of Vs-II to reduce leukocytes adhesion on $ApoE^{-/-}$ mice arteries but also on the recruitment, transmigration and accumulation of M1 macrophages in the lesions [59]. In the same animal model, Sato et al. showed that Vs-I treatment reduces aortic atherosclerotic lesions development due to reductions in intra-plaque inflammation, macrophage infiltration and aortic smooth muscle cells proliferation and plasma glucose level [60]. From a cellular point of view, Vs-I suppressed the lipopolysaccharide (LPS)-induced production of chemokine MCP1 and vascular damage markers, such as VCAM-1 and E-selectin, in human endothelial cells [60]. At the same time, Vs-I was found to reduce M1 pro-inflammatory macrophages differentiation and IL6 release but also oxidized low-density lipoprotein (oxLDL)-induced foam cell formation of macrophages [60]. Of great clinical interest, Vs-I is expressed around Monckeberg's medial calcific sclerosis in human radial arteries [60]. Additionally, the immunomodulatory role of Chr has been reported in a mice model of ulcerative colitis induced by dextran sulfate sodium administration [61,62]. This model decreased Chr expression [62]. Furthermore, Chr treatment, by intracolonic administration, significantly reduced the inflammation and severity of colitis. The anti-inflammatory effects were due to the differentiation of macrophages into M2 anti-inflammatory clones with the consequent reduction of released IL-18 and the increased expression of M2 markers [61,62]. Using the same animal model, Kapoor and colleagues demonstrated that intrarectal Chr treatment reduced colitis severity and inflammation [63]. In parallel, this was associated with a significant decrease in the expression of CD11c, CD40, CD80, CD86 IL6 and IL12p40 in the inflamed colonic mucosa, mesenteric lymph nodes and spleen [63]. In addition, Chr reduces in CD11c positive cells the expression of CD80, CD86 and NF-κB in the spleen and colon, respectively [63]. All these in vivo data demonstrated that Chr has protective properties against intestinal inflammation and exerts the role of immunomodulator for intestinal macrophages and dendritic cells (DCs). These effects were also demonstrated in vitro with macrophages showing that Chr increased the production of anti-inflammatory factors and the M2 differentiation [61]. At the same time, Chr treatment significantly reduced the expression of M1 macrophage markers and the activation of the NF-κB pathway [62]. Additionally, in this case, in vitro experiments with M1 macrophages demonstrated that this peptide could decrease cellular migration, proinflammatory cytokines production and release, and NF-κB phosphorylation [62]. Furthermore, treatment with Chr or a conditioned medium of Chr-treated macrophages M2 induced epithelial cell proliferation and migration but also decreased oxidative stress and pro-inflammatory cytokine production [61]. In addition, the Chr treatment of naïve bone marrow-derived CD11c positive DCs reduced the LPS-induced expression of CD40, CD80, CD86 IL-6 and IL-12p40 [63]. These results were also confirmed in intestinal tissue isolated from patients with ulcerative colitis, demonstrating that Chr expression was down-regulated in these patients compared to healthy controls [62]. Indeed, the mRNA levels of Chr were positively correlated with the mRNA expression of M2 macrophages activation markers and negatively to the expression of collagen, IL-8 and IL-18, but also with M1 activation markers (TLR-4 expression and NF-kB activation) and consequent pro-inflammatory cytokines production [61,62]. In these patients, the reduction of Chr level is also associated with a negative linear relationship with CD11c and CD86 [63]. Moreover, another study confirmed the anti-inflammatory effects of Chr on monocytes. Treatment with this peptide significantly inhibited the transcription of pro-inflammatory factors, such as NF-kB and AP-1, in these cells [64]. Furthermore, Rabbi and colleagues showed that Cts reduced intestinal inflammation and the onset of colitis lesions by a Stat-3 activation [65]. At the same time, the markers of M1 macrophage activation and the colonic levels of pro-inflammatory cytokines, such as IL-6, IL-1β and TNF-α, were significantly decreased by Cts treatment [66]. However, Cts did not influence M2 macrophage markers [66]. Furthermore, these anti-inflammatory effects were confirmed in cellular experiments using macrophages isolated from the peritoneal cavity and the bone marrow, demonstrating that in vitro treatment with Cts significantly decreased the production of the pro-inflammatory cytokines and phosphorylation of Stat-3 [65]. In addition, peritoneal macrophages isolated from naïve mice and treated with Cts and LPS displayed a reduction in the expression and production of pro-inflammatory cytokines blocking the activation of M1 macrophages [66]. In addition, the genetic deletion of Cts induces hypertensive conditions and left ventricular hypertrophy accompanied by significant macrophage infiltration in cardiac tissue and adrenal gland [67]. In this context, the absence of Cts induced an increased level of pro-inflammatory cytokines TNF-α, C-C motif chemokine ligand (CCL)-2, 3, C-X-C motif chemokine ligand (CXCL)-1 and catecholamines but also an elevated inflammation in heart with the up-regulation of cardiac genes, such as *Tnfa*, *Ifng*, *Emr1*, *Itgam*, *Itgax*, *Nos2a*, *IL12b*, *Ccl2* and *Cxcl1* [67]. It is of great interest that the intraperitoneal administration of Cts reversed this phenotype. In addition, macrophage depletion blocked the onset of hypertension in Cts-knockout (KO) mice [67]. Furthermore, bone-marrow transfer of KO animals in wild-type (WT) counterparts induced hypertension and cardiac inflammation, while opposite conditions showed the opposite phenotype [67]. All these data strongly suggest that the anti-hypertensive effects of Cts are partially mediated by an immunosuppressive action of this peptide on macrophages [67]. The role of Cts as an immunomodulator was also explored in the context of atherosclerosis and vascular injury. In fact, in vitro treatment with Cts on endothelial cells significantly reduces the release of TNF-α and vascular damage markers, such as ICAM-1 and VCAM-1, after LPS exposure [68]. At the same time, Cts treatment suppresses inflammatory responses and oxidizes the low-density lipoprotein-induced foam cell formation of human macrophages [68]. Kojima and colleagues demonstrated that Cts injection to $ApoE^{-/-}$ mice significantly reduces macrophage infiltration and the consequent atherosclerotic lesions onset in the aorta but also suppresses aortic smooth muscle cells proliferation and collagen deposition in atheromatous plaques [68]. Furthermore, in vitro experiments with human aortic smooth muscle cells showed that Cts treatment can block collagen-1 and fibronectin expression and migration, proliferation and apoptotic process [68]. Of significant clinical impact, coronary artery disease patients displayed a substantial reduction of plasmatic levels of Cts but an increased expression in coronary atheromatous plaques [68]. In an acute pulmonary embolism in vivo model and cellular experiments with human pulmonary artery endothelial cells, Cts treatment was found to abolish thrombin-induced inflammation blocking TLR-4 expression and p38 phosphorylation, decreasing the consequent acute pulmonary embolism [69]. In conclusion, Vs-I and Chr than Cts are key attenuators of inflammation in different tissue and pathological conditions by reducing immune cell infiltration and inflammatory activation (Figure 2).

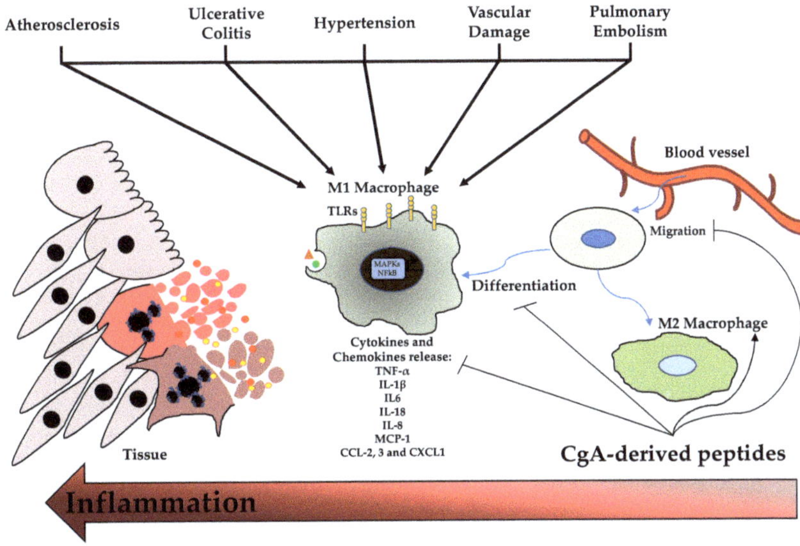

Figure 2. ChromgraninA (CgA)-derived Peptides and Inflammation. Damage or stress stimuli induce the migration, proliferation and activation of macrophages with consequent pro-inflammatory cytokines and chemokines release. This phenomenon generates cell death and inflammation. CgA-derived peptides can reduce M1 polarization and promote M2 anti-inflammatory macrophages differentiation. C-C motif chemokine ligand (CCL); C-X-C motif chemokine ligand (CXCL); interleukin (IL); monocyte chemoattractant protein (MCP); toll-like receptor (TLR); tumor necrosis factor (TNF).

4. Dermaseptins and Anticancer Therapy

4.1. Dermaseptins

Dermaseptins (DRSs) are a class of peptides identified in the skin secretions of several Amazonian tree frogs of the family *Phyllomedusidae*, in particular, the species *Phyllomedusa* [70,71]. The first member of DRS family was isolated from the skin secretion of *P. sauvagei* and named DRS-S1 [72]. This 34-residues-containing peptide has antimicrobial activity against Gram-positive and Gram-negative bacteria, yeast and protozoa without affecting mammalian cells. The second is DRS-B2, isolated from exudates of *P. bicolor* [73–75]. It is also known as adenoregulin due to its capacity to increase the affinity of the agonist toward the receptor of adenosine A1 [76]. DRS-B2, with its 33 amino-acid residues, is considered the most abundant member and the most active peptide of the B family (B for the frog species *P. bicolor*). To date, more than 65 DRSs, listed in the APD3 database (https://aps.unmc.edu/, accessed on 1 June 2022) [26], have been isolated primarily from the skin secretion of South American tree frogs of the 67-member family *Phyllomedusidae* (https://amphibiansoftheworld.amnh.org/, accessed on 1 June 2022). Multiple alignments of the 67 sequences of DRSs listed in APD3 clearly showed that these polycationic peptides, rich in Lys residue, share a signature consisting of a conserved Trp residue at position three and a consensus AA(A/G)KAAL(G/N)A motif in the middle region [70]. Their MIC values are in the low micromolar range for a large panel of microorganisms, comprising bacteria (*S. aureus*, *E. coli*), yeast (*C. albicans*), filamentous fungi (*A. fumigatus*) and protozoa, such as *Leishmania Mexicana*, and show no hemolytic activity against human and rabbit erythrocytes. The mode of action by which DRSs kill these microorganisms follows the "carpet" mechanism [77,78]. These polycationic peptides, destructured in aqueous media, adopt an alpha helix structure upon contact with the host cell plasma membranes and then interact with their negative charges [79–82]. Once bound, the peptide will disrupt membrane permeability and cause the death of the microorganism (Figure 1).

4.2. Dermaseptins and Anticancer Properties

The first two anticancer DRSs peptides were isolated from the South American Amazonian tree frog, *Phyllomedusa bicolor*. These molecules, DRSs-B2 and DRS-B3, were tested in vitro against a human prostatic adenocarcinoma PC-3 cell line, showing an antiproliferative effect with an EC_{50} around 2–3 μM and demonstrating the inhibition of proliferation of more than 90% [83]. In addition, these two peptides also inhibited PC-3 cell colony formation in soft agar and the proliferation, differentiation and capillary formation of endothelial cells [83,84]. Furthermore, DRS-B2 blocks the proliferation and colony formation of several human tumor cell types, such as prostatic adenocarcinoma LNCAP, prostatic carcinoma DU145, mammary carcinoma (MDA-MB2318) cell lines and B-lymphoma lines [83]. These effects were also confirmed in vivo by a cell line PC3 murine xenograft model, showing that DRS-B2 inhibits tumor growth [83]. The anticancer mechanism of action of DRS-B2 was demonstrated by in vitro experiments with tumor PC3 cells. This peptide interacted with tumor cell surface, aggregating and penetrating the cells. Furthermore, it induced the release of cytosolic lactate dehydrogenase, a marker of cytotoxicity and necrosis, but no effects were observed on mitochondrial membrane potential and caspase 3 activations for apoptotic involvement [83]. Concerning the mechanisms of action of DRS-B2, confocal microscopy studies revealed that this peptide rapidly accumulates to cytoplasmic membranes, packed in vesicles and into the nucleus [85]. These effects were also partially mediated by glycosaminoglycans' interaction with DRS-B2 and the consequent structural modification of the peptide with the α-helical domain [85]. Recently, a synthetic hormonotoxin molecule composed of dermaseptin-B2 associated with luteinizing hormone-releasing hormone (LHRH) was tested to improve the peptide's antitumor activity, reducing its peripheral toxicity and lethality. This hormonotoxin displayed an anticancer effect very similar to DRS-B2 both in vitro and in vivo [86]. The LHRH addition to dermaseptin-B2 does not alter the peptide's secondary structure and biological function [86]. On the other hand, double staining flow cytometry analysis showed that this hormonotoxin induced apoptosis instead of a necrotic process caused by DRS-B2 [86]. This different anticancer mechanism of action explains better tolerance and the lower toxicity of the hormonotoxin compared to dermaseptin-B2 [86]. In addition, other biochemical approaches have been used to increase the antitumor activity of DRS, delivering these agents in tumor cells, as seen in DRS-DStomo01 peptide [87]. DRS-DStomo01 was entrapped in chitosan nanoparticles, and the antitumor activity was tested in vitro against HeLa cells. The peptide induces DNA fragmentation and mitochondrial hyperpolarization with consequent cytotoxicity for cancer cells [87]. However, when used in chitosan nanoparticles, DRS-DStomo01 was more active than free peptides [87]. In 2016, two novel members of DRSs family were identified in the skin secretion of the frog *Pachymedusa dacnicolor* and called DRS-PD 1 and 2 [88]. Both peptides were reported to be active against many microorganisms, such as *E. coli*, *S. aureus*, *P. aeruginosa* and *C. albicans*, but with no lytic effects on mammalian red cells [88]. DRS-PD 2 displayed anti-proliferative effects against cancer cell lines, such as H157, PC-3 and U251-MG, within the concentration range of 10^{-9} to 10^{-4} M [88]. This property was also reported for DRS-PD 1 but only for human neuronal glioblastoma U251MG cell line [88]. In addition, these peptides could also inhibit the proliferation of human microvessel endothelial cells with the same concentration range for anticancer activity [88]. Other peptides from the South American orange-legged leaf frog (*Phyllomedusa hypochondrialis*) were identified and called DRS-PH. This peptide was active against several pathogens, such as *E. coli*, *P. aeruginosa*, *S. aureus* and its methicillin-resistant strain (MRSA), *E. faecalis* and *C. albicans*, in a concentration range from 1 μM to 512 μM [89]. Once again, this DRS displayed a broad spectrum of anticancer properties against different cancer cell lines, including MCF-7, H157, U251MG, MDA-MB-435S and PC-3 [89]. Very recently, different studies identified, from the skin of *Phyllomedusa sauvagei*, DRSs-PS type 1, 3 and 4, characterizing their anticancer properties [90–92]. In 2019, Long and coworkers demonstrated that DRS-PS1 has antimicrobial effects against *S. aureus*, *E. coli* and *C. albicans* [90]. Interestingly, DRS-PS 1 showed anti-proliferative effects on human glioblastoma U-251 MG, perturbing cell membrane integrity at the concentration of 0.1 μM [90]. Furthermore, the anticancer action with lower concentrations involves apoptosis activation by mitochondrial-related signal involvement [90]. DRS-PS 3 showed a broad spectrum of antimicrobial activities against several pathogens, such as *S. aureus*, *E. coli* and *C. albicans*, at high concentrations but with reduced cytotoxicity for erythrocytes [91]. However, the synthetic, more cationic and hydrophobic analogues created by replacing acidic amino acids D and E at 5 and 17, respectively, of the DRS-PS3 sequence by lysines (K5/D5, K17/E17-DRS-PS 3) or by replacing two neutral amino acids A10 and G11 with the hydrophobic amino acid leucine (L10/A10, L11/G11-DRS-PS 3) strongly increased their antimicrobial activities against the same pathogens with MIC values of 8 μM or less [91]. On the other hand, both artificial analogues exhibit a more significant hemolytic effect on red blood cells than DRS-PS 3 [91]. Furthermore, these peptides showed anticancer activities against H157, PC3 and HMEC-1 cell lines in the micromolar range but the most active was L10/A10, L11/G11-DRS-PS 3 [91]. Additionally, DRS-PS 4 displayed antimicrobial effects with many pathogens, such as *S. aureus* and MRSA, *E. faecalis*, *E. coli*, *P. aeruginosa* and *C. albicans*, in a range of concentrations from 1 μM to 32 μM, with biofilms eradicating properties of these microorganisms [92]. The antimicrobial mechanism of action is based on the ability of this peptide to permeabilize the bacterial cell membrane [92]. However, the hemolysis activity of DRS-PS 4 was tested using horse red blood cells showing slight effects at antimicrobial concentrations [92]. In addition, the anticancer activity of DRS-PS 4 was also evaluated on several human cell lines, including U251MG, MDA-MB-435S, H157, PC-3 and MCF-7, displaying a dose-dependent inhibitory activity with high cytotoxicity in a concentration range from 10^{-9} to 10^{-4} M [92]. On the other hand, it presents a slight suppressing effect on human microvascular endothelial cells [92]. Very recently, Dong et al. discovered the DRS-PP from frog *Phyllomedusa palliata*. This peptide was active at 2 μM against *E. coli*, *S. aureus* and MRSA, *C. albicans*, *P. aeruginosa*, *E. faecalis* and *K. pneumoniae* [93]. It is of great interest that DRS-PP showed anti-proliferative effects with cytotoxic activities on different cancer cells, such as H157, MCF-7, PC-3 and U251 MG, but no effects on human microvascular endothelial cells [93]. In vivo studies confirmed the anticancer property of this agent; in fact, DRS-PP was tested on a subcutaneous H157 tumor model of nude mice showing significant anti-tumor activity in a dose-dependent manner without hepatopulmonary and toxic side effects [93]. These effects are mediated via disruptive membrane action but exert pro-apoptotic effects induced by mitochondrial and death receptor pathways [93]. Finally, in 2021, DRS-TO was identified in the tiger-striped Leaf Frog, *Phyllomedesa tomopterna*, showing that this peptide was active against *S. aureus* and MRSA, *E. faecalis*, *E. coli*, and *C. albicans* [94]. Additionally, no hemolytic effect was observed on red blood cells, but DRS-TO showed anticancer activity against U251MG, H157 and PC-3 cancer cell lines at higher concentrations [94]. All these data report the great potential of DRSs as anticancer agents and their mechanism of action targeting membrane but also inducing pro-apoptotic effects by mitochondrial dysfunction and death receptor pathways (Figure 3).

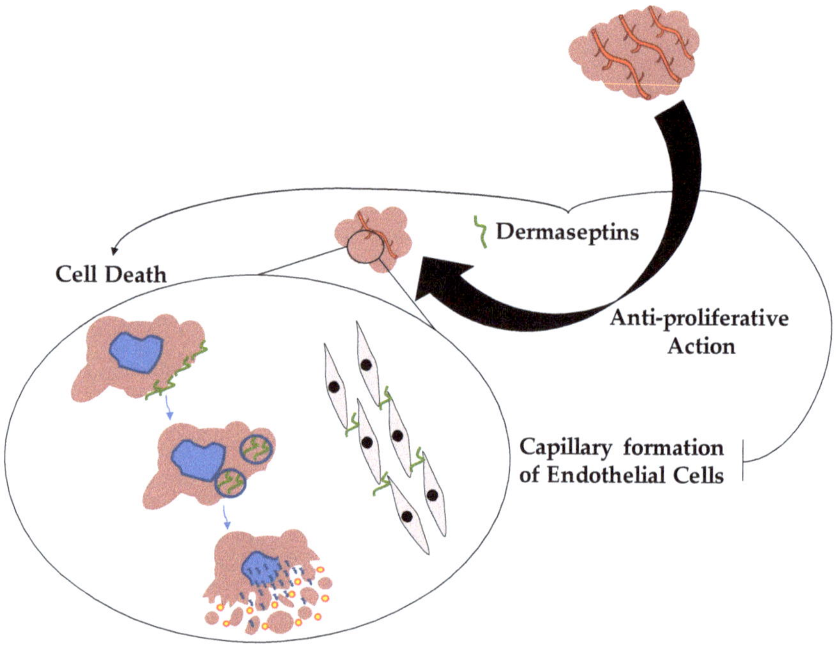

Figure 3. Dermaseptins and Cancer. Dermaseptins act as an anti-proliferative agent against several cancer cells in vitro and in vivo. The anticancer mechanism of action is based on the ability of these MF-AMPs to accumulate in cancer cells, inducing cell death and blocking tumor vascularization.

5. Conclusions and Future Perspectives

In conclusion, the role of MF-AMPs is reported to be crucial in preventing infection. These molecules are essential for the first response to infections and may represent an alternative approach and open future clinical applications for the resolution of antibiotic resistance. At the same time, the endogenous MF-AMPs displayed different effects on different organs and cell types. Among these, they can influence the immune system's inflammatory process and cellular components. In this perspective, CgA-derived peptides, such as Chr and Cts, are the perfect examples of immunomodulation mediated by MF-AMPs. On the other hand, exogenous MF-AMPs produced by different species but not in humans showed several potential therapeutic approaches. As reported in this review, DRSs appear to be excellent anti-proliferative factors with varying models of cancer cells. From a clinical point of view, these peptides may represent elective candidates for future anticancer therapy.

Author Contributions: Writing—initial draft preparation, F.S.; writing—review and editing, M.A. and J.-E.G. All authors have read and agreed to the published version of the manuscript.

Funding: This review was funded by Natural Sciences and Engineering Research Council of Canada (project number RGPIN-2022-03626) to J.-E.G.

Acknowledgments: F.S. thanks "Università Italo-Francese" for the financial support provided in the context of 'Vinci Project 2014' (no. C2-72) during the work realized for his Ph.D. thesis on MF-AMPs. The authors thank Marie-Hélène Metz-Boutigue for the stimulating discussions and her full support for all the above-mentioned research on CgA-derived peptides.

Conflicts of Interest: The authors declare no conflict of interest.

References

1. Riera Romo, M.; Perez-Martinez, D.; Castillo Ferrer, C. Innate immunity in vertebrates: An overview. *Immunology* **2016**, *148*, 125–139. [CrossRef] [PubMed]
2. Auvynet, C.; Rosenstein, Y. Multifunctional host defense peptides: Antimicrobial peptides, the small yet big players in innate and adaptive immunity. *FEBS J.* **2009**, *276*, 6497–6508. [CrossRef] [PubMed]
3. Wang, Z.; Wang, G. APD: The Antimicrobial Peptide Database. *Nucleic Acids Res.* **2004**, *32*, D590–D592. [CrossRef] [PubMed]
4. Brandwein, M.; Bentwich, Z.; Steinberg, D. Endogenous Antimicrobial Peptide Expression in Response to Bacterial Epidermal Colonization. *Front. Immunol.* **2017**, *8*, 1637. [CrossRef]
5. De Mandal, S.; Panda, A.K.; Murugan, C.; Xu, X.; Senthil Kumar, N.; Jin, F. Antimicrobial Peptides: Novel Source and Biological Function with a Special Focus on Entomopathogenic Nematode/Bacterium Symbiotic Complex. *Front. Microbiol.* **2021**, *12*, 555022. [CrossRef]
6. Li, J.; Hu, S.; Jian, W.; Xie, C.; Yang, X. Plant antimicrobial peptides: Structures, functions, and applications. *Bot. Stud.* **2021**, *62*, 5. [CrossRef]
7. Masso-Silva, J.A.; Diamond, G. Antimicrobial peptides from fish. *Pharmaceuticals* **2014**, *7*, 265–310. [CrossRef]
8. Fasina, Y.O.; Obanla, T.; Dosu, G.; Muzquiz, S. Significance of Endogenous Antimicrobial Peptides on the Health of Food Animals. *Front. Vet. Sci.* **2021**, *8*, 585266. [CrossRef]
9. Bartels, E.J.H.; Dekker, D.; Amiche, M. Dermaseptins, Multifunctional Antimicrobial Peptides: A Review of Their Pharmacology, Effectivity, Mechanism of Action, and Possible Future Directions. *Front. Pharmacol.* **2019**, *10*, 1421. [CrossRef]
10. Li, Y. The role of antimicrobial peptides in cardiovascular physiology and disease. *Biochem. Biophys. Res. Commun.* **2009**, *390*, 363–367. [CrossRef]
11. Stuart, B.A.R.; Franitza, A.L.; Lezi, E. Regulatory Roles of Antimicrobial Peptides in the Nervous System: Implications for Neuronal Aging. *Front. Cell Neurosci.* **2022**, *16*, 843790. [CrossRef] [PubMed]
12. Pan, L.L.; Liang, W.; Ren, Z.; Li, C.; Chen, Y.; Niu, W.; Fang, X.; Liu, Y.; Zhang, M.; Diana, J.; et al. Cathelicidin-related antimicrobial peptide protects against ischaemia reperfusion-induced acute kidney injury in mice. *Br. J. Pharmacol.* **2020**, *177*, 2726–2742. [CrossRef] [PubMed]
13. Yeung, A.T.; Gellatly, S.L.; Hancock, R.E. Multifunctional cationic host defence peptides and their clinical applications. *Cell. Mol. Life Sci.* **2011**, *68*, 2161–2176. [CrossRef] [PubMed]
14. Jafari, A.; Babajani, A.; Sarrami Forooshani, R.; Yazdani, M.; Rezaei-Tavirani, M. Clinical Applications and Anticancer Effects of Antimicrobial Peptides: From Bench to Bedside. *Front. Oncol.* **2022**, *12*, 819563. [CrossRef] [PubMed]
15. Antimicrobial Resistance, C. Global burden of bacterial antimicrobial resistance in 2019: A systematic analysis. *Lancet* **2022**, *399*, 629–655. [CrossRef]
16. Beovic, B. The issue of antimicrobial resistance in human medicine. *Int. J. Food Microbiol.* **2006**, *112*, 280–287. [CrossRef]
17. Tagliabue, A.; Rappuoli, R. Changing Priorities in Vaccinology: Antibiotic Resistance Moving to the Top. *Front. Immunol.* **2018**, *9*, 1068. [CrossRef]
18. Rima, M.; Rima, M.; Fajloun, Z.; Sabatier, J.M.; Bechinger, B.; Naas, T. Antimicrobial Peptides: A Potent Alternative to Antibiotics. *Antibiotics* **2021**, *10*, 1095. [CrossRef]
19. Leaper, D.; Wilson, P.; Assadian, O.; Edmiston, C.; Kiernan, M.; Miller, A.; Bond-Smith, G.; Yap, J. The role of antimicrobial sutures in preventing surgical site infection. *Ann. R. Coll. Surg. Engl.* **2017**, *99*, 439–443. [CrossRef]
20. Di Somma, A.; Moretta, A.; Cane, C.; Cirillo, A.; Duilio, A. Antimicrobial and Antibiofilm Peptides. *Biomolecules* **2020**, *10*, 652. [CrossRef]
21. Guryanova, S.V.; Ovchinnikova, T.V. Immunomodulatory and Allergenic Properties of Antimicrobial Peptides. *Int. J. Mol. Sci.* **2022**, *23*, 2499. [CrossRef] [PubMed]
22. Sierra, J.M.; Fuste, E.; Rabanal, F.; Vinuesa, T.; Vinas, M. An overview of antimicrobial peptides and the latest advances in their development. *Expert. Opin. Biol. Ther.* **2017**, *17*, 663–676. [CrossRef] [PubMed]
23. Yeaman, M.R.; Yount, N.Y. Mechanisms of antimicrobial peptide action and resistance. *Pharmacol. Rev.* **2003**, *55*, 27–55. [CrossRef]
24. Moellering, R.C., Jr. Discovering new antimicrobial agents. *Int. J. Antimicrob. Agents* **2011**, *37*, 2–9. [CrossRef]
25. Wang, G.; Li, X.; Wang, Z. APD2: The updated antimicrobial peptide database and its application in peptide design. *Nucleic Acids Res.* **2009**, *37*, D933–D937. [CrossRef]
26. Wang, G.; Li, X.; Wang, Z. APD3: The antimicrobial peptide database as a tool for research and education. *Nucleic Acids Res.* **2016**, *44*, D1087–D1093. [CrossRef]
27. Chung, C.R.; Jhong, J.H.; Wang, Z.; Chen, S.; Wan, Y.; Horng, J.T.; Lee, T.Y. Characterization and Identification of Natural Antimicrobial Peptides on Different Organisms. *Int. J. Mol. Sci.* **2020**, *21*, 986. [CrossRef]
28. Malmsten, M. Antimicrobial peptides. *Ups. J. Med. Sci.* **2014**, *119*, 199–204. [CrossRef]
29. Ribeiro, S.M.; Felicio, M.R.; Boas, E.V.; Goncalves, S.; Costa, F.F.; Samy, R.P.; Santos, N.C.; Franco, O.L. New frontiers for anti-biofilm drug development. *Pharmacol. Ther.* **2016**, *160*, 133–144. [CrossRef]
30. De la Fuente-Nunez, C.; Korolik, V.; Bains, M.; Nguyen, U.; Breidenstein, E.B.; Horsman, S.; Lewenza, S.; Burrows, L.; Hancock, R.E. Inhibition of bacterial biofilm formation and swarming motility by a small synthetic cationic peptide. *Antimicrob. Agents Chemother.* **2012**, *56*, 2696–2704. [CrossRef]

31. Sanchez-Gomez, S.; Martinez-de-Tejada, G. Antimicrobial Peptides as Anti-biofilm Agents in Medical Implants. *Curr. Top. Med. Chem.* **2017**, *17*, 590–603. [CrossRef] [PubMed]
32. Helle, K.B.; Metz-Boutigue, M.H.; Cerra, M.C.; Angelone, T. Chromogranins: From discovery to current times. *Pflugers Arch.* **2018**, *470*, 143–154. [CrossRef] [PubMed]
33. Lugardon, K.; Raffner, R.; Goumon, Y.; Corti, A.; Delmas, A.; Bulet, P.; Aunis, D.; Metz-Boutigue, M.H. Antibacterial and antifungal activities of vasostatin-1, the N-terminal fragment of chromogranin A. *J. Biol. Chem.* **2000**, *275*, 10745–10753. [CrossRef]
34. Zhang, D.; Shooshtarizadeh, P.; Laventie, B.J.; Colin, D.A.; Chich, J.F.; Vidic, J.; de Barry, J.; Chasserot-Golaz, S.; Delalande, F.; Van Dorsselaer, A.; et al. Two chromogranin a-derived peptides induce calcium entry in human neutrophils by calmodulin-regulated calcium independent phospholipase A2. *PLoS ONE* **2009**, *4*, e4501. [CrossRef] [PubMed]
35. Eissa, N.; Hussein, H.; Kermarrec, L.; Ali, A.Y.; Marshall, A.; Metz-Boutigue, M.H.; Hendy, G.N.; Bernstein, C.N.; Ghia, J.E. Chromogranin-A Regulates Macrophage Function and the Apoptotic Pathway in Murine DSS colitis. *J. Mol. Med.* **2018**, *96*, 183–198. [CrossRef] [PubMed]
36. Laguerre, F.; Anouar, Y.; Montero-Hadjadje, M. Chromogranin A in the early steps of the neurosecretory pathway. *IUBMB Life* **2020**, *72*, 524–532. [CrossRef]
37. Pasqua, T.; Corti, A.; Gentile, S.; Pochini, L.; Bianco, M.; Metz-Boutigue, M.H.; Cerra, M.C.; Tota, B.; Angelone, T. Full-length human chromogranin-A cardioactivity: Myocardial, coronary, and stimulus-induced processing evidence in normotensive and hypertensive male rat hearts. *Endocrinology* **2013**, *154*, 3353–3365. [CrossRef] [PubMed]
38. Radek, K.A.; Lopez-Garcia, B.; Hupe, M.; Niesman, I.R.; Elias, P.M.; Taupenot, L.; Mahata, S.K.; O'Connor, D.T.; Gallo, R.L. The neuroendocrine peptide catestatin is a cutaneous antimicrobial and induced in the skin after injury. *J. Investig. Dermatol.* **2008**, *128*, 1525–1534. [CrossRef]
39. Penna, C.; Tullio, F.; Perrelli, M.G.; Mancardi, D.; Pagliaro, P. Cardioprotection against ischemia/reperfusion injury and chromogranin A-derived peptides. *Curr. Med. Chem.* **2012**, *19*, 4074–4085. [CrossRef]
40. Kim, T.; Loh, Y.P. Chromogranin A: A surprising link between granule biogenesis and hypertension. *J. Clin. Investig.* **2005**, *115*, 1711–1713. [CrossRef]
41. Pasqua, T.; Rocca, C.; Spena, A.; Angelone, T.; Cerra, M.C. Modulation of the coronary tone in the expanding scenario of Chromogranin-A and its derived peptides. *Future Med. Chem.* **2019**, *11*, 1501–1511. [CrossRef] [PubMed]
42. Eissa, N.; Hussein, H.; Hendy, G.N.; Bernstein, C.N.; Ghia, J.E. Chromogranin-A and its derived peptides and their pharmacological effects during intestinal inflammation. *Biochem. Pharmacol.* **2018**, *152*, 315–326. [CrossRef] [PubMed]
43. Helle, K.B.; Angeletti, R.H. Chromogranin A: A multipurpose prohormone? *Acta Physiol. Scand.* **1994**, *152*, 1–10. [CrossRef] [PubMed]
44. Koshimizu, H.; Cawley, N.X.; Kim, T.; Yergey, A.L.; Loh, Y.P. Serpinin: A novel chromogranin A-derived, secreted peptide up-regulates protease nexin-1 expression and granule biogenesis in endocrine cells. *Mol. Endocrinol.* **2011**, *25*, 732–744. [CrossRef] [PubMed]
45. Mizuhashi, F.; Koide, K.; Toya, S.; Takahashi, M.; Mizuhashi, R.; Shimomura, H. Levels of the antimicrobial proteins lactoferrin and chromogranin in the saliva of individuals with oral dryness. *J. Prosthet. Dent.* **2015**, *113*, 35–38. [CrossRef]
46. Briolat, J.; Wu, S.D.; Mahata, S.K.; Gonthier, B.; Bagnard, D.; Chasserot-Golaz, S.; Helle, K.B.; Aunis, D.; Metz-Boutigue, M.H. New antimicrobial activity for the catecholamine release-inhibitory peptide from chromogranin A. *Cell. Mol. Life Sci.* **2005**, *62*, 377–385. [CrossRef]
47. Lugardon, K.; Chasserot-Golaz, S.; Kieffer, A.E.; Maget-Dana, R.; Nullans, G.; Kieffer, B.; Aunis, D.; Metz-Boutigue, M.H. Structural and biological characterization of chromofungin, the antifungal chromogranin A-(47-66)-derived peptide. *J. Biol. Chem.* **2001**, *276*, 35875–35882. [CrossRef]
48. Lugardon, K.; Chasserot-Golaz, S.; Kieffer, A.E.; Maget-Dana, R.; Nullans, G.; Kieffer, B.; Aunis, D.; Metz-Boutigue, M.H. Structural and biological characterization of chromofungin, the antifungal chromogranin A (47-66)-derived peptide. *Ann. N. Y. Acad. Sci.* **2002**, *971*, 359–361. [CrossRef]
49. Aardal, S.; Helle, K.B.; Elsayed, S.; Reed, R.K.; Serck-Hanssen, G. Vasostatins, comprising the N-terminal domain of chromogranin A, suppress tension in isolated human blood vessel segments. *J. Neuroendocrinol.* **1993**, *5*, 405–412. [CrossRef]
50. Metz-Boutigue, M.H.; Goumon, Y.; Strub, J.M.; Lugardon, K.; Aunis, D. Antimicrobial chromogranins and proenkephalin-A-derived peptides: Antibacterial and antifungal activities of chromogranins and proenkephalin-A-derived peptides. *Ann. N. Y. Acad. Sci.* **2003**, *992*, 168–178. [CrossRef]
51. Metz-Boutigue, M.H.; Kieffer, A.E.; Goumon, Y.; Aunis, D. Innate immunity: Involvement of new neuropeptides. *Trends Microbiol.* **2003**, *11*, 585–592. [CrossRef] [PubMed]
52. Park, H.S.; Lee, S.C.; Cardenas, M.E.; Heitman, J. Calcium-Calmodulin-Calcineurin Signaling: A Globally Conserved Virulence Cascade in Eukaryotic Microbial Pathogens. *Cell Host Microbe* **2019**, *26*, 453–462. [CrossRef] [PubMed]
53. Mahata, S.K.; O'Connor, D.T.; Mahata, M.; Yoo, S.H.; Taupenot, L.; Wu, H.; Gill, B.M.; Parmer, R.J. Novel autocrine feedback control of catecholamine release. A discrete chromogranin a fragment is a noncompetitive nicotinic cholinergic antagonist. *J. Clin. Investig.* **1997**, *100*, 1623–1633. [CrossRef] [PubMed]
54. Taylor, C.V.; Taupenot, L.; Mahata, S.K.; Mahata, M.; Wu, H.; Yasothornsrikul, S.; Toneff, T.; Caporale, C.; Jiang, Q.; Parmer, R.J.; et al. Formation of the catecholamine release-inhibitory peptide catestatin from chromogranin A. Determination of proteolytic cleavage sites in hormone storage granules. *J. Biol. Chem.* **2000**, *275*, 22905–22915. [CrossRef]

55. Jean-Francois, F.; Khemtemourian, L.; Odaert, B.; Castano, S.; Grelard, A.; Manigand, C.; Bathany, K.; Metz-Boutigue, M.H.; Dufourc, E.J. Variability in secondary structure of the antimicrobial peptide Cateslytin in powder, solution, DPC micelles and at the air-water interface. *Eur. Biophys. J.* **2007**, *36*, 1019–1027. [CrossRef]
56. Jean-Francois, F.; Castano, S.; Desbat, B.; Odaert, B.; Roux, M.; Metz-Boutigue, M.H.; Dufourc, E.J. Aggregation of cateslytin beta-sheets on negatively charged lipids promotes rigid membrane domains. A new mode of action for antimicrobial peptides? *Biochemistry* **2008**, *47*, 6394–6402. [CrossRef]
57. Jean-Francois, F.; Elezgaray, J.; Berson, P.; Vacher, P.; Dufourc, E.J. Pore formation induced by an antimicrobial peptide: Electrostatic effects. *Biophys. J.* **2008**, *95*, 5748–5756. [CrossRef]
58. Jean-Francois, F.; Desbat, B.; Dufourc, E.J. Selectivity of cateslytin for fungi: The role of acidic lipid-ergosterol membrane fluidity in antimicrobial action. *FASEB J.* **2009**, *23*, 3692–3701. [CrossRef]
59. Xiong, W.; Wang, X.; Dai, D.; Zhang, B.; Lu, L.; Tao, R. The anti-inflammatory vasostatin-2 attenuates atherosclerosis in ApoE(−/−) mice and inhibits monocyte/macrophage recruitment. *Thromb. Haemost.* **2017**, *117*, 401–414. [CrossRef]
60. Sato, Y.; Watanabe, R.; Uchiyama, N.; Ozawa, N.; Takahashi, Y.; Shirai, R.; Sato, K.; Mori, Y.; Matsuyama, T.; Ishibashi-Ueda, H.; et al. Inhibitory effects of vasostatin-1 against atherogenesis. *Clin. Sci.* **2018**, *132*, 2493–2507. [CrossRef]
61. Eissa, N.; Hussein, H.; Kermarrec, L.; Grover, J.; Metz-Boutigue, M.E.; Bernstein, C.N.; Ghia, J.E. Chromofungin Ameliorates the Progression of Colitis by Regulating Alternatively Activated Macrophages. *Front. Immunol.* **2017**, *8*, 1131. [CrossRef] [PubMed]
62. Eissa, N.; Hussein, H.; Kermarrec, L.; Elgazzar, O.; Metz-Boutigue, M.H.; Bernstein, C.N.; Ghia, J.E. Chromofungin (CHR: CHGA47-66) is downregulated in persons with active ulcerative colitis and suppresses pro-inflammatory macrophage function through the inhibition of NF-kappaB signaling. *Biochem. Pharmacol.* **2017**, *145*, 102–113. [CrossRef] [PubMed]
63. Kapoor, K.; Eissa, N.; Tshikudi, D.; Bernstein, C.N.; Ghia, J.E. Impact of intrarectal chromofungin treatment on dendritic cells-related markers in different immune compartments in colonic inflammatory conditions. *World J. Gastroenterol.* **2021**, *27*, 8138–8155. [CrossRef] [PubMed]
64. Schneider, F.; Marban, C.; Ajob, G.; Helle, S.; Guillot, M.; Launoy, A.; Maestraggi, Q.; Scavello, F.; Rohr, O.; Metz-Boutigue, M.H. In Trauma Patients, the Occurrence of Early-Onset Nosocomial Infections Is Associated with Increased Plasma Concentrations of Chromogranin A. *Shock* **2018**, *49*, 522–528. [CrossRef]
65. Rabbi, M.F.; Labis, B.; Metz-Boutigue, M.H.; Bernstein, C.N.; Ghia, J.E. Catestatin decreases macrophage function in two mouse models of experimental colitis. *Biochem. Pharmacol.* **2014**, *89*, 386–398. [CrossRef]
66. Rabbi, M.F.; Eissa, N.; Munyaka, P.M.; Kermarrec, L.; Elgazzar, O.; Khafipour, E.; Bernstein, C.N.; Ghia, J.E. Reactivation of Intestinal Inflammation Is Suppressed by Catestatin in a Murine Model of Colitis via M1 Macrophages and Not the Gut Microbiota. *Front. Immunol.* **2017**, *8*, 985. [CrossRef]
67. Ying, W.; Tang, K.; Avolio, E.; Schilling, J.M.; Pasqua, T.; Liu, M.A.; Cheng, H.; Gao, H.; Zhang, J.; Mahata, S.; et al. Immunosuppression of Macrophages Underlies the Cardioprotective Effects of CST (Catestatin). *Hypertension* **2021**, *77*, 1670–1682. [CrossRef]
68. Kojima, M.; Ozawa, N.; Mori, Y.; Takahashi, Y.; Watanabe-Kominato, K.; Shirai, R.; Watanabe, R.; Sato, K.; Matsuyama, T.A.; Ishibashi-Ueda, H.; et al. Catestatin Prevents Macrophage-Driven Atherosclerosis but Not Arterial Injury-Induced Neointimal Hyperplasia. *Thromb. Haemost.* **2018**, *118*, 182–194. [CrossRef]
69. Chen, H.; Liu, D.; Ge, L.; Wang, T.; Ma, Z.; Han, Y.; Duan, Y.; Xu, X.; Liu, W.; Yuan, J.; et al. Catestatin prevents endothelial inflammation and promotes thrombus resolution in acute pulmonary embolism in mice. *Biosci. Rep.* **2019**, *39*, BSR20192236. [CrossRef]
70. Amiche, M.; Ladram, A.; Nicolas, P. A consistent nomenclature of antimicrobial peptides isolated from frogs of the subfamily Phyllomedusinae. *Peptides* **2008**, *29*, 2074–2082. [CrossRef]
71. Nicolas, P.; El Amri, C. The dermaseptin superfamily: A gene-based combinatorial library of antimicrobial peptides. *Biochim. Biophys. Acta* **2009**, *1788*, 1537–1550. [CrossRef]
72. Mor, A.; Nguyen, V.H.; Delfour, A.; Miglioresamour, D.; Nicolas, P. Isolation, Amino-Acid-Sequence, and Synthesis of Dermaseptin, a Novel Antimicrobial Peptide of Amphibian Skin. *Biochemistry* **1991**, *30*, 8824–8830. [CrossRef] [PubMed]
73. Amiche, M.; Ducancel, F.; Lajeunesse, E.; Boulain, J.C.; Menez, A.; Nicolas, P. Molecular cloning of a cDNA encoding the precursor of adenoregulin from frog skin. Relationships with the vertebrate defensive peptides, dermaseptins. *Biochem. Biophys. Res. Commun.* **1993**, *191*, 983–990. [CrossRef]
74. Amiche, M.; Ducancel, F.; Mor, A.; Boulain, J.C.; Menez, A.; Nicolas, P. Precursors of vertebrate peptide antibiotics dermaseptin b and adenoregulin have extensive sequence identities with precursors of opioid peptides dermorphin, dermenkephalin, and deltorphins. *J. Biol. Chem.* **1994**, *269*, 17847–17852. [CrossRef]
75. Mor, A.; Amiche, M.; Nicolas, P. Structure, synthesis, and activity of dermaseptin b, a novel vertebrate defensive peptide from frog skin: Relationship with adenoregulin. *Biochemistry* **1994**, *33*, 6642–6650. [CrossRef] [PubMed]
76. Daly, J.W.; Caceres, J.; Moni, R.W.; Gusovsky, F.; Moos, M.; Seamon, K.B.; Milton, K.; Myers, C.W. Frog Secretions and Hunting Magic in the Upper Amazon—Identification of a Peptide That Interacts with an Adenosine Receptor. *Proc. Natl. Acad. Sci. USA* **1992**, *89*, 10960–10963. [CrossRef]
77. Pouny, Y.; Rapaport, D.; Mor, A.; Nicolas, P.; Shai, Y. Interaction of antimicrobial dermaseptin and its fluorescently labeled analogues with phospholipid membranes. *Biochemistry* **1992**, *31*, 12416–12423. [CrossRef]

78. Shai, Y. Mechanism of the binding, insertion and destabilization of phospholipid bilayer membranes by alpha-helical antimicrobial and cell non-selective membrane-lytic peptides. *Biochim. Biophys. Acta* **1999**, *1462*, 55–70. [CrossRef]
79. Bechinger, B. The structure, dynamics and orientation of antimicrobial peptides in membranes by multidimensional solid-state NMR spectroscopy. *Biochim. Biophys. Acta* **1999**, *1462*, 157–183. [CrossRef]
80. Zasloff, M. Antimicrobial peptides in health and disease. *N. Engl. J. Med.* **2002**, *347*, 1199–1200. [CrossRef]
81. Lequin, O.; Bruston, F.; Convert, O.; Chassaing, G.; Nicolas, P. Helical structure of dermaseptin B2 in a membrane-mimetic environment. *Biochemistry* **2003**, *42*, 10311–10323. [CrossRef] [PubMed]
82. Amiche, M.; Galanth, C. Dermaseptins as models for the elucidation of membrane-acting helical amphipathic antimicrobial peptides. *Curr. Pharm. Biotechnol.* **2011**, *12*, 1184–1193. [CrossRef] [PubMed]
83. Van Zoggel, H.; Carpentier, G.; Dos Santos, C.; Hamma-Kourbali, Y.; Courty, J.; Amiche, M.; Delbe, J. Antitumor and angiostatic activities of the antimicrobial peptide dermaseptin B2. *PLoS ONE* **2012**, *7*, e44351. [CrossRef]
84. Van Zoggel, H.; Hamma-Kourbali, Y.; Galanth, C.; Ladram, A.; Nicolas, P.; Courty, J.; Amiche, M.; Delbe, J. Antitumor and angiostatic peptides from frog skin secretions. *Amino Acids* **2012**, *42*, 385–395. [CrossRef] [PubMed]
85. Dos Santos, C.; Hamadat, S.; Le Saux, K.; Newton, C.; Mazouni, M.; Zargarian, L.; Miro-Padovani, M.; Zadigue, P.; Delbe, J.; Hamma-Kourbali, Y.; et al. Studies of the antitumor mechanism of action of dermaseptin B2, a multifunctional cationic antimicrobial peptide, reveal a partial implication of cell surface glycosaminoglycans. *PLoS ONE* **2017**, *12*, e0182926. [CrossRef]
86. Couty, M.; Dusaud, M.; Miro-Padovani, M.; Zhang, L.; Zadigue, P.; Zargarian, L.; Lequin, O.; de la Taille, A.; Delbe, J.; Hamma-Kourbali, Y.; et al. Antitumor Activity and Mechanism of Action of Hormonotoxin, an LHRH Analog Conjugated to Dermaseptin-B2, a Multifunctional Antimicrobial Peptide. *Int. J. Mol. Sci.* **2021**, *22*, 11303. [CrossRef]
87. Medeiros, K.A.; Joanitti, G.A.; Silva, L.P. Chitosan nanoparticles for dermaseptin peptide delivery toward tumor cells in vitro. *Anticancer Drugs* **2014**, *25*, 323–331. [CrossRef]
88. Shi, D.; Hou, X.; Wang, L.; Gao, Y.; Wu, D.; Xi, X.; Zhou, M.; Kwok, H.F.; Duan, J.; Chen, T.; et al. Two Novel Dermaseptin-Like Antimicrobial Peptides with Anticancer Activities from the Skin Secretion of Pachymedusa dacnicolor. *Toxins* **2016**, *8*, 144. [CrossRef]
89. Huang, L.; Chen, D.; Wang, L.; Lin, C.; Ma, C.; Xi, X.; Chen, T.; Shaw, C.; Zhou, M. Dermaseptin-PH: A Novel Peptide with Antimicrobial and Anticancer Activities from the Skin Secretion of the South American Orange-Legged Leaf Frog, *Pithecopus (Phyllomedusa) hypochondrialis*. *Molecules* **2017**, *22*, 1805. [CrossRef]
90. Long, Q.; Li, L.; Wang, H.; Li, M.; Wang, L.; Zhou, M.; Su, Q.; Chen, T.; Wu, Y. Novel peptide dermaseptin-PS1 exhibits anticancer activity via induction of intrinsic apoptosis signalling. *J. Cell Mol. Med.* **2019**, *23*, 1300–1312. [CrossRef]
91. Tan, Y.; Chen, X.; Ma, C.; Xi, X.; Wang, L.; Zhou, M.; Burrows, J.F.; Kwok, H.F.; Chen, T. Biological Activities of Cationicity-Enhanced and Hydrophobicity-Optimized Analogues of an Antimicrobial Peptide, Dermaseptin-PS3, from the Skin Secretion of *Phyllomedusa sauvagii*. *Toxins* **2018**, *10*, 320. [CrossRef] [PubMed]
92. Chen, D.; Zhou, X.; Chen, X.; Huang, L.; Xi, X.; Ma, C.; Zhou, M.; Wang, L.; Chen, T. Evaluating the Bioactivity of a Novel Antimicrobial and Anticancer Peptide, Dermaseptin-PS4(Der-PS4), from the Skin Secretion of *Phyllomedusa sauvagii*. *Molecules* **2019**, *24*, 2974. [CrossRef] [PubMed]
93. Dong, Z.; Hu, H.; Yu, X.; Tan, L.; Ma, C.; Xi, X.; Li, L.; Wang, L.; Zhou, M.; Chen, T.; et al. Novel Frog Skin-Derived Peptide Dermaseptin-PP for Lung Cancer Treatment: In vitro/vivo Evaluation and Anti-tumor Mechanisms Study. *Front. Chem.* **2020**, *8*, 476. [CrossRef]
94. Chen, Z.; Xi, X.; Lu, Y.; Hu, H.; Dong, Z.; Ma, C.; Wang, L.; Zhou, M.; Chen, T.; Du, S.; et al. In vitro activities of a novel antimicrobial peptide isolated from phyllomedusa tomopterna. *Microb. Pathog.* **2021**, *153*, 104795. [CrossRef] [PubMed]

pharmaceutics

MDPI

Review

Chromogranin A and Its Fragments in the Critically Ill: An Expanding Domain of Interest for Better Care

Francis Schneider [1,2,*], Raphaël Clère-Jehl [1], Francesco Scavello [2,3], Thierry Lavigne [1,2,4], Angelo Corti [5], Tommaso Angelone [6], Youssef Haïkel [2,7] and Philippe Lavalle [2]

1 Médecine Intensive—Réanimation, Hôpital de Hautepierre, Hôpitaux Universitaires de Strasbourg, Faculté de Médecine, FMTS at Unistra, 67085 Strasbourg, France
2 Biomaterials and Bioengeneering, UMR_S1121, FMTS at Unistra, 67085 Strasbourg, France
3 IRCCS Humanitas Research Hospital, 20089 Rozzano, MI, Italy
4 Hygiène Hospitalière et Médecine Préventive, Pôle de Santé Publique, Hôpitaux Universitaires de Strasbourg, 67091 Strasbourg, France
5 San Raffaele Scientific Institute, Division of Experimental Oncology, Vita-Salute San Raffaele University, Via Olgettina 58, 20132 Milan, MI, Italy
6 Laboratory of Cellular and Molecular Cardiac Pathophysiology, Department of Biology, Ecology, and Earth Science, University of Calabria, 87036 Rende, CS, Italy
7 Nouvel Hôpital Civil, Hôpitaux Universitaires de Strasbourg, Faculté de Chirurgie dentaire, Université de Strasbourg, 67091 Strasbourg, France
* Correspondence: francis.schneider@chru-strasbourg.fr; Tel.: +33-388127906

Abstract: Life-threatening diseases challenge immunity with a release of chromogranins. This report focuses on Chromogranin A (CGA) and some of its derived peptides in critically ill patients, with attention paid to their potential to become biomarkers of severity and actors of defense. First, we studied whether circulating CGA may be a biomarker of outcome in non-selected critically ill patients: CGA concentrations were reliably associated with short-term death, systemic inflammation, and multiple organ failure. Additionally, when studying Vasostatin-I, the major N-terminal fragment of CGA, we noted its reliable prognostic value as early as admission if associated with age and lactate. In trauma patients, CGA concentrations heralded the occurrence of care-related infections. This was associated with an in vitro inhibitor impact of Chromofungin on both NF-kappa B- and API-transcriptional activities. Secondly, in life-threatening disease-induced oxidative stress, the multimerization of Vasostatin-I occurs with the loss of its anti-microbial properties ex vivo. In vivo, a 4%-concentration of non-oxidized albumin infusion reversed multimerization with a decrease in care-related infections. Finally, in vitro Catestatin impacted the polymorphonuclear cells-Ca++-dependent, calmodulin–regulated iPLA2 pathway by releasing immunity-related proteins. Furthermore, human Cateslytin, the active domain of Catestatin, helped destroy S. aureus: this prompted the creation of synthetic D-stereoisomer of CGA-derived peptides against superbugs for the protection of implanted devices. In conclusion, CGA consideration in the critically ill is only starting, but it offers interesting perspectives for improved outcomes.

Keywords: albumin; biomaterials; Catestatin; Chromogranin A; Chromofungin; critically ill; outcome; prognosis; superbugs; Vasostatin-I

Citation: Schneider, F.; Clère-Jehl, R.; Scavello, F.; Lavigne, T.; Corti, A.; Angelone, T.; Haïkel, Y.; Lavalle, P. Chromogranin A and Its Fragments in the Critically Ill: An Expanding Domain of Interest for Better Care. *Pharmaceutics* **2022**, *14*, 2178. https://doi.org/10.3390/pharmaceutics14102178

Academic Editor: Rakesh Tiwari

Received: 18 August 2022
Accepted: 9 October 2022
Published: 12 October 2022

1. Introduction

Chromogranins proteins had attracted the interest of scientists and clinicians since the mid-sixties when Blaschko H et al. [1] and Helle KB [2] released the first two manuscripts stored in the PubMed database on this topic. Since then, no less than 9891 papers have been published on chromogranins, of which, however, only 19 are indicated as concerning data recorded in patients admitted in critical care wards (Figure 1). The interest in chromogranins as mediators of life-stressing diseases on human health, although thus only at an early stage, has been slowly growing over the last decade. Fifteen years ago, our study group was among the first to consider CGA as a possible object of examination in critical illness. Yet, this area of assessment is complicated. First, in vivo studies require testing large cohorts of patients with clinical phenotypes as similar as possible. This necessitates extensive multicenter studies when clinically relevant hypotheses have been validated in feasibility studies. For instance, now that we have identified Vasostatin-I (VS-I) as a myocardial depressant factor [3], it becomes interesting to gather data to comprehend its role in depressed cardiac contractility in acute hypokinetic shock. Second, clinicians must test in vivo clear-cut null hypotheses, which may be very difficult in multiple organ failure patients. As an example of such difficulty, systemic inflammation in response to strong pro-inflammatory triggers is frequent in intensive care units (ICUs), but in vivo studies require the assessment of chromogranins at the same stage of a given disease mandatorily. In each setting, pharmacokinetics, pharmacodynamics, and the volume of distribution of the proteins of interest are not currently established; and this is independent of the availability of adequate techniques of measurements that may vary according to the characteristics of both the test used and the physiological liquids from which samples have been harvested. Finally, having decided to assess the presence of these proteins, adequate biological sampling techniques must be chosen. This sometimes requires the in-house development of biological assays [4].

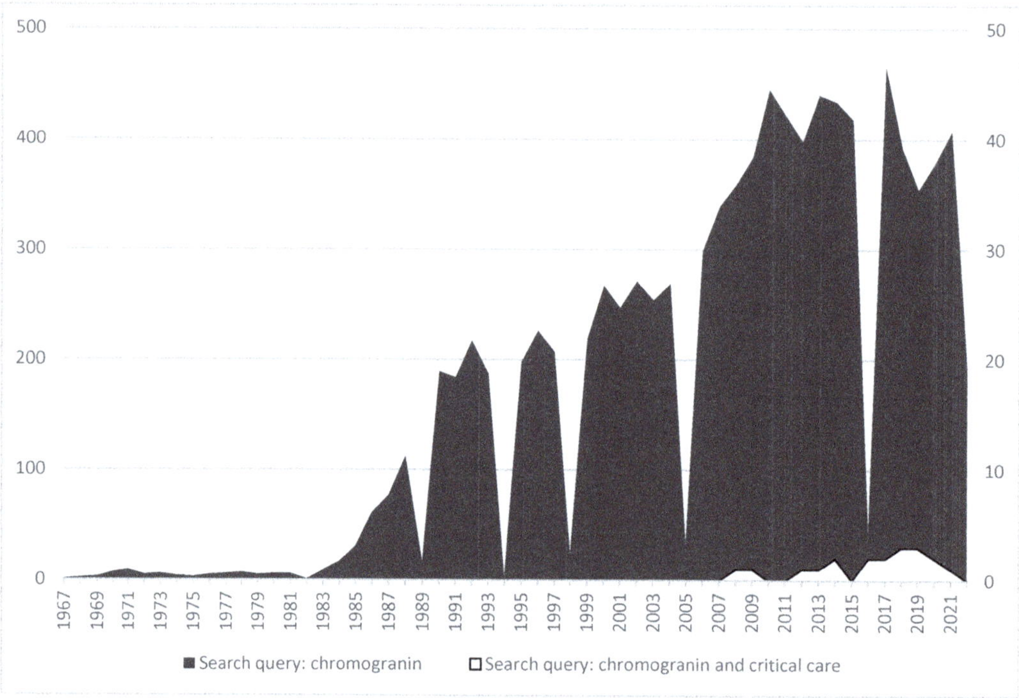

Figure 1. Number of annual papers issued in the PubMed database according to the query from 1967 to 2022. Please note that the scale of measurement is ten times larger for the query "Chromogranin" (left vertical axis) than for the query "chromogranin and critical care" (right vertical axis).

Patients admitted to an ICU for a disease with organ failure are at risk of short-term death. In this setting, in response to the initial assault on their integrity, they develop a systemic inflammatory response to maintain vital organs functioning and optimal tissue repair. This response is temporary and well-balanced between pro- and anti-inflammatory mediators; it results in recovery, provided the initial trigger of the disease does not overwhelm physiological responses. Following initial local injury, the organism develops (i) a hormonal, metabolic response, (ii) an immunological response with a neurological component, and (iii) a hemodynamic response called "acute circulatory failure" or shock [5,6]. During that period, the organism rapidly releases chromogranins not only from adrenal medulla stockpiles but also from the diffuse neuroendocrine cell islets existing throughout the organism. From blood concentration assessments, physicians can indirectly assess the balance between synthesis, release, and clearance of the molecules while also considering the possible impact of extra-renal purification techniques.

Bearing these precautions in mind, we report our experience on CGA (and related peptides) in critically ill and experimental conditions related to acute and severe illnesses.

2. CGA and Vasostatin-I (CGA1-76) Are Biomarkers of Severity in the Critically Ill

In 2007, we assessed whether CGA could become a biomarker of severity in life-threatening diseases on hospital admission. Our primary aim was to improve the triage of intermediate-severity patients to select the most adapted care for them. At that time, the agreed practice was to assess severity in ICU patients with the Simplified Acute Physiology Score 2 (SAPS II) based on a European/North American multicenter study [7]. This score provides an accurate estimate of the risk of death in the ICU without having to specify a primary diagnosis. It proved efficient notwithstanding that since it was available only at hour 24 of admission, it could not become a tool for early triage, whereas a rapidly available and performant biological test would.

The idea of assessing CGA was prompted by the fact that life-threatening diseases trigger a physiological response to an aggressive challenge to health with a triple component: (i) a hormonal, metabolic response; (ii) an immunological response, and (iii) a hemodynamic response—all of which are also linked by a neural immunity balance [6]. Following damage to the body's integrity, a patient will activate the release of various hormones, such as adrenalin and cortisol, as well as many others [8]. The adrenocortical response is similar in all mammals: afferent impulses from receptors implanted in the injury site stimulate the secretion of hypothalamic-releasing factors, further stimulating the pituitary gland. Consequently, the adrenal cortex releases cortisol and the adrenal medulla adrenaline together with chromogranins that are stored and co-released by the chromaffin cells. A feasibility study confirmed the hypothesis that patients with acute organ failure present at admission increased serum concentrations of CGA as a probable component of the early hormonal "fight-or-flight" adaptive response to stress [9]. Importantly, circulating CGA concentrations increased in association with systemic inflammation rather than with infection and the SAPS II, suggesting a possible link with either survival or one of the components of the score. These data align with what happens in humans even in the context of a stressing injury less severe than trauma, thus indicating that the slightest injury to the skin integrity also results in a signal of stress with the release of chromogranin-derived anti-microbial peptides (AMP) [10]. In a confirmation study [11], we reported that CGA concentrations correlate better with systemic inflammation (assessed by both C-reactive protein and procalcitonin) than with infection. Yet, CGA levels reached the highest values in septic shock. In addition, admission CGA values were equivalent to the SAPS II in predicting 28-day mortality. This offered the attractive opportunity for assessing outcomes as early as a few hours after admission in patients free of any other cause of CGA increase.

Further insight into the use of CGA as a biomarker of triage required an extension of the study size. We wondered whether a single admission dosage of VS-I (CGA-1-76), the major N-terminal fragment, would enable greater accuracy than a single CGA test. In a pilot prospective observational study performed in third-level French ICUs, we demonstrated that admission concentrations of VS-I were increased when compared with the values recorded in controls, and this was even more so when the shock was present [12]: VS-I concentrations above 3.97 ng/mL were indicative of poor outcome. Furthermore, including arterial lactate and age in the prediction model improved the reliability of the assessment, making it significantly better than the SAPS II as early as 4 h after admission.

In a further study, we investigated whether CGA could be a marker of the severity of the initial challenging disease as regards its impact on morbidity. Care-related infections are a major issue in critical care for many reasons (costs, bed- blocking and emergence of antibiotics-resistant superbugs). We chose the study phenotype of "multiple trauma" in patients previously in good health. This model appeared more suitable for investigating the impact of a basic injury on health, given that older patients with comorbidities were excluded. The study model, by definition, included a "two-hit" challenge to health: first, the trauma itself and, shortly thereafter, the surgical procedure, which represents a second hit for CGA release. In such patients, we noted an altered plasmatic CGA response revealing a potential mechanism for an association with nosocomial infection [13]. Over several days from admission, CGA concentrations increased compared with control values, but they also leveled off at a relatively high level associated with acute renal failure. Importantly, admission values of CGA significantly increased in those multiple trauma patients who subsequently developed a nosocomial infection: a concentration of 67 ng/mL predicted this occurrence with a sensitivity of 100% and a specificity of 70%, leading to a negative likelihood ratio of almost zero. Recently, we confirmed that admission CGA achieved similar performance for predicting nosocomial infections in COVID patients requiring oxygenation support [14]. This suggests that CGA may reflect not only the triggering disease but also the neuro-hormonal response by the neuroendocrine tissues. In multiple trauma, we finally investigated the ability of VS-I to modulate the innate response of monocytes that are called on to upgrade the patient's defense [13]. Acting as a cell-penetrating peptide, the CGA47-70 fragment, not however including either its scrambled or its prolonged isoforms, was able to downregulate both NF-kappa B and AP-1: this suggests an indirect anti-inflammatory pathway potentially entailing a risk of temporary immune deficiency. These data support the possible occurrence of some forms of care-related infections when proteases of CGA are upregulated.

3. Fine-Tuned Albumin Infusion Modifies CGA-Derived Peptides Multimers In Vitro and Impacts on Nosocomial Infection Occurrence

Care-related infections are a matter of worry in the ICU: they increase the costs of treatment and length of stay and kill, on average, 10 to 15% of the patients. Finally, acute stress and anti-microbial treatments are also responsible for the emergence of transmittable-resistant microbes. Despite hygiene, antibiotics stewardship, and precautions, these infections will occur in many patients with systemic inflammatory conditions at the acute phase of a disease. Oxidative stress induces damage to proteins within the circulation and beyond. The mechanism of oxidative stress undoubtedly affects many proteins, including those belonging to innate defense. In several patients, we recorded multimers of interest at the acute phase of the disease, including multimers of granins (see Figures 2 and 3). These multimers persist longer in the circulation of those patients with the highest and longest rate of infusion of norepinephrine for shock. Human serum albumin (HSA) displays properties for the care of critically ill patients. Among others, HSA provides an opportunity to restore in vitro the native status of proteins, as reported previously by our group [15]. However, this effect has never been reported in ICU patients that are prone to develop either colonization or infection. We decided to perform a pilot study on critically ill patients at risk of nosocomial infection to test whether therapeutic non-oxidized HSA would prevent such infections [16]. This study included a biochemical analysis of the interactions of the CGA-derived peptide VS-I, for which the anti-microbial activity is related to its non-oxidative state [15]. The results were that: (i) in vivo, therapeutic HSA significantly lessens both colonization and infection occurrences in patients with shock; (ii) this was possible provided therapeutic HSA is prescribed as a continuous low dose infusion of 4% HSA. We showed, in addition, that, in vitro, both natural and recombinant VS-I develop biochemical interactions with several natural and synthetic isoforms of albumin (HSA, bovine serum albumin, therapeutic HSA) via the hydrophobic domain of VS-I17-40 (which includes the disulfide bridge C17-C38). This allows the oxidation of VS-I to be reversed, rendering it more efficient as an anti-microbial protein even in tissues where the pH decreases at 6 in microcirculation during shock. We deduced that the rate of therapeutic HSA infusion is essential in vivo when seeking to restore the physiologic activities of defense. A prospective multicenter open-label randomized trial confirmed this data in septic shock patients for which continuously infused 4% therapeutic HSA over the first week of shock decreased nosocomial infection by two-thirds when compared with the intermittent infusion of similar doses of 20% HSA [17]. These results are noteworthy because they explain the discrepancies existing in meta-analyses on the benefit of therapeutic HSA: many studies postulate the lack of efficiency of therapeutic albumin in septic conditions, whereas others have found significant improvements in restricted populations [18,19]. In fact, protocols of infusion and amounts of therapeutic albumin differ from one study to another, and so do the isoforms of albumin tested, which explains why physicians do not achieve the goal of defense reversion. Our final proposition is that therapeutic albumin should be: (i) chosen as 4% albumin with a high potential of antioxidant activity [20]; (ii) infused continuously at a rate of 10–12 mL/kg/24 h over 5 days to limit the risk of care-related infections.

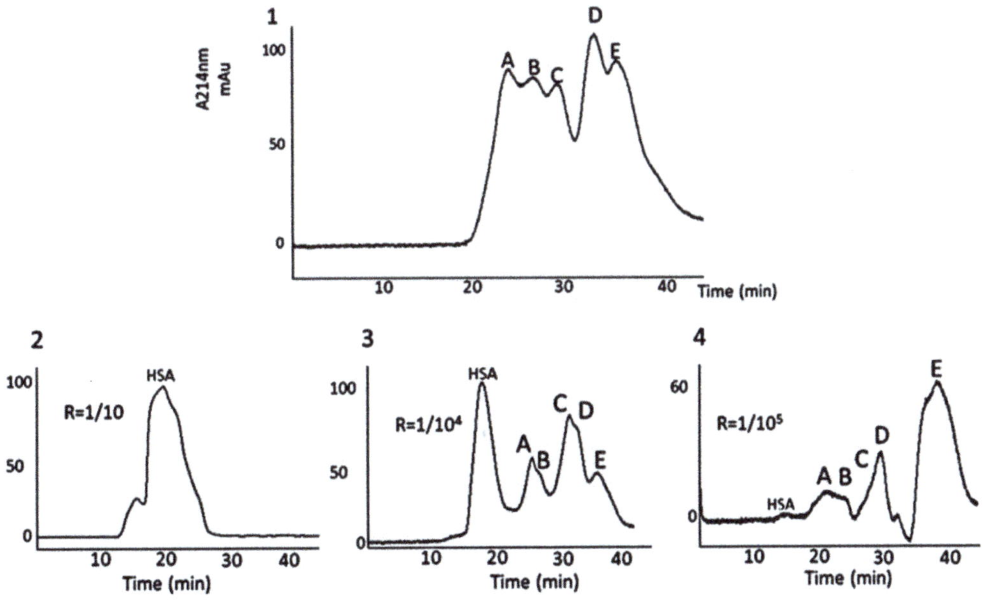

Figure 2. Gel filtration HPLC of purified and oxidized VS-I samples to evaluate the impact of different concentrations of therapeutic human serum albumin (HSA) on multimers of VS-I. For the four graphs, the X-axis corresponds to the elution time (expressed in min), which is linearly related to the molecular weight of the components included in the peaks. The Y-axis corresponds to absorbance expressed in milliUnits of absorbance. Whether in vitro or in vivo, the oxidation of VS-I leads to multimerization, as shown on the first chromatogram (upper graph, numbered 1). The monomeric form of the peptide corresponds to peak E, while peaks A-D correspond to VS-I multimers. When adding fresh, non-oxidized therapeutic HSA at a molar albumin/VS-I ratio (R) from 1/10 to $1/10^5$ (as shown on chromatograms 2–4), the release of the monomeric VS-I increases (see changes in the amplitude of peak E). This counterintuitive result explains the possible release of monomers of VS-I with the restoration of its anti-microbial properties [15].

4. In Vitro, CGA-Derived Peptides Modify the Immunological Activities of Specific Cells Belonging to Defense

CGA has a prohormone function; numerous cleavage products of this protein display activity in the domain of defense [21,22]. Therefore, we speculated that the CGA-derived-peptides Chromofungin (CHR, CGA47-66) and Catestatin (CAT, CGA344-364) might impact the functioning of human cells involved in immunity. We harvested polymorphonuclear cells (PMNs) and monocytes from healthy controls. We exposed them ex vivo to the peptides to assess the consequences as they may occur in vivo during systemic inflammation.

We first performed experiments on PMNs [23]. We demonstrated that, after intracellular penetration, both CHR and CAT provoked a rapid and synergic Ca++- entry with no lytic effect on the cells. This occurred provided free calcium was available in the extracellular space. It was concentration-dependent and in the range of concentrations relevant for clinical effects. We also explored the impact of scrambled isomers and amino-acid substitutions. In doing so, we identified the need for a perfect respect of the chemical primary structure of both CAT and CHR to obtain the expected pharmacodynamical impact, indicating that a precise mechanism of both cell entry and action is required for physiological effects. The Ca++- entry evoked by CHR and CAT is consistent with Ca++-selective store-operated calcium channel activity. Once inside the cells, CAT and CHR interact with calmodulin, thereby allowing the release of lysophospholipids by membrane-bound iPLA2 and subsequent store-operated calcium channels. As an ultimate result of intracellular Ca++ concentration increase, the PMNs release secretions, among which we isolated factors involved in innate immunity such as lactotransferrin, neutrophil gelatinase, lysozyme, S100 A, and S100B calcium-binding proteins. These data point to a role of CAT and CHR in Ca++ signaling outside the chromaffin cells, with an impact on the activation of PMNs through a mechanism not related to a cellular membrane-bound receptor but in line with a cell-penetrating peptide activity. This action explains how the neuro-hormonal response to stress may trigger a rapid-onset effective enhanced pro-inflammatory PMNs-related mechanism of defense in vivo in any vascularized tissue where an insult occurs. It is also of note that CAT has been reported to act via the nicotinic acetylcholine receptor (nAChR), a classical surface receptor, which can also participate in anti-inflammatory responses through neural immunity regulation [6].

Our group also explored monocytes. We assessed a possible effect of the CGA47-70-derived peptide (which includes CHR) detected in the plasma of multiple trauma patients who are prone to develop care-related infections [13]. This molecule entered the monocyte progressively over 5 to 15 min with at least two intracytoplasmic localizations, one of which was detected in the perinuclear region. In further cellular investigations, including luciferase assays, we showed that CGA47-66 could inhibit both NF-kappa B and AP-1, which play a role in amplifying and perpetuating the inflammatory processes in vivo. Such activities suggest an anti-inflammatory potential for this peptide with a risk of deleterious imbalance of innate immunity.

5. CGA-Derived Peptides as Actors against Superbugs

Antibiotic-resistant microbes (bacteria, fungi, and yeasts) are detected increasingly in samples harvested from ICU patients, and they trigger significant morbidity when comorbidities are present. There is, therefore, a need for better tools to cure patients, in addition to the discovery of new anti-microbial drugs.

Based on the observation that bacterial host cells genetically engineered to express CGA for industrial production are dying upon the induction of CGA expression, our group hypothesized that chromogranins impact bacterial survival (see in [22]). Following the HPLC of the protein material secreted by chromaffin cells, it was concluded that several CGA-derived peptides and CGA itself, proenkephalin-A, and free ubiquitin participated partially in the struggle for survival after a stressing challenge. A summary of the antibacterial, antifungal, and anti-malaria properties of some of these chromaffin cell-derived peptides was given in a recent review [22]. Interestingly, while these properties are related to the CGA-derived molecules, they may sometimes be enhanced through synergic associations with either therapeutic albumin or commercially available anti-microbial drugs. Indeed, since AMPs interact with cell membranes, they represent candidates to potentiate anti-microbial drugs, as shown for some molecules marketed for ICU patients [24,25].

Our data also suggest that, sometimes, infectious diseases occur when these AMPs fail to fulfill their missions in oxidative stressing conditions.

Recently, our group also reported that CTS (CGA344-364), but not Cateslytin (CTL, CGA352-366), interacts with circulating albumin, which underlines the important role of the C-terminal part of CTS for the binding process [26]. This interaction improves the anti-microbial activity of this peptide against C. albicans at a concentration of 4 μM, demonstrating a synergistic effect [26].

Finally, human intervention on some of the L-isomers of these molecules may also modify their potency in an attempt to coat medical implants [24,27]. In recent studies, we have tested natural peptides against superbugs carried by ICU patients. It emerged that some natural and some synthetic peptides, though not all, recovered significant bactericidal activity in vitro when associated with or chemically modified. Investigations further showed that AMPs act as either cell-penetrating peptides destabilizing the cell wall of the microbes or as intracellular molecular actors interacting with calmodulin with subsequent limitation of the rate of activity of calmodulin-activated enzymes, some of which play a role in hyphal growth. These data prompted the conclusion that biomaterials intended for human implantation may benefit from the coating by such molecules or from the latter being sprayed (on wounds, for instance). Indeed, contaminations of medical devices and surgical sites continue to contribute to significant hospital morbidity. We have, therefore, successfully moved to functionalize bioprostheses with biomaterials designed to combat biofilm-associated infections. Interestingly, a self-killing approach with the bacterial-controlled release of AMPs has been reported [28]. Lastly, CGA-derived peptides were included in hydrogels to prevent oral cavity infection [29]. CAT was functionalized with polyarginine and hyaluronic acid on a silver platform: this, in addition to CAT's anti-microbial activity, strongly limited the local production of inflammatory cytokines. This is a matter of interest in relation to buccal implantation [30], with potential as regards devices requiring transient implantation in ICU patients.

6. Enigma and Future for CGA and Its Derived Peptides in the Critically Ill

Based on available data, one first challenge is understanding the mechanisms and relevance of the multimerization of CGA and some of its derived peptides in acute stressing diseases. As shown in Figure 3, according to time from admission to the recovery of shock, we have studied by Western blotting analyses the plasma of septic shock patients with antibodies directed against VS-I. As indicated, we have observed time-dependent changes in the processing of VS-I-tagged multimers in the bloodstream during the first days of ICU stay. These data indicate time-dependent changes in the processing of molecules, which supports characterizing chromogranins (CGA, CGA-related peptides, and chromogranins B and C) as acute phase proteins. The multimers rapidly diminished and, in a second phase, progressively vanished from circulation at the time when the treatment by catecholamines was possible to stop. Whether multimerization is just a modality of transport for CGA from chromaffin cells to distant targets in tissues or a mechanism of protection of CGA from enzymatic processing within circulation in this setting remains unresolved and merits close understanding to avoid interrupting a physiological process intended to protect CGA from immediate endogenous processing. On the one hand, the pharmacological manipulation of multimers with low doses of therapeutic albumin enables the release of monomeric molecules if these molecules are already oxidized [16]. On the other hand, the release of CAT, which has no such disulfide bridge available, requires another pharmacological approach, possibly by limiting the upregulation of inducible enzymes responsible for the processing of CGA. In clinical settings with systemic inflammation, such intervention has been recommended for vascular iNO-synthase inhibition. Finally, there is another challenge in relation to ICU patients as far as health stress is concerned: would CGA predict the risk of re-admission to an ICU for an improving patient when he/she leaves the ICU for an intermediate care facility? This issue has never been reported to date but would be of significant interest in situations of overcrowded ICUs.

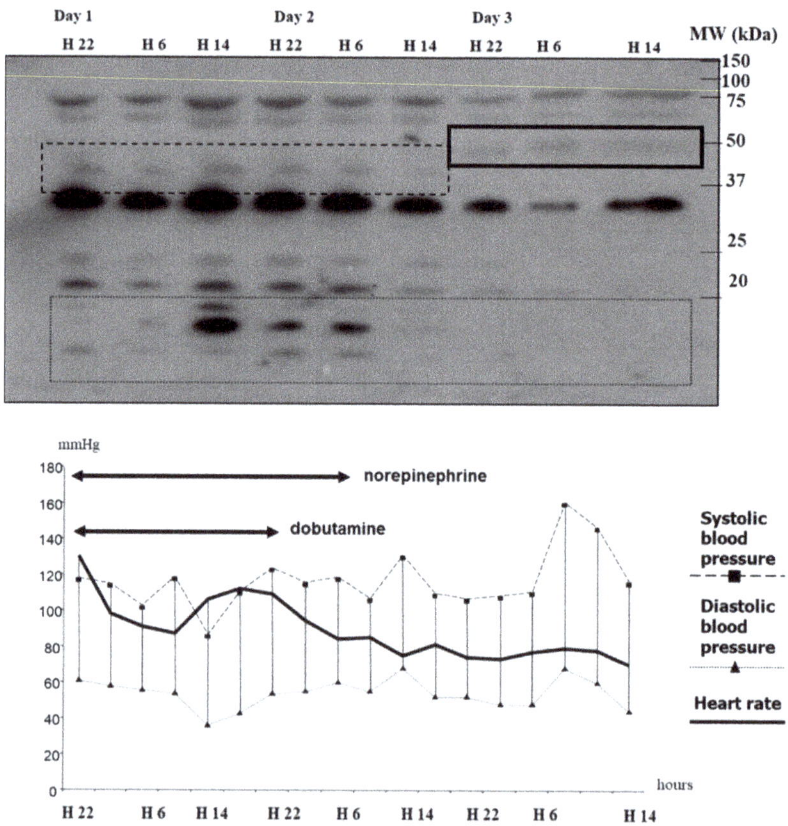

Figure 3. Typical Western blotting analyses of plasma samples to evaluate the time-dependent changes in extracellular processed VS-I (**top** panel), and simultaneous hemodynamic profile (**bottom** panel) during a human septic shock. Time (Day X) runs over 3 days from admission (Day 1, hour 10:00 pm (H22)) to the end of the third day (Day 3, 02.00 pm (H14) and is represented on the X-axis (top and bottom lines). Plasma samples harvested by 8 h, from admission until Day 3, were immediately centrifugated (4000 rounds/min at 4 °C), and SDS-PAGE electrophoresis followed by electrophoretic blotting with immunological detection of VS-I was performed in standard conditions. The anti-VS-I antibodies were a generous gift of Pr A. Corti, Milan, Italy. Periods of norepinephrine and dobutamine infusion are represented as bold arrows from admission until weaning. Please note that the processing of chromogranin A is notably changing around Day 2 at 2:00 (H14) when the patient's circulatory status no longer requires the two vasopressors. The global immunoblots are decreasing in intensity, and both small and large molecule multimerization is decaying (top panel). This phenomenon is contemporary to the decrease in doses of vasopressors (norepinephrine and dobutamine) and corresponds to circulatory failure recovery. Altogether, these data explain (i) why dosages of any of these proteins must be performed at a similar time window of the disease if a proper interpretation of their role is to be considered; (ii) that a pharmacological intervention must be scheduled at a moment when it can be efficient. Thus, our data on 4% albumin-impact on multimerization show that such an intervention must start as early as possible after admission, and it has no sense after the 5th day of disease onset [16]. Please remember that the apparent molecular weight (MW) of full-length CGA is around 70–75 kDa, and that of VS-I is approximately 18 kDa, which explains the immunoblotting of multimers on the top panel. Boxes (solid line, dashed line . . .) focus on processed molecules of interest, tagged with VS-I-antibodies: a careful identification must specify whether some multimers are not just large monomeric, full-length CGA molecules containing the VS-I domain.

A second issue concerns the possible use of CGA-derived peptides with defense properties in helping to keep implantable medical devices immune to infectious attacks once implanted. Such a development would not only be of interest to surgeons who implant several categories of prostheses but would also represent a significant breakthrough for critically ill patients. Although these proteins are endogenous peptides with few detectable side effects at nano- or micro-molar concentrations, convenient therapeutic use in humans has never been found. No data exist on in vivo consequences of the infusion of VS-I to test vasoactivity in humans. However, some experimental ex vivo studies have described the effect of VS-I on human vessels and in experimental animals [31–34]. Although this peptide has proven capable of ex vivo fungicidal and anti-bacterial activity even in multidrug-resistant microbes, it has never been tested in vivo as an anti-microbial drug. No ethical issues have been reported as explaining such a situation. Still, one can reasonably imagine how large the amount of VS-I required to produce persistent physiological effects on humans would be. One additional explanation is that the efficiency and the cost/effectiveness ratio of the drug used as a single anti-infectious agent would be questionable. Nonetheless, it does not seem unreasonable to test this approach to prevent colonization by superbugs in critical patients as they frequently display certain forms of immune suppression [35] or even as local adjuvant treatment. Duration of the peptide availability for long implantation is another limit for such use. Our group has recently proposed that implantable devices can be coated whenever possible to lessen the risk of care-related infections [27,28]. We have succeeded in incorporating CAT—and its active core CTL—by linking them to materials through a spacer, which is cleavable by enzymes from bacterial strains prone to colonizing intravascular prostheses. This opens the perspective of a new defense strategy for fighting infection of implants: the innovative component of the strategy is that the availability of the AMP is longer and diminishes only if the microbe is present with the required enzyme for the release of the coated peptide. For example, we successfully used the endo-protease Glu-C produced by S. aureus, although it had been previously shown that hyaluronidase from both S. aureus and yeast also works [27,28,30]. To make the strategy even more efficient, we have also tested D-stereoisomers of some CGA-derived peptides: these isomers proved very stable against enzymatic proteolysis. The preliminary results encourage further investigation. First, the dimeric form of CTL linked by three polyethylene glycols substantially enhances the anti-bacterial activity against S. aureus. In contrast, dimerization was not required to ensure better destruction of C. albicans. Second, the D- CTL peptide displays interesting, enhanced activity against some Gram-negative superbugs relevant as far as ICU patients' infections are considered. Third, the non-toxic peptide DOPA5T- CTL can be employed as a "self-killing strategy" regarding S. aureus having certain protease activities; in addition, once released, the anti-microbial CGA-derived peptides still boost local immunity in dendritic cells and CD14 cells as well in tissues where a high concentration of microbes would justify a sustained release of the host defense peptide. These results indicate that medical use is foreseeable with some synthetic peptides such as CTL or VS-I in the near future, provided that technical and scientific progress enables the perfect impregnation of the implant.

A third issue arose recently: COVID became a significant threat in the hospital as far as its critical forms are concerned. De Lorenzo et al. asserted that CGA concentrations could predict death [36]. They showed that dying COVID patients demonstrated higher CGA levels on admission than survivors. Indeed, in our differently designed study of COVID patients admitted for acute respiratory failure and hypoxemia, admission plasma CGA concentrations instead predicted the occurrence of morbidity rather than mortality, which was better forecast by the CAT/CGA ratio [14]. Our study suggested that the stressing challenge of COVID was probably not hypoxemia itself—in line with its effect in vitro—since matched ICU control patients without hypoxemia did display levels of CGA similar to those of hypoxic COVID patients [37]. Because standard inflammation parameters did also not correlate with either CGA or CAT, we also examined the possibility that long-lasting circulating concentrations of CAT could interfere with systemic inflammation. According to a previous study by our group, such concentrations physiologically boost, through a cell-penetrating-peptide mechanism, the release of pro-inflammatory molecules by PMNs, molecules which are currently recognized as biomarkers of severity in COVID [23]. In addition to this mechanism of inflammation, circulating CAT will also be available to engage in a molecular receptor-linked action on nAChR. There are strong scientific arguments to prove that CAT is a non-competitive inhibitor of this receptor, which provides a new and better comprehension of imbalanced neural regulation of innate immunity in critically ill patients [38]. This activity of CAT will be of major significance regarding all critically ill patients in explaining morbidity linked to the failure to arrive at a proper balance between pro- and anti-inflammatory pathways after life-threatening stress. Indeed, the action of CAT on nAChR provides an explanation not only for the occurrence of care-related infection but also for long-lasting multiple organ failure-associated myopathy [39].

Finally, there are other relevant issues regarding critically ill patients in connection with the in vivo processing of CGA. Among the difficult matters to explore is the involvement of CGA and its derivatives in acute cardiac diseases, whereas its involvement in chronic heart diseases has already been investigated [40–45]. Acute cardiac failure is far from exceptional in acute-onset conditions such as shock. No compelling reason exists to exclude a possible role for full-length CGA or selected CGA-derived peptides in transient cardiomyopathy or some severe arrhythmias.

7. Conclusions

In recent years, progress has been achieved in comprehending the biological significance of full-length CGA and selective CGA-derived peptides measured in the bloodstream of critically ill patients. These molecules will undoubtedly benefit from specific assessments in larger groups of seriously ill patients with selected clinical phenotypes. Such a prerequisite is mandatory to understand how well-conserved molecules can contribute to understanding pathophysiological conditions that are still unexplained and, therefore, not efficiently treated. For further experiments, we propose testing the implications of CGA and CGA-derived peptides in acute heart diseases and in acute neurological abnormalities that are time-dependently observed in the critically ill. In addition, given the presence of other granins, these molecules will also be investigated in a second step. If our hypotheses turn out to be correct, we believe this will open new insights into the care of the critically ill with multiple organ failure. Linking our efforts in clinical settings and ex vivo experiments will result in better use of endogenous CGA-derived peptides for diagnostic tools and synthetic CGA-derived peptides for implants in the critically ill.

Author Contributions: Conceptualization, F.S. (Francis Schneider); investigation, F.S. (Francis Schneider), R.C.-J., F.S. (Francesco Scavello) and T.L.; resources F.S. (Francis Schneider), Y.H. and P.L.; writing—original draft preparation, F.S. (Francis Schneider) and R.C.-J.; writing-review and editing, F.S. (Francis Schneider), A.C., T.A. and Y.H. All authors have read and agreed to the published version of the manuscript.

Funding: This research received no external funding.

Institutional Review Board Statement: Not applicable.

Informed Consent Statement: Not applicable.

Data Availability Statement: Not applicable.

Acknowledgments: This research program was possible with the help (notably through constant discussions) and the logistical support of other European research groups interested in these molecules highly conserved during evolution. We also thank INSERM U1121 and Hôpitaux Universitaires de Strasbourg, both at Strasbourg University, France, for their support.

Conflicts of Interest: The authors declare no conflict of interest related to this paper.

References

1. Blaschko, H.; Comline, R.S.; Schneider, F.H.; Silver, M.; Smith, A.D. Secretion of a chromaffin granule protein, chromogranin, from the adrenal gland after splanchnic stimulation. *Nature* **1967**, *215*, 58–59. [CrossRef]
2. Helle, K.B. The chromogranin if the adrenal medulla: A high-density lipoprotein. *Biochem. J.* **1968**, *109*, 43–44. [CrossRef]
3. Schneider, F.; Castelain, V.; Herbrecht, J.E.; Hellé, S.; Metz-Boutigue, M.H. Adrenal gland-released Vasostatin-Iis a myocardial depressant factor. *Br. J. Clin. Pharmacol.* **2020**, *86*, 825–828. [CrossRef]
4. Chung, H.; Corti, A.; Crippa, L.; Schneider, F.; Metz-Boutigue, M.H.; Garnero, P. Development of an immunoassay for the derived-peptide of chromogranin A Vasostatin-I (1-76): Assessment of severity in patients with sepsis. *Biomarkers* **2012**, *17*, 430–434. [CrossRef] [PubMed]
5. Lord, J.M.; Midwinter, M.J.; Chen, Y.-F.; Belli, A.; Brohi, K.; Kovacs, E.J.; Koenderman, L.; Kubes, P.; Lilford, R.J. The systemic immune response to trauma: An overview of pathophysiology and treatment. *Lancet* **2014**, *18*, 1455–1465. [CrossRef]
6. Pavlov, V.A.; Tracey, K.J. Neural regulation of immunity: Molecular mechanisms and clinical translation. *Nat. Neurosci.* **2017**, *2*, 156–166. [CrossRef] [PubMed]
7. Le Gall, J.R.; Lemeshow, S.; Saulnier, F. A new simplified acute physiology score (SAPS II) based on a European/North American multicenter study. *JAMA* **1993**, *270*, 2957–2963. [CrossRef]
8. Cannon, W.B. The emergency function of the adrenal medulla in pain and the major emotions. *Am. J. Physiol.* **1914**, *33*, 356–372. [CrossRef]
9. Zhang, D.; Lavaux, T.; Voegeli, A.-C.; Lavigne, T.; Castelain, V.; Meyer, N.; Sapin, R.; Aunis, D.; Metz-Boutigue, M.-H.; Schneider, F. Prognostic value of chromogranin A at admission in critically ill patients: A cohort study in a medical intensive care unit. *Clin. Chem.* **2008**, *54*, 1497–1503. [CrossRef]
10. Tasiemski, A.; Hammad, H.; Vandenbulcke, F.; Breton, C.; Bilfinger, T.J.; Pestel, J.; Salzet, M. Presence of chromogranin-derived antimicrobial peptides in plasma during coronary artery bypass surgery and evidence of an immune origin of these peptides. *Blood* **2002**, *100*, 553–559. [CrossRef]
11. Zhang, D.; Lavaux, T.; Sapin, R.; Lavigne, T.; Castelain, V.; Aunis, M.; Metz-Boutigue, M.-H.; Schneider, F. Serum concentration of chromogranin A at admission: An early biomarker of severity in critically ill patients. *Ann. Med.* **2009**, *41*, 38–44. [CrossRef] [PubMed]
12. Schneider, F.; Bach, C.; Chung, H.; Crippa, L.; Lavaux, T.; Bollaert, P.-E.; Wolff, M.; Corti, A.; Launoy, A.; Delabranche, X.; et al. Vasostatin-I, a chromogranin A-derived peptide, in non-selected critically ill patients: Distribution, kinetics, and prognostic significance. *Intensiv. Care Med.* **2012**, *38*, 1514–1522. [CrossRef] [PubMed]
13. Schneider, F.; Marban, C.; Ajob, G.; Helle, S.; Guillot, M.; Launoy, A.; Maestraggi, Q.; Scavello, F.; Rohr, O.; Metz-Boutigue, M.-H. In trauma patients, the occurrence of early-onset nosocomial infections is associated with increased plasma concentrations of Chromogranin A. *Shock* **2018**, *49*, 522–528. [CrossRef]
14. Schneider, F.; Le Borgne, P.; Herbrecht, J.-E.; Danion, F.; Solis, M.; Hellé, S.; Betscha, C.; Clere-Jehl, R.; Lefebvre, F.; Castelain, V.; et al. Assessment of plasma Catestatin in COVID-19 reveals a hitherto unknown inflammatory activity with impact on morbidity-mortality. *Front. Immunol.* 2022, *in press*. [CrossRef]
15. Lugardon, K.; Raffner, R.; Goumon, Y.; Corti, A.; Delmas, A.; Bulet, P.; Aunis, D.; Metz-Boutigue, M.H. Antibacterial and antifungal activities of Vasostatin-I, the N-terminal fragment of chromogranin A. *J. Biol. Chem.* **2000**, *275*, 10745–10753. [CrossRef] [PubMed]
16. Schneider, F.; Dureau, A.-F.; Hellé, S.; Betscha, C.; Senger, B.; Cremel, G.; Boulmedais, F.; Strub, J.-M.; Corti, A.; Meyer, N.; et al. A pilot study on continuous infusion of 4% Albumin in critically ill patients: Impact on nosocomial infection via a reduction mechanism for oxidized substrates. *Crit. Care Explor.* **2019**, *1*, e0044. [CrossRef]
17. Schneider, F.; Castelain, V.; Morel, G.; Dureau, A.-F.; Poidevin, A.; Ludes, P.-O.; Fabacher, T.; Senger, B.; Meyer, N.; Metz-Boutigue, M.-H. Continuous 4 percent Albumin versus intermittent 20 percent Albumin in adults with septic shock: A prospective, pHSAe IV, open-label randomized trial. *Am. J. Intern. Med.* **2020**, *8*, 89–100. [CrossRef]
18. China, L.; Freemantle, N.; Forrest, E.; Kallis, Y.; Ryder, S.D.; Wright, G.; Portal, A.J.; Salles, N.B.; Gilroy, D.W.; O'Brien, A. A randomized trial of albumin infusions in hospitalized patients with cirrhosis. *N. Eng. J. Med.* **2021**, *384*, 808–817. [CrossRef]
19. Caraceni, P.; Riggio, O.; Angeli, P. ANSWER Study Investigators. Long-term albumin administration in decompensated cirrhosis: An open-label randomized trial. *Lancet* **2018**, *391*, 2417–2429, Erratum in *Lancet* **2018**, *392*, 386. [CrossRef]
20. Plantier, J.L.; Duretz, V.; Devos, V.; Urbain, R.; Jorieux, S. Comparison of antioxidant properties of different therapeutic albumin preparations. *Biologicals* **2016**, *44*, 226–233. [CrossRef]
21. Shooshtarizadeh, P.; Zhang, D.; Chich, J.-F.; Gasnier, C.; Schneider, F.; Haïkel, Y.; Aunis, D.; Metz-Boutigue, M.-H. The antimicrobial peptides derived from chromogranin/secretogranin family, new actors of innate immunity. *Regul. Pept.* **2010**, *165*, 102–110. [CrossRef] [PubMed]
22. Scavello, F.; Kharouf, H.; Lavalle, P.; Haïkel, Y.; Schneider, F.; Metz-Boutigue, M.H. The antimicrobial peptides secreted by the chromaffin cells of the adrenal medulla link the neuroendocrine and immune systems: From basic to clinical studies. *Front. Immunol.* **2022**, *13*, 977175. [CrossRef] [PubMed]
23. Zhang, D.; Shooshtarizadeh, P.; Laventie, B.-J.; Colin, D.A.; Chich, J.-F.; Vidic, J.; de Barry, J.; Chasserot-Golaz, S.; Delalande, F.; Van Dorsselaer, A.; et al. Two chromogranin a-derived peptides induce calcium entry in human neutrophils by calmodulin-regulated calcium independent phospholipase A2. *PLoS ONE* **2009**, *4*, e4501. [CrossRef] [PubMed]

24. Zaet, A.; Dartevelle, P.; Daouad, F.; Ehlinger, C.; Quilès, F.; Francius, G.; Boehler, C.; Bergthold, C.; Frisch, B.; Prévost, G.; et al. D-Cateslytin, a new antimicrobial peptide with therapeutic potential. *Sci. Rep.* **2017**, *7*, 15199. [CrossRef]
25. Dartevelle, P.; Ehlinger, C.; Zaet, A.; Boehler, C.; Rabineau, M.; Westermann, B.; Strub, J.-M.; Cianférani, S.; Haïkel, Y.; Metz-Boutigue, M.-H.; et al. D-Cateslytin: A new antifungal agent for the treatment of oral Candida albicans associated infections. *Sci. Rep.* **2018**, *8*, 9235. [CrossRef]
26. Mancino, D.; Kharouf, N.; Scavello, F.; Hellé, S.; Salloum-Yared, F.; Mutschler, A.; Mathieu, E.; Lavalle, P.; Metz-Boutigue, M.-H.; Haïkel, Y. The Catestatin-derived peptides are new actors to fight the development of oral candidosis. *Int. J. Mol. Sci.* **2022**, *23*, 2066. [CrossRef]
27. Cado, G.; Aslam, R.; Séon, L.; Garnier, T.; Fabre, R.; Parat, A.; Chassepot, A.; Voegel, J.-C.; Senger, B.; Schneider, F.; et al. Self-defensive biomaterial coating against bacteria and yeasts: Polysaccharide multilayer film with embedded antimicrobial peptide. *Adv. Funct. Mater.* **2013**, *23*, 4801–4809. [CrossRef]
28. Scavello, F.; Mutschler, A.; Hellé, S.; Schneider, F.; Chasserot-Golaz, S.; Strub, J.-M.; Cianferani, S.; Haikel, Y.; Metz-Boutigue, M.-H. Catestatin in innate immunity and Cateslytin-derived peptides against superbugs. *Sci. Rep.* **2021**, *11*, 15615. [CrossRef]
29. Mateescu, M.; Baixe, S.; Garnier, T.; Jierry, L.; Ball, V.; Haikel, Y.; Metz-Boutigue, M.H.; Nardin, M.; Schaaf, P.; Etienne, O.; et al. Antibacterial Peptide-Based Gel for Prevention of Medical Implanted-Device Infection. *PLoS ONE* **2015**, *10*, e0145143. [CrossRef]
30. Özçelik, H.; Vrana, N.E.; Gudima, A.; Riabov, V.; Gratchev, A.; Haïkel, Y.; Metz-Boutigue, M.-H.; Carradò, A.; Faerber, J.; Roland, T.; et al. Harnessing the multifunctionality in nature: A bioactive agent release system with self-antimicrobial and immunomodulatory properties. *Adv. Healthc. Mater.* **2015**, *4*, 2026–2036. [CrossRef] [PubMed]
31. Aardal, S.; Helle, K.B. The vasoinhibitory activity of bovine chromogranin A fragment (vasostatin) and its independence of extracellular calcium in isolated segments of human blood vessels. *Regul. Pept.* **1992**, *41*, 9–18. [CrossRef]
32. Aardal, S.; Helle, K.B.; Elsayed, S.; Reed, R.K.; Serck-Hanssen, G. Vasostatins, comprising the N-terminal domain of chromogranin A, suppress tension in isolated human blood vessel segments. *J. Neuroendocrinol.* **1993**, *5*, 405–412. [CrossRef] [PubMed]
33. Brekke, J.F.; Osol, G.J.; Helle, K.B. N-terminal chromogranin-derived peptides as dilators of bovine coronary resistance arteries. *Regul. Pept.* **2002**, *105*, 93–100. [CrossRef]
34. Roatta, S.; Passatore, M.; Novello, M.; Colombo, B.; Dondossola, E.; Mohammed, M.; Losano, G.; Corti, A.; Helle, K.B. The chromogranin A-derived N-terminal peptide vasostatin-I: In vivo effects on cardiovascular variables in the rabbit. *Regul. Pept.* **2011**, *168*, 10–20. [CrossRef] [PubMed]
35. Pfortmueller, C.A.; Meisel, C.; Fux, M.; Schefold, J.C. Assessment of immune organ dysfunction in critical illness: Utility of innate immune response markers. *Intensiv. Care Med. Exp.* **2017**, *5*, 49. [CrossRef]
36. De Lorenzo, R.; Sciorati, C.; Ramirez, G.A.; Colombo, B.; Lorè, N.I.; Capobianco, A.; Tresoldi, C.; Cirillo, D.M.; Ciceri, F.; Corti, A.; et al. Chromogranin A plasma levels predict mortality in COVID-19. *PLoS ONE* **2022**, *17*, e0267235. [CrossRef]
37. Dry, K.L.; Phillips, J.H.; Dart, A.M. Catecholamine release from bovine adrenal chromaffin cells during anoxia or metabolic inhibition. *Circ. Res.* **1991**, *69*, 466–474. [CrossRef]
38. Mahata, S.K.; O'Connor, D.T.; Mahata, M.; Yoo, S.H.; Taupenot, L.; Wu, H.; Gill, B.M.; Parmer, R.J. Novel autocrine feedback control of catecholamine release. A discrete chromogranin a fragment is a non-competitive nicotinic cholinergic antagonist. *J. Clin. Investig.* **1997**, *100*, 1623–1633. [CrossRef]
39. Rudolf, R.; Straka, T. Nicotinic acetylcholine receptor at vertebrate motor endplates: Endocytosis, recycling, and degradation. *Neurosci. Lett.* **2019**, *711*, 134434. [CrossRef]
40. Tota, B.; Angelone, T.; Cerra, M.C. The surging role of Chromogranin A in cardiovascular homeostasis. *Front. Chem.* **2014**, *2*, 64. [CrossRef]
41. Mahata, S.K.; Corti, A. Chromogranin A and its fragments in cardiovascular, immunometabolic, and cancer regulation. *Ann. N. Y. Acad. Sci.* **2019**, *1455*, 34–58. [CrossRef]
42. Filice, E.; Pasqua, T.; Quintieri, A.M.; Cantafio, P.; Scavello, F.; Amodio, N.; Cerra, M.C.; Marban, C.; Schneider, F.; Metz-Boutigue, M.-H.; et al. Chromofungin, CGA47-66-derived peptide, produces basal cardiac effects and postconditioning cardioprotective action during ischemia/reperfusion injury. *Peptides* **2015**, *71*, 40–48. [CrossRef] [PubMed]
43. Rocca, C.; De Bartolo, A.; Grande, F.; Rizzuti, B.; Pasqua, T.; Giordano, F.; Granieri, M.C.; Occhiuzzi, M.A.; Garofalo, A.; Amodio, N.; et al. Cateslytin abrogates lipopolysaccharide-induced cardiomyocyte injury by reducing inflammation and oxidative stress through toll like receptor 4 interaction. *Int. Immunopharmacol.* **2021**, *94*, 107487. [CrossRef] [PubMed]
44. Angelone, T.; Mazza, R.; Cerra, M.C. Chromogranin-A: A multifaceted cardiovascular role in health and disease. *Curr. Med. Chem.* **2012**, *19*, 4042–4050. [CrossRef] [PubMed]
45. Tarantino, N.; Santoro, F.; Di Biase, L.; Di Terlizzi, V.; Vitale, E.; Barone, R.; Della Rocca, D.G.; Cruz, N.S.D.L.D.L.; Di Biase, M.; Brunetti, N.D. Chromogranin-A serum levels in patients with Tako-tsubo syndrome and ST elevation acute myocardial infarction. *Int. J. Cardiol.* **2020**, *320*, 12–17. [CrossRef]

pharmaceutics

Review

Catestatin: Antimicrobial Functions and Potential Therapeutics

Suborno Jati [1,†], Sumana Mahata [2,†], Soumita Das [3], Saurabh Chatterjee [4] and Sushil K. Mahata [2,5,*,‡]

1 Department of Chemistry and Biochemistry, University of California San Diego, La Jolla, CA 92093, USA; sjati@ucsd.edu
2 Department of Medicine, University of California San Diego, La Jolla, CA 92093, USA; sumahata@health.ucsd.edu
3 Department of Biomedical and Nutritional Science, University of Massachusetts Lowell, Lowell, MA 01854, USA; soumita_das@uml.edu
4 Department of Medicine, University of California Irvine, Irvine, CA 92697, USA; saurabhc@hs.uci.edu
5 VA San Diego Healthcare System, 3350 La Jolla Village Drive, San Diego, CA 92161, USA
* Correspondence: smahata@health.ucsd.edu; Tel.: +1-(858)-552-8585 (ext. 2637)
† These authors contributed equally to this work.
‡ Dedicated to Marie Helene Metz-Boutigue for establishing catestatin as an antimicrobial and a cell permeable peptide.

Abstract: The rapid increase in drug-resistant and multidrug-resistant infections poses a serious challenge to antimicrobial therapies, and has created a global health crisis. Since antimicrobial peptides (AMPs) have escaped bacterial resistance throughout evolution, AMPs are a category of potential alternatives for antibiotic-resistant "superbugs". The Chromogranin A (CgA)-derived peptide Catestatin (CST: $hCgA_{352-372}$; $bCgA_{344-364}$) was initially identified in 1997 as an acute nicotinic-cholinergic antagonist. Subsequently, CST was established as a pleiotropic hormone. In 2005, it was reported that N-terminal 15 amino acids of bovine CST ($bCST_{1-15}$ aka cateslytin) exert antibacterial, antifungal, and antiyeast effects without showing any hemolytic effects. In 2017, D-$bCST_{1-15}$ (where L-amino acids were changed to D-amino acids) was shown to exert very effective antimicrobial effects against various bacterial strains. Beyond antimicrobial effects, D-$bCST_{1-15}$ potentiated (additive/synergistic) antibacterial effects of cefotaxime, amoxicillin, and methicillin. Furthermore, D-$bCST_{1-15}$ neither triggered bacterial resistance nor elicited cytokine release. The present review will highlight the antimicrobial effects of CST, $bCST_{1-15}$ (aka cateslytin), D-$bCST_{1-15}$, and human variants of CST (Gly364Ser-CST and Pro370Leu-CST); evolutionary conservation of CST in mammals; and their potential as a therapy for antibiotic-resistant "superbugs".

Keywords: Chromogranin A; catestatin; gut microbiome; antimicrobial peptide; cell permeable peptide

Citation: Jati, S.; Mahata, S.; Das, S.; Chatterjee, S.; Mahata, S.K. Catestatin: Antimicrobial Functions and Potential Therapeutics. *Pharmaceutics* **2023**, *15*, 1550. https://doi.org/10.3390/pharmaceutics15051550

Academic Editors: Scavello Francesco, Jean-Eric Ghia and Amiche Mohamed

Received: 1 April 2023
Revised: 9 May 2023
Accepted: 14 May 2023
Published: 20 May 2023

1. Introduction

Microbial infections in critically ill patients are a global threat. With failing host defense, the use of antibiotics has taken the place for containment of those infections. Nevertheless, the microbes have also evolved with time to develop resistance against those drugs. This arm's race between antibiotic drugs and pathogens had led to the rise of multi-drug-resistant microbes, called "superbugs", which emphasizes the urgent need to develop new modes of treatment. Since antimicrobial peptides (AMPs) have evaded bacterial resistance for millions of years of evolution [1], AMPs could be a potential solution for antibiotic resistant "superbugs". Chromogranin A (CgA), the acidic and secretory pro-protein [2,3], is proteolytically cleaved to generate several biologically important peptides, including Catestatin (CST: $hCgA_{352-372}$) [4–12]. The 21 amino acid peptide CST (human: $S_{352}SMKLSFRARAYGFRGPGPQL_{372}$; bovine: $R_{344}SMRLSFRARGYGFRGPGLQL_{364}$) was identified in 1997 as a physiologic brake in catecholamine secretion, which acts by non-competitive inhibition of nicotinic-cholinergic signaling [5,13–18]. CST is now established as a pleiotropic peptide [6,19,20]

The non-hemolytic antimicrobial (bacteria, fungus, and yeast) effects of CST ($bCST_{1–15}$ aka cateslytin) were first reported in 2005, where CST was shown to act by penetrating fungal and yeast cell membranes [21]. The study showed that less than 10 μM peptide was required to kill the bacteria. Antibacterial activities were also reported for the two human variants of CST ($G_{364}S$-CST and $P_{370}L$-CST) [22] with minimal inhibitory concentration (MIC) of 1–20 μM [21]. Later, D-$bCST_{1–15}$ (where L-amino acids were changed to D-amino acids: $r_{344}smrlsfrargygfr_{358}$) was shown to exert more potent antibacterial (both Gram-positive and Gram-negative bacteria) effects than natural CST [23]. The present review will focus on the antimicrobial effects of CST, with special emphasis on the mechanisms underlying its antibacterial effects, therapeutic potential, and evolutionary conservation.

2. Antibacterial Effects of CST

2.1. Inhibition of Bacterial Growth by CST

The group of Metz-Boutigue first demonstrated the antibacterial activity of CST. Her group used $bCST_{344–358}$ (coining the term cateslytin to describe this antimicrobial effect) to reveal the inhibition of growth of the Gram-positive and Gram-negative bacteria [21]. The minimal inhibitory concentrations (MICs) of CST ($bCgA_{344–358}$, $hCgA_{352–372}$, $Gly_{364}Ser$-CST and $Pro_{370}Leu$-CST) for Gram-positive bacteria (*Micrococcus luteus, Bacillus megaterium,* Group A *Streptococcus, S. aureus* ATCC 25923, *S. aureus* ATCC 49775, *S. aureus* S1 MRSA, *S. aureus* S1 MSSA, and *S. aureus* DmprF) range from 0.8 μM to >100 μM (Figure 1) [21,24]. The minimal concentration with 100% inhibition (MIC_{100}) for Gram-positive bacteria range from 2 μM to >100 μM. The MIC of CST was higher (8 μM to 50 μM) for Gram-negative bacteria (*Escherichia coli* D22, *E. coli* 029, and *Pseudomonas aeruginosa*) compared to Gram-positive bacteria (Figure 1). Likewise, the MIC_{100} of CST was higher (15 μM to 150 μM) for Gram-negative bacteria compared to Gram-positive bacteria (Figure 1). The higher MIC and MIC_{100} values of CST for Gram-negative bacteria are consistent with the presence of extra outer membrane containing lipopolysaccharide (LPS) [25,26]. Beyond the extra-thick cell membrane, Gram-negative bacteria also release exotoxins such as tetanus [27] and cholera toxins [28] that worsen prognosis.

Bacterial species	WT-CST $bCgA_{344-358}$ MIC (μM)	WT-CST $bCgA_{344-358}$ MIC_{100} (μM)	WT-CST $hCgA_{352-372}$ MIC (μM)	WT-CST $hCgA_{352-372}$ MIC_{100} (μM)	G364S-CST $hCgA_{352-372}$ MIC (μM)	G364S-CST $hCgA_{352-372}$ MIC_{100} (μM)	P370L-CST $hCgA_{352-372}$ MIC (μM)	P370L-CST $hCgA_{352-372}$ MIC_{100} (μM)
Gram-positive bacteria *Micrococcus luteus*	0.8	2	5	15	1	2	2	10
Bacillus megaterium	0.8	2	-	-	-	-	-	-
Group A *Streptococcus*			75	75		30		30-50
Staphylococcus aureus ATCC 25923	40		>100	>100	>100	>100		
Staphylococcus aureus ATCC 49775	37		125					
Staphylococcus aureus S1 (MRSA)	37		130					
Staphylococcus aureus S1 (MSSA)	45		150					
Staphylococcus aureus *ΔmprF*			20	30	5-10	10	5	5
Gram-negative bacteria *Escherichia coli D22*	8	15	15	150	10	40	20	100
Escherichia coli O29			30	30	20	20-30	10	5-10
Pseudomonas aeruginosa			50	50				

Figure 1. Effects of wild-type (WT)-CST and natural human variants of CST (Gly364Ser and Pro370Leu) on the growth of Gram-positive and Gram-negative bacteria showing minimal inhibitory concentration (MIC) and lethal concentration (MIC_{100}) of CST.

D-$bCST_{1–15}$ was reported to exert more effective antimicrobial effects against various bacterial strains than L-$bCST_{1–15}$ [23]. In addition to its antimicrobial effects, D-$bCST_{1–15}$ was reported to potentiate (additive/synergistic) the antibacterial effects of cefotaxime, amoxicillin, and methicillin [23]. Furthermore, it has been shown that D-bCST$_{1-15}$ neither triggered bacterial resistance nor elicited cytokine release [23]. In addition, D-bCST$_{1-15}$ was reported to be more resistant to degradation by secreted bacterial protease than L-bCST$_{1-15}$ [23]. Thus, it was suggested that D-bCST$_{1-15}$ can be used as a monotherapy or as a combination therapy with currently prescribed antibiotics to counteract various diseases associated with bacterial infection [23].

2.2. Composition of Bacterial Membranes

While antibiotics target specific cellular activities (e.g., synthesis of DNA, protein, or cell wall), AMPs target the LPS layer of the cell membrane. Extensive studies have been conducted to learn the composition of the bacterial membrane. The bacterial cytoplasmic membrane consists of zwitterionic phospholipids (phosphatidylcholine, phosphatidylethanolamine, sphingomyelin, etc.) and anionic phospholipids (phosphatidyl serine, phosphatidyl glycerol, etc.), providing them with a negative charge [29–32]. In contrast, besides the cytoplasmic membrane, Gram-negative bacteria contain an additional strong electronegative LPS-containing thick outer membrane [25,26]. Furthermore, the peptidoglycan layer on the outer side of the cytoplasmic membrane is much thicker in Gram-positive bacteria compared to Gram-negative bacteria (20–80 nm versus ~10 nm) [33,34]. The peptidoglycan layer in Gram-positive bacteria is connected by electronegative wall lipoteichoic acids and anchored on the phospholipid bilayer by electronegative lipoteichoic acids [35]. In contrast, in Gram-negative bacteria, the LPS forms the major lipid component of the outer leaflet of the outer membrane [35].

2.3. Secondary Structure of CST Explains the Antibacterial Effects of CST

Based on their secondary structure, AMPs are generally categorized into four groups: (i) α-helical AMPs, (ii) β-sheet AMPs, (iii) extended AMPs, and cationic loop AMPs [36]. Homology modeling followed by molecular dynamics simulation of bovine CST (bCgA$_{342-370}$) performed in a water shell led to a β-strand-loop-β-strand structure. Molecular dynamics and computer simulations of human CST$_{1-21}$ revealed the following: R_{10}, A_{11}, and Y_{12} contribute to a 3_{10} helix [37]. In contrast, F_7, R_8, A_9, F_{14}, R_{15}, G_{16}, P_{17}, and G_{18} contribute to the antiparallel β-sheet [37]. The mechanism of the antibacterial action of CST$_{1-21}$ could start by interacting with negatively charged moieties such as LPS in the outer membranes of Gram-negative bacteria and lipoteichoic acid in the wall of Gram-positive bacteria. The primary structure of CST reveals that CST contains cationic and hydrophobic residues and adopt a β-sheet secondary structure via intermolecular forces [38]. This folding structure would facilitate CST to fold into an amphiphilic conformation with positively charged (polar) and hydrophobic (nonpolar) faces (Figure 2). The presence of a great number of positively charged residues (5 in bCST and 4 in hCST) will allow CST to interact preferentially with negatively charged bacterial membranes [1,39]. Since the hydrophilic and hydrophobic amino acids of CST are structurally segregated, it will provide solubility of CST in both aqueous and lipid-rich environments, as suggested for other AMPs [40]. In addition, positively charged amino acids in CST formed amphipathic structures, as evidenced by separated hydrophobic and hydrophilic surface domains [39,41] (Figure 2). When the concentration of CST would exceed a certain critical concentration, the cell membrane would form pores, leading to content leakage, cell lysis, and finally death. Since cyclization of peptide has been reported to induce high antimicrobial activity [39,41], it is reasonable to assume that cyclization of CST would markedly improve the antibacterial activity of CST.

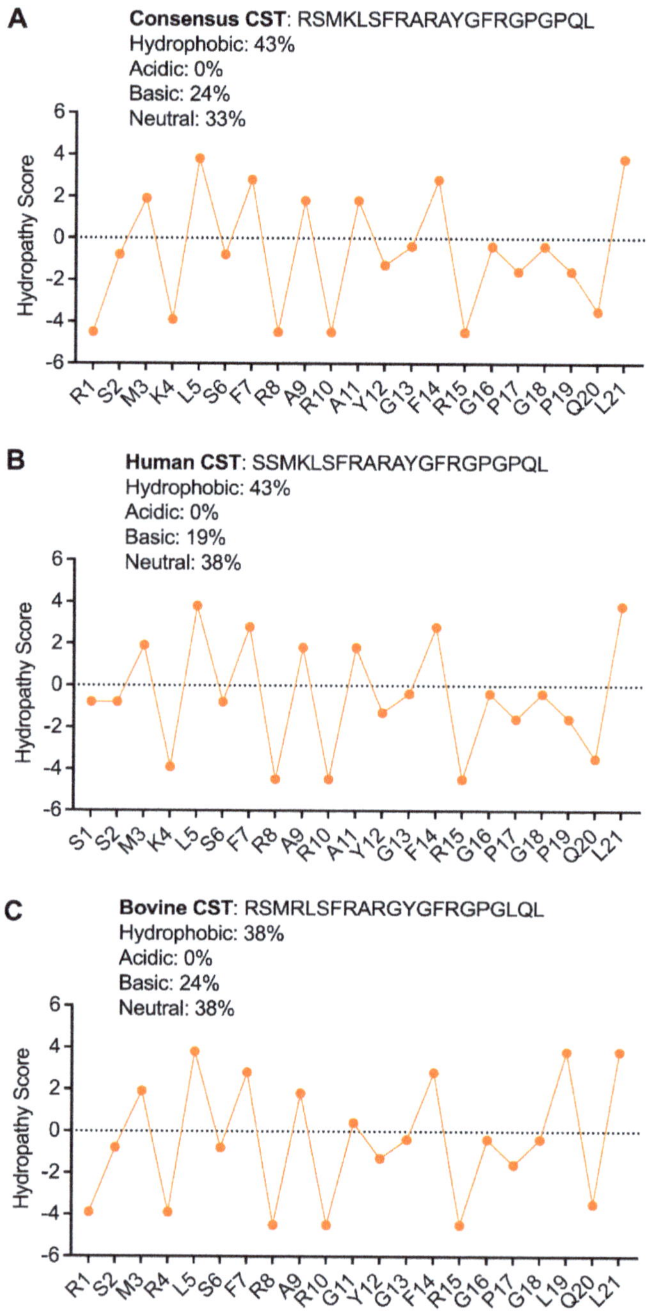

Figure 2. Hydropathy profiles of (**A**) Consensus CST, (**B**) Human CST, and (**C**) Bovine CST. The values are plotted based on the parameters used from Kyte and Doolittle, 1982. The values above zero represents the hydrophobic property of the amino acids that might contribute to the hydrophobic core of the peptide. The values below zero represent the hydrophilic property of the amino acids, which are instrumental in interaction with other protein factors.

Metz-Boutigue's group has shown that $bCgA_{344-358}$ is unstructured in solution but is converted to an antiparallel β-structure and forms aggregates at the surface of negatively charged bacterial membranes [42]. As for catecholamine secretion [15], arginine residues were found to be crucial for binding to negatively charged lipids [42,43]. They proposed that the phase boundary defects caused by zones of different rigidity and thickness lead to permeability induction and peptide crossing through the bacterial membrane [42]. The fact that CST penetrates through the bacterial wall was shown by measuring the optical density of the released β-galactosidase from ML-35p [24]. Electron microscopical studies of *E. coli* ML-35p confirmed that CST rapidly disrupts the *E. coli* membrane, with visible membrane blebbing compared to untreated cells within 10 min [24].

2.4. CST as a Potential Therapy for Bacterial Diseases

AMPs derived from CgA display antimicrobial activities by lytic effects at micromolar range against Gram-positive bacteria, filamentous fungi, and yeasts. Interestingly, CST-derived peptides can kill "superbugs" and more particularly *S. aureus* [44]. Considering the actions of CST on *E. coli*, it could be useful as a therapeutic target for the Gram-negative bacteria that cause many serious infections, including Cholera [45], *E. coli* [46], *Yersinia* [47], *Campylobacter* [48], *Legionella* [49], *Salmonella* [50], *Klebsiella* [51], *Pseudomonas* [52], *Francisella tularensis* [53], *Salmonella typhi* [54], and microbes associated with drug resistance [55–57], and CST might be used as a therapeutic target for the above diseases.

3. Antifungal and Anti-Yeast Effects of CST

3.1. Inhibition of Growth of Fungus and Yeast by CST

Fungal infections are common on the surface of skin, nails, or mucous membranes (superficial or mucocutaneous), underneath the skin (subcutaneous), or in the lungs, brain, or heart (deep infection). Deep fungal infections include Histoplasmosis [58], Coccidioidomycosis (Valley fever) [59], Blastomycosis [60], Aspergillosis [61], Candidal urinary tract infection [62], invasive candidiasis [63,64], *Pneumocystis* pneumonia [65], Mucormycosis [66,67], and Cryptococcosis [68,69]. It is becoming increasingly evident that resistance to antifungal therapy is on the rise [70,71], which calls for the development of alternative therapy for these infections. Host-defense peptides are emerging as new promising candidates to counteract antifungal resistance [72]. To this end, Metz-Boutigue's group tested the effects of CST on the growth of fungus and yeasts. They found MIC values of CST or its human variants ranging from 0.2 μM to 75 μM against a host of fungal species (*Neurospora crassa, Aspergillus fumigatus, A. niger, Nectria haematococca, Fusarium culmorum, F. oxysporum, Trichophyton mentagrophytes*, and *T. rubrum*) [21,24] (Figure 3). The MIC_{100} values of CST or its human variants against the above fungal species ranged from 0.8 μM to 100 μM [21,24] (Figure 3). CST and its human variants also displayed similar inhibitory effects on the growth of yeasts, with MIC ranging from 1.2 μM to >240 μM (Figure 3) [21,24]. The MIC_{100} of CST and its variants against the above yeasts ranged from 6 μM to 75 μM [21,24] (Figure 4). Similar to the effects of retro-inverso (RI)-CST (Amino-lqpGpGrfGyararfslkmss-carboxyl, with the CST sequence reversed from carboxyl → amino, and chirality was inversed from L → D) on catecholamine secretion [73], D-CST exhibited comparable inhibitory effects on the growth of yeast compared to L-CST with MIC ranged from 2 μM to 9.6 μM [23]. D-CST was also uncovered to be resistant to proteolytic digestion [23,44,74]. Akin to L-CST, D-CST can also be used to develop therapies for drug-resistant microbial infection [75].

Phylum Fungal species	WT-CST $bCgA_{344-358}$ MIC (μM)	WT-CST $bCgA_{344-358}$ MIC_{100} (μM)	WT-CST $hCgA_{352-372}$ MIC (μM)	WT-CST $hCgA_{352-372}$ MIC_{100} (μM)	G364S-CST $hCgA_{352-372}$ MIC (μM)	G364S-CST $hCgA_{352-372}$ MIC_{100} (μM)	P370L-CST $hCgA_{352-372}$ MIC (μM)	P370L-CST $hCgA_{352-372}$ MIC_{100} (μM)
Ascomycota *Neurospora crassa*	1.2	3.2	20	50	3	5	3	10
Aspergillus fumigatus	10	80	20-80	30	10-20	80	10-20	100
Aspergillus niger			20	30	20	30	10	20
Nectria haematococca	0.2	0.8						
Fusarium culmorum	2	8						
Fusarium oxysporum	6	10						
Trichophyton mentagrophytes	4	20						
Trichophyton rubrum			75	75	20		30	

Figure 3. Effects of WT-CST and natural human variants of CST (Gly364Ser and Pro370Leu) on the growth of fungal species showing MIC and MIC_{100} of CST.

Yeast species	WT-L-CST $bCgA_{344-358}$ MIC (μM)	WT-L-CST $bCgA_{344-358}$ MIC_{100} (μM)	WT-D-CST $bCgA_{344-358}$ MIC (μM)	WT-L-CST $bCgA_{344-364}$ MIC (μM)	WT-CST $hCgA_{352-372}$ MIC (μM)	WT-CST $hCgA_{352-372}$ MIC_{100} (μM)	G364S-CST $hCgA_{352-372}$ MIC_{100} (μM)	P370L-CST $hCgA_{352-372}$ MIC_{100} (μM)
Candida albicans ATCC 14053						50-75	30	30
Candida albicans	1.2	8						
Candida albicans "S"	7.9		5.5	30	>240			
Candida albicans "R"	9.6		9.6	50	>240			
Candida tropicalis	1.8	10						
Candida tropicalis "S"	9.8		8.1	50	>240			
Candida tropicalis "R"	2.0		2.0	20	>240			
Candida glabrata	8	30						
Candida glabrata "S"	38.2	13.4						
Candida glabrata "R"	61.4	15.0		>100	>240			
Candida neoformans	1.4	6		>100	>240			

Figure 4. Effects of WT-CST and natural human variants of CST (Gly364Ser and Pro370Leu) on the growth of yeast species showing MIC and MIC_{100} of CST.

3.2. Mechanisms Underlying the Antifungal and Antiyeast Activities of CST

The composition of fungal cell membranes is similar to that of bacteria, comprising zwitterionic and anionic phospholipids. In contrast, the fungal cell wall is composed of chitin, glucan, ergosterol, and mannoprotein, which reside on the surface of the cytoplasmic membrane. Because of the negative charge on the cytoplasmic membrane, CST could exert its anti-fungal activities in a similar way to its antibacterial activity. Metz-Boutigue's group used confocal laser microscopy to analyze the interaction of the synthetic rhodamine-labeled cateslytin ($bCgA_{344-358R}$) with fungal (*A. fumigatus*) and yeast (*C. albicans*) membranes [21]. Rhodaminated cateslytin (1 μM) was detected in the inner compartment after 2 min of incubation, implicating rapid and efficient penetration through the cell wall [21]. Using time-lapse video microscopy of fungal growth, they have shown that rhodaminated cateslytin blocked the growth and development of nascent fungus [21]. Penetration of rhodaminated cateslytin takes place at both ends of the small fungi (three cells and expressing a slow growth rate) as compared to larger fungi with a higher growth rate where penetration takes place at one end [21]. Sequence homology of the well-known cell-permeable peptide penetratin with CST representing seven mammalian orders (Primates, Rodentia, Artiodactyla, Perissodactyla, Carnivora, Cetacea, and Monotremata) revealed 63.64% to 75% similarity, which should qualify CST as a cell-permeable peptide (Figure 5).

Common Name	Scientific Name	Order	Sequence	Identity	Similarity
Penetratin			R Q I K I W F Q N R R		
Consensus mammalian CST			R S M K L S F R A R	40%	70%
Human	*Homo sapiens*	Primates	M K L S F R A R	37.5%	75%
House mouse	*Mus musculus*	Rodentia	R S M K L S F R T R	40%	70%
Cattle	*Bos taurus*	Artiodactyla	R S M R L S F R A R	30%	70%
Pig	*Sus scrofa*	Artiodactyla	R S M K L S F R	37.5%	75%
Horse	*Equus caballus*	Perissodactyla	R S M K L S F R A R	40%	70%
Walrus	*Odobanus rosmarus divergens*	Carnivora	R S M K L S F R A R	40%	70%
Killer whale	*Orcinus orca*	Cetacea	R A M K L S F R A R	40%	70%
Platypus	*Ornithorhynchus anatinua*	Monotremata	R S M K L S F K T H K	27%	64%
Identical: Black; Similar: Blue; Not Similar: Red					

Figure 5. Homology between cell permeable peptide penetratin and CST in seven mammalian orders.

4. CST Regulation of Gut Microbiota

4.1. Microbiomes in Colonic Mucosa versus Feces

Recent studies have identified a larger role of gut microbiota in gut-immune homeostasis and in intestinal pathology. The human intestinal microbiota is dominated by five phyla: high-abundant (>80%) (1) Bacillota (aka Firmicutes) and (2) Bacteroidota; less-abundant (3) Actinomycetota (aka Actinobacteria), (4) Pseudomonadota (aka Proteobacteri), and (5) Verrucomicrobiota [76] as compared to four phyla in mice: high-abundant Bacteroidota, Bacillota, Deferribacterota, and Pseudomonadota [77]; and (4) low-abundant Actinomycetota and Verrucomicrobiota compared to humans. In mouse colonic mucosa samples, 19 phyla were identified [78] (Figure 6). Although CST failed to alter bacterial populations in the four high-abundant phyla, it altered colonic mucosa-associated bacterial community composition at lower taxonomic levels, including orders Bacteroidales, Clostridiales, and YS2, and Families Chitinophagaceae, Clostridiaceae, Coriobacteriaceae, Pseudomonadaceae, Rikenellaceae, and Ruminococcaceae [78]. While CST increased the relative abundance of Bacteroidota, it caused a marked decrease in the Bacillota population (Figure 6). *Bacteroides* and *Parabacteroides* species, representing ~25% of the colonic microbiota, transform simple and complex sugars into volatile short-chain fatty acids (SCFAs), such as acetate, butyrate, and propionate [79–81], which are absorbed in the colon as a nutrient [82,83]. In addition to colonic nutrients, SCFAs are well established for their roles in accelerating gut transit time via the release of serotonin [84,85]. SCFAs also release glucagon-like peptide 1 from the enteroendocrine L-cells [86–88] and improve insulin sensitivity [89–91]. *Bacteroides thetaiotaomicron* produces significant amounts of glycosylhydrolases, which prevent obesity [92]. Other *Bacteroides* species are also reported to prevent obesity and increase insulin sensitivity [93,94]. Furthermore, *Bacteroides fragilis* produces zwitterionic polysaccharide, which activates $CD4^+$ T cells to produce interleukin 10 (IL-10). IL-10 plays crucial roles in preventing abscess formation and other unchecked inflammatory responses [95,96]. The functional correlation between different CST mutants across species and their respective microbiota has remained elusive.

	WT Mucosal samples	WT+Sal Fecal samples	WT+CST Fecal samples	CST-KO+Sal Fecal samples	CST-KO+CST Fecal samples	WT+FMT-CST-KO Fecal samples	CST-KO+FMT-WT Fecal samples
Major Phyla (>1%)	Bacillota (aka Firmicutes)	Bacillota	Bacillota ↓	Bacillota	Bacillota ↓↓	Bacillota	Bacillota
	Bacteroidota	Bacteroidota	Bacteroidota ↑↑	Bacteroidota	Bacteroidota ↑	Bacteroidota ↑	Bacteroidota ↓
	Deferribacterota	Deferribacterota	Deferribacterota	Deferribacterota	Deferribacterota	Deferribacterota	Deferribacterota
	Pseudomonadota (aka Proteobacteria)	Pseudomonadota	Pseudomonadota ↑	Pseudomonadota	Pseudomonadota ↑	Pseudomonadota	Pseudomonadota
Minor Phyla (<1%)	Actinomycetota (aka Actinobacteria)	Actinomycetota	Actinomycetota	Actinomycetota	Actinomycetota	Actinomycetota ↑	Actinomycetota
	Parcubacteria (aka OD1)	Parcubacteria	Parcubacteria ↑	Parcubacteria	Parcubacteria	Parcubacteria ↑	Parcubacteria
	Saccharibacteria (aka TM7)	Saccharibacteria			Saccharibacteria	Saccharibacteria	Saccharibacteria
		Campylobacterota	Campylobacterota	Campylobacterota	Campylobacterota	Campylobacterota	Campylobacterota
	Cyanobacteria	Cyanobacteria	Cyanobacteria	Cyanobacteria	Cyanobacteria	Cyanobacteria	Cyanobacteria
		Desulfobacterota	Desulfobacterota ↑	Desulfobacterota	Desulfobacterota	Desulfobacterota	Desulfobacterota
	Fibrobacterota	Fibrobacterota			Fibrobacterota	Fibrobacterota	Fibrobacterota
	Mycoplasmatota (aka Tenericutes)	Mycoplasmatota			Mycoplasmatota	Mycoplasmatota	Mycoplasmatota
	Verrucomicrobiota	Verrucomicrobiota	Verrucomicrobiota	Verrucomicrobiota ↓	Verrucomicrobiota ↑↑	Verrucomicrobiota ↓	Verrucomicrobiota

Figure 6. Abundance of bacterial species in mucosal and fecal samples in WT and CST-KO mice in presence or absence of CST as well as after fecal microbial transplant. Green arrows indicate CST effects; red arrows indicate FMT effects.

4.2. Microbiomes in CST Knockout (CST-KO) Mice and Inflammation

CST knockout (CST-KO) mice were generated in 2018 and are: insulin-resistant on a normal chow diet [97], hyperadrenergic [98], hypertensive [98], and with a leaky gut [99]. The microbiome in CST-KO mice was found to be quite different in composition than its WT littermates [99]. Microbial richness revealed a significant decrease in CST-KO compared to WT mice [100]. (Figure 7). Surprisingly, Verrucomicrobiota population was very low in CST-KO mice, indicating low levels of *Akkermansia* species. Since *A. muciniphila* modulates obesity by regulating metabolism and energy homeostasis to improve insulin sensitivity and glucose homeostasis [101], low Verrucomicrobiota population possibly contributed to the insulin resistance reported for CST-KO mice [97].

4.3. Alteration of Diversity and Composition of the Microbiota in the CST-KO after Supplementation with CST

Decreased amplicon sequence variants and abundance-based coverage estimator indices in CST-KO mice were restored after supplementation with CST for 15 days [100]. Akin to richness scores, supplementation of CST-KO mice with CST increased the diversity index as assessed by Shanon's *H* and inverted Simpson's index [100]. At the phylum level, CST decreased Bacillota phylum and increased Bacteroidota, Patescibacteria, Desulfobacterota, Verrucomicrobiota, and Proteobacteria in both CST-KO and WT mice [100]. In contrast, CST increased *Alistipes*, *Akkermansia*, and *Roseburia* genera only in CST-KO mice [100].

Consensus: RSMKLSFRARAYGFRGPGPQL

Common Name	Scientific Name	Order	CgA Domain	Sequence (10, 20)	
Human	*Homo sapiens*	Primates	CgA352-372	SSMKLSFRARAYGFRGPGPQL	21
Bonobo	*Pan paniscus*	Primates	CgA354-374	SSMKLSFRARAYGFRGPGPQL	21
Chimpanzee	*Pan troglodytes*	Primates	CgA354-374	SSMKLSFRARAYGFRGPGPQL	21
Western low-land gorilla	*Gorilla gorilla gorilla*	Primates	CgA354-374	SSMKLSFRARAYGFRGPGPQL	21
Sumatran orangutan	*Pongo abelii*	Primates	CgA352-372	SSMKLPFRARAYGFRGPGPQL	21
Silvery gibbon	*Hylobates moloch*	Primates	CgA353-373	SSMKLSFRARAYGFRGPGPQL	21
Black snub-nosed monkey	*Rhinopithecus bieti*	Primates	CgA348-368	RSMKLSFRARAYGFRGPGPQL	21
Crab-eating macaque	*Macaca fascicularis*	Primates	CgA353-373	RSMKLSFRARAYGFRGPGPQL	21
Golden snub-nosed monkey	*Rhinopithecus roxellana*	Primates	CgA348-368	RSMKLSFRARAYGFRGPGPQL	21
Green monkey	*Chlorocebus sabaeus*	Primates	CgA352-372	RSMKLSFRARAYGFRGPGPQL	21
Night monkey	*Aotus nancymaae*	Primates	CgA351-371	HSMKLSFQARAYGFRGPGPQL	21
Old world monkey	*Colobus angolensis palliatus*	Primates	CgA350-370	RSMKLSFRARAYGFRGPGPQL	21
Olive baboon	*Papio anubis*	Primates	CgA355-375	RSMKLSFRARAYGFRGPGPQL	21
Pig-tailed macaque	*Macaca nemestrina*	Primates	CgA353-373	RSMKLSFRARAYGFRGPGPQL	21
Red colobus	*Piliocolobus tephrosceles*	Primates	CgA353-373	RSMKLSFRARAYGFRGPGPQL	21
Sooty mangabey	*Cercocebus atys*	Primates	CgA354-374	GSMKLSFRARAYGFRGPGPQL	21
Rhesus monkey	*Macaca mulatta*	Primates	CgA350-370	RSMKLSFRARAYGFRGPGPQL	21
Tufted capuchin	*Sapajus apella*	Primates	CgA351-371	HSMKLSFRGRAYGFRGPGPQL	21
Black rat	*Rattus rattus*	Rodentia	CgA360-380	RSMKLSFRTRAYGFRDPGPQL	21
Brown rat	*Rattus norvegicus*	Rodentia	CgA366-386	RSMKLSFRARAYGFRDPGPQL	21
Golden hamster	*Mesocricetus auratus*	Rodentia	CgA335-355	RSMKLSFRARAYGFRDPGPQP	21
Grasshopper mouse	*Onychomys torridus*	Rodentia	CgA347-367	RSMKLPFRARAYGFRDPGPQL	21
House mouse	*Mus musculus*	Rodentia	CgA364-384	RSMKLSFRTRAYGFRDPGPQL	21
Multimammate mouse	*Mastomys coucha*	Rodentia	CgA356-376	RSMKLSFRARAYGFRDPGPQL	21
Ryuku mouse	*Mus carol*	Rodentia	CgA359-379	RSMKLSFRTRAYGFRDPGPQL	21
Shrew mouse	*Mus pahari*	Rodentia	CgA347-367	RSMKLSFRTRAYGFRDPGPQL	21
European rabbit	*Oryctolagus cuniculus*	Lagomorpha	CgA342-362	RSMKLSFQPRAYGFRGPGPQL	21
Alpaca	*Vicugna pacos*	Artiodactyla	CgA348-368	RSMKLSFRARAYGFRGPGLQL	21
Arabian camel	*Camelus dromedarius*	Artiodactyla	CgA347-367	HSMKLSFRARAYGFRGPGLQL	21
Cattle	*Bos taurus*	Artodactyla	CgA344-364	RSMRLSFRARGYGFRGPGLQL	21
Goat	*Capra hircus*	Artiodactyla	CgA338-358	RSMRLSFQARGYGFRGPGLQL	21
Pig	*Sus scrofa*	Artiodactyla	CgA343-363	RSMKLSFRAPAYGFRGPGLQL	21
Sheep	*Ovis aries*	Artiodactyla	CgA347-367	RSMRLSFQARGYGFRGPGLQL	21
Donkey	*Equus asinus*	Perissodactyla	CgA343-363	RSMKLSFRARAYGFRGPGLQL	21
Horse	*Equus caballus*	Perissodactyla	CgA343-363	RSMKLSFRARAYGFRGPGLQL	21
Southern white rhinoceros	*Ceratotherium simum*	Perissodactyla	CgA347-367	RSMKLSFRARAYGFRGPGLQL	21
California sea lion	*Zalophus californianus*	Carnivora	CgA344-364	RSMRLSFRARAYDFRGPGLQL	21
Cat	*Felis catus*	Carnivora	CgA349-369	SSMKLSFRARAYDFRGPGLQL	21
Cheetah	*Acinonyx jubatus*	Carnivora	CgA348-368	SSMKLSFRARAYDFRGPGLQL	21
Dog	*Canis lupus*	Carnivora	CgA343-363	HSMKLSFRARAYDFRGPGLPL	21
Giant panda	*Ailuropoda melanoleuca*	Carnivora	CgA346-366	RSMKLSFRARAYDFRGPGLQL	21
Grizzly bear	*Ursus arctos horribilis*	Carnivora	CgA347-367	RSMKLSFRARAYDFRGPGLQL	21
Harbor seal	*Phoca vitulina*	Carnivora	CgA348-368	RSMKLSFRARAYDFRGPGLQL	21
Leopard	*Panthera pardus*	Carnivora	CgA348-368	SSMKLSFRARAYDFRGPGLQL	21
Mongooses	*Suricata suricatta*	Carnivora	CgA350-370	RSMKLAFRARAYDFRGPGLQL	21
Monk seal	*Neomonachus schauinslandi*	Carnivora	CgA344-364	RSMKLSFRARAYDFRGPGLQL	21
Northern fur seal	*Callorhinus ursinus*	Carnivora	CgA343-363	RSMRLSFRARAYDFRGPGLQL	21
Walrus	*Odobenus rosmarus divergens*	Carnivora	CgA340-360	RSMKLSFRARAYDFRGPGLQL	21
Weddell seal	*Leptonychotes weddellii*	Carnivora	CgA347-367	RSMKLSFRARAYDFRGPGLQL	21
Dolphin	*Lipotes vexillifer*	Cetacea	CgA339-359	RAMKLSFRARGYDFSGPGLPL	21
Killer whale	*Orcinus orca*	Cetacea	CgA340-360	RAMKLSFRARGYDFSGPGLQL	21
Sperm whale	*Physeter catodon*	Cetacea	CgA355-375	RAMKLSFRARGYDFRGPGLQL	21
Platypus	*Ornithorhynchus anatinua*	Monotremata	CgA370-390	RSMKLSFKTHKYDFGEAGRPK	21

Figure 7. Homology of CST sequence in 53 mammalian species belonging to eight orders. CST sequences were aligned using the MUSCLE method provided by SnapGene software from the following mammalian species: human (Homo sapiens: NM_001275), bonobo (*Pan paniscus*: XP_008956465.1), chimpanzee (*Pan troglodytes*: PNI97600.1), western low-land gorilla (*Gorilla gorilla gorilla*: XM_019009788.2), Sumatran orangutan (*Pongo abelii*: XM_002825045.3), silvery gibbon (*Hylobates moloch*: XP_031990963.1), black snub-nosed monkey (*Rhinopithecus bieti*: XM_017857899.1), crab-eating macaque (*Macaca fascicularis*: XP_045252830.1), golden snub-nosed monkey (*Rhinopithecus roxellana*: XM_010384506.1), green monkey (*Chlorocebus sabaeus*: XM_007987644.2), night monkey (*Aotus nancymaae*: XM_012455409.1), old world monkey (*Colobus angolensis palliates*: XM_011949380.1), olive baboon (*Papio Anubis*: XM_031667888.1), pig-tailed macaque (*Macaca nemestrina*: XM_011717182.1), red colobus (*Piliocolobus tephrosceles*:

XM_023205512.3), Sooty mangabey (*Cercocebus atys*: XM_012083744.1), rhesus monkey (*Macaca mulatta*: NM_001278450.1), tufted capuchin (*Sapajus apella*: XM_032287580.1), black rat (*Rattus rattus*: XM_032908276.1), brown rat (*Rattus norvegicus*: XM_032908276.1), golden hamster (*Mesocricetus auratus*: XM_005068386.4), grasshopper mouse (*Onychomys torridus*: XM_036206345.1), house mouse (*Mus musculus*: NM_007693.2), multimammate mouse (*Mastomys coucha*: XM_031357132.1), Ryuku mouse (*Mus caroli*: XM_021179357.1), Shrew mouse (*Mus pahari*: XM_021202342.2), European rabbit (*Oryctolagus cuniculus*: XM_051826432.1), alpaca (*Vicugna pacos*: XP_031534667.1), Arabian camel (*Camelus dromedarius*: XM_031454226.1), cattle (*Bos taurus*: NM_181005.2), goat (*Capra hircus*: XM_018066172.1), pig (*Sus scrofa*: NP_001157477.2), sheep (*Ovis aries*: XP_004018008.3), donkey (*Equus asinus*: XP_014687627.1), horse (*Equus caballus*: NP_001075283.2), southern white rhinoceros (*Ceratotherium simum*: XP_004434274.1), California sea lion (*Zalophus californianus*: XP_027424506.2), cat (*Felis catus*: XP_023111743.1), cheetah (*Acinonyx jubatus*: XP_026922275.1), dog (*Canis lupus*: XP_038528993.1), giant panda (*Ailuropoda melanoleuca*: XP_019660005.1), grizzly bear (*Ursus arctos horribilis*: XP_048075839.1), harbor seal (*Phoca vitulina*: XP_032261715.1), leopard (*Panthera pardus*: XP_019317643.2), mongooses (*Suricata suricatta*: XP_029807749.1), monk seal (*Neomonachus* schauinslandi: XP_021535325.1), Northern fur seal (*Callorhinus ursinus*: XP_025726236.1), walrus (*Odobenus rosmarus divergens*: XP_004394547.1), weddell-seal (*Leptonychotes weddellii*: XP_030873380.1), dolphin (*Lipotes vexillifer*: XP_007454783.1), killer whale (*Orcinus orca*: XP_004262400.1), sperm whale (*Physeter catodon*: XP_023986851.1), and platypus (*Ornithorhynchus anatinua*: XP_039767777.1). Yellow shows an amino acid match between species.

4.4. Restoration of Microbial Dysbiosis in CST-KO Mice after Fecal Microbial Transplant (FMT) from WT Donor Mice

FMT is now established as an effective therapeutic modality in the treatment of the following diseases: (i) antibiotic-refractory recurrent *Clostridium difficile* colitis with a success rate of up to 95% [102–105], (ii) constipation, (iii) irritable bowel syndrome, and (iv) inflammatory bowel disease [106–108]. Therefore, attempts were made recently to assess whether gut microbial population in mice can be reversed by reciprocal FMT. WT mice that received FMT from the CST-KO mice ($WT^{FMT\text{-}CST\text{-}KO}$) encompassed a reduction of Clostridia and *Akkermansia* [109], which are linked to metabolic disorders and insulin resistance [110,111] and a marked increase in the Proteobacteria population, which are associated with active inflammatory bowel disease (IBD) states [112,113]. Of note, CST-KO mice are insulin-resistant on a normal chow diet [97]. In contrast, insulin-resistant CST-KO mice that received FMT from the WT mice ($CST\text{-}KO^{FMT\text{-}WT}$) showed an increase in richness, a notable reduction of *Staphylococcus*, and an increase in the butyrate-producing *Intestinimonas* [109] (Figure 6). Butyrate, taken up directly by colonocytes, serves not only as a direct source of energy that contributes directly to a healthy gut, but also acts as a signaling molecule that affects many factors such as satiety, secretion of hormones, and glucose metabolism [114–116]. Furthermore, reduced levels of butyrate are strongly associated with IBD and metabolic disorders [117,118]. Butyrate has also been shown to restore gut barrier integrity [119], modulates regulatory T cell function [120–122], and regulates certain serine proteases [123,124].

5. Catestatin and Innate Immunity

The first indication for the role of CST in innate immunity came from a study in rats where intravenous administration of CST was shown to reduce pressor responses by electrical stimulation [125]. The hypotensive effect of CST was revealed to be mediated at least in part by profuse histamine release (by ~21-fold) and action at the H_1 receptor [125]. The in vivo studies were later confirmed in peritoneal and pleural mast cells where CST caused dose-dependent release of histamine utilizing signaling pathways established for wasp venom peptide mastoparan and other amphiphilic cationic neuropeptides (the peptidergic pathway) [126]. This pathway is in sharp contrast to the nicotinic-cholinergic pathway used by CST to induce catecholamine secretion from chromaffin cells [5]. Subsequent studies uncover the following: (i) release of immunoreactive CST-containing peptides from human stimulated polymorphonuclear neutrophils [21]; (ii) detection of CST in mouse peritoneal macrophages by Western blots [98]; (iii) detection of CST in human monocytes and monocyte-derived macrophages by Western blots [127]; (iv) blockade of lipopolysaccharide (LPS)-induced increase in expression of tumor necrosis factor alpha [127]; (v) decreased expression of proinflammatory cytokines by CST in plasma and heart [98]; (vi) inhibition of infiltration of macrophages in obese liver [97]; (vii) degranulation of primary mast cells from human peripheral blood [128]; and (viii) low plasma CST in fatal COVID-19 patients [129]. These findings implicate CST as an immunomodulatory peptide. Since receptor-ligand interactions are an essential driver of host-immune response [130], it is important to examine if CST can bind with a receptor on immune cells and regulate their polarization and function in host defense.

6. Evolutionary Conservation and Selection Pressure on CST in Mammals

6.1. Homology of CST in Mammals

Sequence alignment of CST in 53 mammalian species belonging to eight orders revealed >80% homology in 52 species, except in Platypus (lowest in the mammalian phylogenetic tree) where the homology with the primates (highest in the mammalian phylogenetic tree) was >58% (Figure 7), indicating that CST is highly conserved in mammals. The homology of individual amino acids is summarized in Figure 8. Aromatic amino acids such as phenylalanine, tyrosine and tryptophan are reported to exhibit a rigid, planar structure and possess added stability due to the π-electron cloud situated above and below the plane of the aromatic ring [131–133]. Therefore, F_7, Y_{12}, F_{14} conserved residues in CST can undergo aromatic–aromatic interactions such as hydrogen bonding coupled with attractive, non-covalent, dipole, and van der Walls interactions, and also pi-stacking of the benzene rings [134–137]. These interactions, in turn, can stabilize the overall structure of CST, as reported earlier for other proteins [138–141]. Analysis of the energetics of protein analyses revealed that the packing of non-polar groups in the protein interior is favorable owing to the favorable enthalpy of van der Walls interactions [142]. Therefore, it is reasonable to assume that van der Walls interactions of the aromatic amino acids (F_7, Y_{12}, F_{14}) in association with van der Walls interactions of apolar (L_5, G_{18}) amino acids provided a global stability for CST [143,144]. In the course of evolution, with the change in interacting partners across species, we see a significant reduction in the conservation of charged residue. Interestingly, for the maintenance of structural framework, a 100% conservation of hydrophobic amino acid is maintained across the species through the mammalian evolutionary ladder.

6.2. Single Nucleotide Polymorphisms (SNPs) in the CST Domain of Mammals

Four non-synonymous SNPs have been identified in CST domain of CgA: $Gly_{364}Ser$ (US, Indian, and Japanese populations) [22,145,146], $Gly_{367}Val$ (only in Indian populations) [145], $Pro_{370}Leu$ (US and Indian populations) [22], and $Arg_{374}Gln$ (US populations only) [22]. $Pro_{370}Leu$-CST has the highest potency of inhibiting catecholamine secretion and desensitizing catecholamine secretion, followed by WT-CST and Gly_{364}-Ser-CST [22]. As a sharp contrast to catecholamine secretion [22], Gly364Ser was reported to be two-times more effective than Pro370Leu in exerting antibacterial activities [21].

R	S	M	K	L	S	F
1 h352	2 h353	3 h354	4 h355	5 h356	6 h357	7 h358
73%	>94%	100%	>90%	100%	>92%	100%
R	A	R	A	Y	G	F
8 h359	9 h360	10 h361	11 h362	12 h363	13 h364	14 h365
>92%	>86%	>96%	>86%	100%	>66%	100%
R	G	P	G	P	Q	L
15 h366	16 h367	17 h368	18 h369	19 h370	20 h371	21 h372
>94%	>83%	>98%	100%	>58%	>94%	>96%

Consensus sequence   Amino Acid number   Homology

Figure 8. Homology of the individual amino acid in catestatin sequence in 53 mammalian species belonging to seven orders.

7. Conclusions

(i) High conservation of CST in mammals: Alignment of CST sequences from 53 mammalian species belonging to eight orders revealed that CST sequence is highly conserved (>90% in 90% species) in mammals: Five (~24%) amino acids (M_3, L_5, F_7, F_{14}, and G_{18}) are 100% conserved; nine (~43%) amino acids (S_2, K_4, S_6, R_8, R_{10}, R_{15}, P_{17}, Q_{20} and L_{21}) are 90–96% conserved; and three (~14%) amino acids (A_9, A_{11}, and G_{16}) are >80% conserved. The least conserved sequences are G_{13} (>66%) and P_{19} (>58%), where human variants of CST were reported for G_{13} ($G_{13}S$) and P_{19} ($P_{19}L$), indicating that natural selection pressures still exist on those two amino acids [147–150].

(ii) CST as an immunomodulatory peptide: Existing literature (expression of CST in innate immune cells [21,98,127,151], inhibition of macrophage infiltration in tissues [97–99], decreased expression of pro-inflammatory cytokines by CST [97,98], and low plasma CST in fatal COVID-19 patients [129]) implicate CST as an immunomodulatory peptide.

(iii) CST as an antimicrobial peptide: Prominent effects of CST in the low micromolar range on inhibition of growth of Gram-positive and Gram-negative bacteria, fungi, and yeast establish CST as an antimicrobial peptide [21].

(iv) D-bCST$_{1–15}$ as a potential therapy for microbial infection: D-bCST$_{1–15}$ could be used as a monotherapy or as a combination therapy with cefotaxime, amoxicillin, and methicillin against the "superbugs" because it has more effective antibacterial activity compared to L-bCST$_{1–15}$, penetration through the bacterial cell wall, resistance to bacterial proteases, undetectable susceptibility to resistance, and potentiation/synergic action of commonly prescribed antibiotics [23].

(v) CST as a cell permeable peptide: Penetration of CST (pI 12.03–12.48) in bacteria, fungus, yeast, and neutrophils [21,152], coupled with 70–75% homology with cell penetrating peptide Penetratin (pI 12.62), rightfully qualify CST as a cell permeable peptide.

(vi) Gut microbiome-mediated improvement in insulin sensitivity by CST: The increased ratio of Bacilotta to Bacteroidota, together with low levels of Verrucomicrobiota (e.g., *Akkermansia* spp.) in CST-KO mice [100], not only explains insulin resistance in CST-KO mice [97] but also implicates that CST is necessary for the maintenance of insulin sensitivity. A decreased ratio of *Bacilotta* to *Bacteroidota* coupled with increased abundance of *Verrucomicrobiota* after supplementation of CST-KO mice with CST [100] confirm that CST is necessary and sufficient to increase insulin sensitivity by modulating gut microbiota. Decreased population of *Akkermansia* and increased population of Proteobacteria in $WT^{FMT\text{-}CST\text{-}KO}$ coupled with increased population of butyrate producing *Intestimonas* in $CST\text{-}KO^{FMT\text{-}WT}$ [109] further substantiates regulation of obesity and insulin resistance by CST [97] via regulation of gut microbial population [100,109].

(vii) Improvement in antimicrobial effect of CST by cyclization: Based on the existing literature [39,41], we propose that cyclization of CST would markedly improve the antibacterial activity of CST.

Author Contributions: S.K.M. conceived the idea and wrote major portion of the manuscript. S.M. made all the figures. S.J., S.M., S.D., S.C. and S.K.M. performed literature search, interpreted data, and wrote the manuscript. All authors have read and agreed to the published version of the manuscript.

Funding: This research was supported by grants from the National Institutes of Health (1 R21 AG072487-01 and 1 R21 AG080246-01 to S.K.M.). S.J. is supported by AFTD Holloway Postdoctoral Fellowship (Award #2020-02).

Institutional Review Board Statement: Not applicable.

Informed Consent Statement: Not applicable.

Data Availability Statement: Not applicable.

Conflicts of Interest: S.K.M. is a co-inventor of a patent on CST regulation of obesity.

References

1. Zasloff, M. Antimicrobial peptides of multicellular organisms. *Nature* **2002**, *415*, 389–395. [CrossRef] [PubMed]
2. Winkler, H.; Fischer-Colbrie, R. The Chromogranins A and B: The first 25 years and future perspectives. *Neuroscience* **1992**, *49*, 497–528. [CrossRef] [PubMed]
3. Montero-Hadjadje, M.; Vaingankar, S.; Elias, S.; Tostivint, H.; Mahata, S.K.; Anouar, Y. Chromogranins A and B and secretogranin II: Evolutionary and functional aspects. *Acta Physiol. (Oxf.)* **2008**, *192*, 309–324. [CrossRef] [PubMed]
4. Bartolomucci, A.; Possenti, R.; Mahata, S.K.; Fischer-Colbrie, R.; Loh, Y.P.; Salton, S.R. The extended granin family: Structure, function, and biomedical implications. *Endocr. Rev.* **2011**, *32*, 755–797. [CrossRef]
5. Mahata, S.K.; O'Connor, D.T.; Mahata, M.; Yoo, S.H.; Taupenot, L.; Wu, H.; Gill, B.M.; Parmer, R.J. Novel autocrine feedback control of catecholamine release. A discrete Chromogranin A fragment is a noncompetitive nicotinic cholinergic antagonist. *J. Clin. Investig.* **1997**, *100*, 1623–1633. [CrossRef]
6. Mahata, S.K.; Corti, A. Chromogranin A and its fragments in cardiovascular, immunometabolic, and cancer regulation. *Ann. N. Y. Acad. Sci.* **2019**, *1455*, 34–58. [CrossRef]
7. Taylor, C.V.; Taupenot, L.; Mahata, S.K.; Mahata, M.; Wu, H.; Yasothornsrikul, S.; Toneff, T.; Caporale, C.; Jiang, Q.; Parmer, R.J.; et al. Formation of the catecholamine release-inhibitory peptide catestatin from Chromogranin A. Determination of proteolytic cleavage sites in hormone storage granules. *J. Biol. Chem.* **2000**, *275*, 22905–22915. [CrossRef]
8. Parmer, R.J.; Mahata, M.; Gong, Y.; Mahata, S.K.; Jiang, Q.; O'Connor, D.T.; Xi, X.-P.; Miles, L.A. Processing of Chromogranin A by plasmin provides a novel mechanism for regulating catecholamine secretion. *J. Clin. Investig.* **2000**, *106*, 907–915. [CrossRef]
9. Jiang, Q.; Taupenot, L.; Mahata, S.K.; Mahata, M.; O'Connor, D.T.; Miles, L.A.; Parmer, R.J. Proteolytic cleavage of Chromogranin A (CgA) by plasmin: Selective liberation of a specific bioactive CgA fragment that regulates catecholamine release. *J. Biol. Chem.* **2001**, *276*, 25022–25029. [CrossRef]
10. Lee, J.C.; Taylor, C.V.; Gaucher, S.P.; Toneff, T.; Taupenot, L.; Yasothornsrikul, S.; Mahata, S.K.; Sei, C.; Parmer, R.J.; Neveu, J.M.; et al. Primary sequence characterization of catestatin intermediates and peptides defines proteolytic cleavage sites utilized for converting Chromogranin A into active catestatin secreted from neuroendocrine chromaffin cells. *Biochemistry* **2003**, *42*, 6938–6946. [CrossRef]
11. Biswas, N.; Vaingankar, S.M.; Mahata, M.; Das, M.; Gayen, J.R.; Taupenot, L.; Torpey, J.W.; O'Connor, D.T.; Mahata, S.K. Proteolytic cleavage of human Chromogranin A containing naturally occurring catestatin variants: Differential processing at catestatin region by plasmin. *Endocrinology* **2008**, *149*, 749–757. [CrossRef] [PubMed]

12. Biswas, N.; Rodriguez-Flores, J.L.; Courel, M.; Gayen, J.R.; Vaingankar, S.M.; Mahata, M.; Torpey, J.W.; Taupenot, L.; O'Connor, D.T.; Mahata, S.K. Cathepsin L Co-Localizes with Chromogranin A in Chromaffin Vesicles to Generate Active Peptides. *Endocrinology* **2009**, *150*, 3547–3557. [CrossRef] [PubMed]
13. Mahata, S.K.; Mahata, M.; Parmer, R.J.; O'Connor, D.T. Desensitization of catecholamine release: The novel catecholamine release-inhibitory peptide catestatin (Chromogranin $A_{344–364}$) acts at the receptor to prevent nicotinic cholinergic tolerance. *J. Biol. Chem.* **1999**, *274*, 2920–2928. [CrossRef] [PubMed]
14. Taupenot, L.; Mahata, S.K.; Mahata, M.; Parmer, R.J.; O'Connor, D.T. Interaction of the catecholamine release-inhibitory peptide catestatin (human Chromogranin A(352–372)) with the chromaffin cell surface and Torpedo electroplax: Implications for nicotinic cholinergic antagonism. *Regul. Pept.* **2000**, *95*, 9–17. [CrossRef] [PubMed]
15. Mahata, S.K.; Mahata, M.; Wakade, A.R.; O'Connor, D.T. Primary structure and function of the catecholamine release inhibitory peptide catestatin (Chromogranin A344–364): Identification of amino acid residues crucial for activity. *Mol. Endocrinol.* **2000**, *14*, 1525–1535. [PubMed]
16. Preece, N.E.; Nguyen, M.; Mahata, M.; Mahata, S.K.; Mahapatra, N.R.; Tsigelny, I.; O'Connor, D.T. Conformational preferences and activities of peptides from the catecholamine release-inhibitory (catestatin) region of Chromogranin A. *Regul. Pept.* **2004**, *118*, 75–87. [CrossRef]
17. Mahata, S.K. Catestatin—The catecholamine release inhibitory peptide: A structural and functional overview. *Curr. Med. Chem. Immun. Endoc. Metab. Agents* **2004**, *4*, 221–234. [CrossRef]
18. Mahapatra, N.R.; Mahata, M.; Mahata, S.K.; O'Connor, D.T. The Chromogranin A fragment catestatin: Specificity, potency and mechanism to inhibit exocytotic secretion of multiple catecholamine storage vesicle co-transmitters. *J. Hypertens.* **2006**, *24*, 895–904. [CrossRef]
19. Mahata, S.K.; Mahata, M.; Fung, M.M.; O'Connor, D.T. Catestatin: A multifunctional peptide from Chromogranin A. *Regul. Pept.* **2010**, *162*, 33–43. [CrossRef]
20. Mahata, S.K.; Kiranmayi, M.; Mahapatra, N.R. Catestatin: A Master Regulator of Cardiovascular Functions. *Curr. Med. Chem.* **2018**, *25*, 1352–1374. [CrossRef]
21. Briolat, J.; Wu, S.D.; Mahata, S.K.; Gonthier, B.; Bagnard, D.; Chasserot-Golaz, S.; Helle, K.B.; Aunis, D.; Metz-Boutigue, M.H. New antimicrobial activity for the catecholamine release-inhibitory peptide from Chromogranin A. *Cell. Mol. Life Sci.* **2005**, *62*, 377–385. [CrossRef] [PubMed]
22. Wen, G.; Mahata, S.K.; Cadman, P.; Mahata, M.; Ghosh, S.; Mahapatra, N.R.; Rao, F.; Stridsberg, M.; Smith, D.W.; Mahboubi, P.; et al. Both rare and common polymorphisms contribute functional variation at CHGA, a regulator of catecholamine physiology. *Am. J. Hum. Genet.* **2004**, *74*, 197–207. [CrossRef] [PubMed]
23. Zaet, A.; Dartevelle, P.; Daouad, F.; Ehlinger, C.; Quiles, F.; Francius, G.; Boehler, C.; Bergthold, C.; Frisch, B.; Prevost, G.; et al. D-Cateslytin, a new antimicrobial peptide with therapeutic potential. *Sci. Rep.* **2017**, *7*, 15199. [CrossRef]
24. Radek, K.A.; Lopez-Garcia, B.; Hupe, M.; Niesman, I.R.; Elias, P.M.; Taupenot, L.; Mahata, S.K.; O'Connor, D.T.; Gallo, R.L. The neuroendocrine peptide catestatin is a cutaneous antimicrobial and induced in the skin after injury. *J. Investig. Dermatol.* **2008**, *128*, 1525–1534. [CrossRef] [PubMed]
25. Scheffers, D.J.; Pinho, M.G. Bacterial cell wall synthesis: New insights from localization studies. *Microbiol. Mol. Biol. Rev.* **2005**, *69*, 585–607. [CrossRef] [PubMed]
26. Brown, A.R.; Gordon, R.A.; Hyland, S.N.; Siegrist, M.S.; Grimes, C.L. Chemical Biology Tools for Examining the Bacterial Cell Wall. *Cell. Chem. Biol.* **2020**, *27*, 1052–1062. [CrossRef]
27. Lopez-Siles, M.; Corral-Lugo, A.; McConnell, M.J. Vaccines for multidrug resistant Gram negative bacteria: Lessons from the past for guiding future success. *FEMS Microbiol. Rev.* **2021**, *45*, fuaa054. [CrossRef]
28. Welch, R.A. Pore-forming cytolysins of gram-negative bacteria. *Mol. Microbiol.* **1991**, *5*, 521–528. [CrossRef]
29. Mahlapuu, M.; Hakansson, J.; Ringstad, L.; Bjorn, C. Antimicrobial Peptides: An Emerging Category of Therapeutic Agents. *Front. Cell. Infect. Microbiol.* **2016**, *6*, 194. [CrossRef]
30. Lee, T.H.; Hirst, D.J.; Aguilar, M.I. New insights into the molecular mechanisms of biomembrane structural changes and interactions by optical biosensor technology. *Biochim. Biophys. Acta* **2015**, *1848*, 1868–1885. [CrossRef]
31. Wadhwani, P.; Epand, R.F.; Heidenreich, N.; Burck, J.; Ulrich, A.S.; Epand, R.M. Membrane-active peptides and the clustering of anionic lipids. *Biophys. J.* **2012**, *103*, 265–274. [CrossRef] [PubMed]
32. Bogdanov, M.; Pyrshev, K.; Yesylevskyy, S.; Ryabichko, S.; Boiko, V.; Ivanchenko, P.; Kiyamova, R.; Guan, Z.; Ramseyer, C.; Dowhan, W. Phospholipid distribution in the cytoplasmic membrane of Gram-negative bacteria is highly asymmetric, dynamic, and cell shape-dependent. *Sci. Adv.* **2020**, *6*, eaaz6333. [CrossRef] [PubMed]
33. Vollmer, W.; Seligman, S.J. Architecture of peptidoglycan: More data and more models. *Trends Microbiol.* **2010**, *18*, 59–66. [CrossRef] [PubMed]
34. Rojas, E.R.; Billings, G.; Odermatt, P.D.; Auer, G.K.; Zhu, L.; Miguel, A.; Chang, F.; Weibel, D.B.; Theriot, J.A.; Huang, K.C. The outer membrane is an essential load-bearing element in Gram-negative bacteria. *Nature* **2018**, *559*, 617–621. [CrossRef] [PubMed]
35. Percy, M.G.; Grundling, A. Lipoteichoic acid synthesis and function in gram-positive bacteria. *Annu. Rev. Microbiol.* **2014**, *68*, 81–100. [CrossRef]
36. Bowdish, D.M.; Davidson, D.J.; Hancock, R.E. A re-evaluation of the role of host defence peptides in mammalian immunity. *Curr. Protein. Pept. Sci.* **2005**, *6*, 35–51. [CrossRef]

37. Sahu, B.S.; Mohan, J.; Sahu, G.; Singh, P.K.; Sonawane, P.J.; Sasi, B.K.; Allu, P.K.; Maji, S.K.; Bera, A.K.; Senapati, S.; et al. Molecular interactions of the physiological anti-hypertensive peptide catestatin with the neuronal nicotinic acetylcholine receptor. *J. Cell Sci.* **2012**, *125*, 2323–2337. [CrossRef]
38. Tsigelny, I.; Mahata, S.K.; Taupenot, L.; Preece, N.E.; Mahata, M.; Khan, I.; Parmer, R.J.; O'Connor, D.T. Mechanism of action of Chromogranin A on catecholamine release: Molecular modeling of the catestatin region reveals a b-strand/loop/b-strand structure secured by hydrophobic interactions and predictive of activity. *Regul. Pept.* **1998**, *77*, 43–53. [CrossRef] [PubMed]
39. Dathe, M.; Nikolenko, H.; Klose, J.; Bienert, M. Cyclization increases the antimicrobial activity and selectivity of arginine- and tryptophan-containing hexapeptides. *Biochemistry* **2004**, *43*, 9140–9150. [CrossRef]
40. Lazzaro, B.P.; Zasloff, M.; Rolff, J. Antimicrobial peptides: Application informed by evolution. *Science* **2020**, *368*, eaau5480. [CrossRef]
41. Wessolowski, A.; Bienert, M.; Dathe, M. Antimicrobial activity of arginine- and tryptophan-rich hexapeptides: The effects of aromatic clusters, D-amino acid substitution and cyclization. *J. Pept. Res.* **2004**, *64*, 159–169. [CrossRef] [PubMed]
42. Jean-Francois, F.; Castano, S.; Desbat, B.; Odaert, B.; Roux, M.; Metz-Boutigue, M.H.; Dufourc, E.J. Aggregation of cateslytin beta-sheets on negatively charged lipids promotes rigid membrane domains. A new mode of action for antimicrobial peptides? *Biochemistry* **2008**, *47*, 6394–6402. [CrossRef] [PubMed]
43. Epand, R.M.; Epand, R.F. Lipid domains in bacterial membranes and the action of antimicrobial agents. *Biochim. Biophys. Acta* **2009**, *1788*, 289–294. [CrossRef]
44. Scavello, F.; Mutschler, A.; Helle, S.; Schneider, F.; Chasserot-Golaz, S.; Strub, J.M.; Cianferani, S.; Haikel, Y.; Metz-Boutigue, M.H. Catestatin in innate immunity and Cateslytin-derived peptides against superbugs. *Sci. Rep.* **2021**, *11*, 15615. [CrossRef] [PubMed]
45. Costa, T.R.; Felisberto-Rodrigues, C.; Meir, A.; Prevost, M.S.; Redzej, A.; Trokter, M.; Waksman, G. Secretion systems in Gram-negative bacteria: Structural and mechanistic insights. *Nat. Rev. Microbiol.* **2015**, *13*, 343–359. [CrossRef] [PubMed]
46. Bonten, M.; Johnson, J.R.; van den Biggelaar, A.H.J.; Georgalis, L.; Geurtsen, J.; de Palacios, P.I.; Gravenstein, S.; Verstraeten, T.; Hermans, P.; Poolman, J.T. Epidemiology of Escherichia coli Bacteremia: A Systematic Literature Review. *Clin. Infect. Dis.* **2021**, *72*, 1211–1219. [CrossRef] [PubMed]
47. Pechous, R.D.; Sivaraman, V.; Stasulli, N.M.; Goldman, W.E. Pneumonic Plague: The Darker Side of Yersinia pestis. *Trends Microbiol.* **2016**, *24*, 190–197. [CrossRef]
48. Kim, S.H.; Chelliah, R.; Ramakrishnan, S.R.; Perumal, A.S.; Bang, W.S.; Rubab, M.; Daliri, E.B.; Barathikannan, K.; Elahi, F.; Park, E.; et al. Review on Stress Tolerance in Campylobacter jejuni. *Front. Cell. Infect. Microbiol.* **2020**, *10*, 596570. [CrossRef]
49. Cunha, B.A.; Burillo, A.; Bouza, E. Legionnaires' disease. *Lancet* **2016**, *387*, 376–385. [CrossRef]
50. Coburn, B.; Grassl, G.A.; Finlay, B.B. Salmonella, the host and disease: A brief review. *Immunol. Cell Biol.* **2007**, *85*, 112–118. [CrossRef]
51. Lee, G.C.; Burgess, D.S. Treatment of Klebsiella pneumoniae carbapenemase (KPC) infections: A review of published case series and case reports. *Ann. Clin. Microbiol. Antimicrob.* **2012**, *11*, 32. [CrossRef] [PubMed]
52. Capatina, D.; Feier, B.; Hosu, O.; Tertis, M.; Cristea, C. Analytical methods for the characterization and diagnosis of infection with *Pseudomonas aeruginosa*: A critical review. *Anal. Chim. Acta* **2022**, *1204*, 339696. [CrossRef] [PubMed]
53. Hennebique, A.; Boisset, S.; Maurin, M. Tularemia as a waterborne disease: A review. *Emerg. Microbes Infect.* **2019**, *8*, 1027–1042. [CrossRef] [PubMed]
54. Butler, T. Treatment of typhoid fever in the 21st century: Promises and shortcomings. *Clin. Microbiol. Infect.* **2011**, *17*, 959–963. [CrossRef]
55. Mazzariol, A.; Bazaj, A.; Cornaglia, G. Multi-drug-resistant Gram-negative bacteria causing urinary tract infections: A review. *J. Chemother.* **2017**, *29*, 2–9. [CrossRef]
56. Blair, J.M.; Richmond, G.E.; Piddock, L.J. Multidrug efflux pumps in Gram-negative bacteria and their role in antibiotic resistance. *Future Microbiol.* **2014**, *9*, 1165–1177. [CrossRef]
57. Vasoo, S.; Barreto, J.N.; Tosh, P.K. Emerging issues in gram-negative bacterial resistance: An update for the practicing clinician. *Mayo Clin. Proc.* **2015**, *90*, 395–403. [CrossRef]
58. Sizemore, T.C. Rheumatologic manifestations of histoplasmosis: A review. *Rheumatol. Int.* **2013**, *33*, 2963–2965. [CrossRef]
59. Jude, C.M.; Nayak, N.B.; Patel, M.K.; Deshmukh, M.; Batra, P. Pulmonary coccidioidomycosis: Pictorial review of chest radiographic and CT findings. *Radiographics* **2014**, *34*, 912–925. [CrossRef]
60. Linder, K.A.; Kauffman, C.A.; Miceli, M.H. Blastomycosis: A Review of Mycological and Clinical Aspects. *J. Fungi* **2023**, *9*, 117. [CrossRef]
61. Mahdavinia, M.; Grammer, L.C. Management of allergic bronchopulmonary aspergillosis: A review and update. *Ther. Adv. Respir. Dis.* **2012**, *6*, 173–187. [CrossRef]
62. Griffith, N.; Danziger, L. Candida auris Urinary Tract Infections and Possible Treatment. *Antibiotics* **2020**, *9*, 898. [CrossRef]
63. Ben-Ami, R. Treatment of Invasive Candidiasis: A Narrative Review. *J. Fungi* **2018**, *4*, 97. [CrossRef]
64. McCarty, T.P.; Pappas, P.G. Invasive Candidiasis. *Infect. Dis. Clin. N. Am.* **2016**, *30*, 103–124. [CrossRef]
65. de Boer, M.G.; de Fijter, J.W.; Kroon, F.P. Outbreaks and clustering of *Pneumocystis pneumonia* in kidney transplant recipients: A systematic review. *Med. Mycol.* **2011**, *49*, 673–680.
66. Jeong, W.; Keighley, C.; Wolfe, R.; Lee, W.L.; Slavin, M.A.; Kong, D.C.M.; Chen, S.C. The epidemiology and clinical manifestations of mucormycosis: A systematic review and meta-analysis of case reports. *Clin. Microbiol. Infect.* **2019**, *25*, 26–34. [CrossRef]

67. Nair, A.G.; Dave, T.V. Transcutaneous retrobulbar injection of amphotericin B in rhino-orbital-cerebral mucormycosis: A review. *Orbit* **2022**, *41*, 275–286. [CrossRef]
68. Setianingrum, F.; Rautemaa-Richardson, R.; Denning, D.W. *Pulmonary cryptococcosis*: A review of pathobiology and clinical aspects. *Med. Mycol.* **2019**, *57*, 133–150. [CrossRef]
69. Montoya, M.C.; Magwene, P.M.; Perfect, J.R. Associations between Cryptococcus Genotypes, Phenotypes, and Clinical Parameters of Human Disease: A Review. *J. Fungi* **2021**, *7*, 260. [CrossRef]
70. Bermas, A.; Geddes-McAlister, J. Combatting the evolution of antifungal resistance in *Cryptococcus neoformans*. *Mol. Microbiol.* **2020**, *114*, 721–734. [CrossRef]
71. Wall, G.; Lopez-Ribot, J.L. Current Antimycotics, New Prospects, and Future Approaches to Antifungal Therapy. *Antibiotics* **2020**, *9*, 445. [CrossRef] [PubMed]
72. Buda De Cesare, G.; Cristy, S.A.; Garsin, D.A.; Lorenz, M.C. Antimicrobial Peptides: A New Frontier in Antifungal Therapy. *mBio* **2020**, *11*, e02123-20. [CrossRef] [PubMed]
73. Biswas, N.; Gayen, J.; Mahata, M.; Su, Y.; Mahata, S.K.; O'Connor, D.T. Novel peptide isomer strategy for stable inhibition of catecholamine release: Application to hypertension. *Hypertension* **2012**, *60*, 1552–1559. [CrossRef] [PubMed]
74. Aslam, R.; Marban, C.; Corazzol, C.; Jehl, F.; Delalande, F.; Van Dorsselaer, A.; Prevost, G.; Haikel, Y.; Taddei, C.; Schneider, F.; et al. Cateslytin, a Chromogranin A derived peptide is active against *Staphylococcus aureus* and resistant to degradation by its proteases. *PLoS ONE* **2013**, *8*, e68993. [CrossRef] [PubMed]
75. Mancino, D.; Kharouf, N.; Scavello, F.; Helle, S.; Salloum-Yared, F.; Mutschler, A.; Mathieu, E.; Lavalle, P.; Metz-Boutigue, M.H.; Haikel, Y. The Catestatin-Derived Peptides Are New Actors to Fight the Development of Oral Candidosis. *Int. J. Mol. Sci.* **2022**, *23*, 2066. [CrossRef]
76. Rios-Covian, D.; Salazar, N.; Gueimonde, M.; de Los Reyes-Gavilan, C.G. Shaping the Metabolism of Intestinal Bacteroides Population through Diet to Improve Human Health. *Front. Microbiol.* **2017**, *8*, 376. [CrossRef]
77. Rabbi, M.F.; Eissa, N.; Munyaka, P.M.; Kermarrec, L.; Elgazzar, O.; Khafipour, E.; Bernstein, C.N.; Ghia, J.E. Reactivation of Intestinal Inflammation Is Suppressed by Catestatin in a Murine Model of Colitis via M1 Macrophages and Not the Gut Microbiota. *Front. Immunol.* **2017**, *8*, 985. [CrossRef]
78. Rabbi, M.F.; Munyaka, P.M.; Eissa, N.; Metz-Boutigue, M.H.; Khafipour, E.; Ghia, J.E. Human Catestatin Alters Gut Microbiota Composition in Mice. *Front. Microbiol.* **2016**, *7*, 2151. [CrossRef]
79. Horvath, T.D.; Ihekweazu, F.D.; Haidacher, S.J.; Ruan, W.; Engevik, K.A.; Fultz, R.; Hoch, K.M.; Luna, R.A.; Oezguen, N.; Spinler, J.K.; et al. *Bacteroides ovatus* colonization influences the abundance of intestinal short chain fatty acids and neurotransmitters. *iScience* **2022**, *25*, 104158. [CrossRef]
80. Fernandez-Julia, P.J.; Munoz-Munoz, J.; van Sinderen, D. A comprehensive review on the impact of beta-glucan metabolism by Bacteroides and *Bifidobacterium* species as members of the gut microbiota. *Int. J. Biol. Macromol.* **2021**, *181*, 877–889. [CrossRef]
81. Bornet, E.; Westermann, A.J. The ambivalent role of Bacteroides in enteric infections. *Trends Microbiol.* **2022**, *30*, 104–108. [CrossRef] [PubMed]
82. Wong, J.M.; de Souza, R.; Kendall, C.W.; Emam, A.; Jenkins, D.J. Colonic health: Fermentation and short chain fatty acids. *J. Clin. Gastroenterol.* **2006**, *40*, 235–243. [CrossRef]
83. Cummings, J.H.; Macfarlane, G.T. Role of intestinal bacteria in nutrient metabolism. *JPEN J. Parenter. Enteral Nutr.* **1997**, *21*, 357–365. [CrossRef]
84. Reigstad, C.S.; Salmonson, C.E.; Rainey, J.F., 3rd; Szurszewski, J.H.; Linden, D.R.; Sonnenburg, J.L.; Farrugia, G.; Kashyap, P.C. Gut microbes promote colonic serotonin production through an effect of short-chain fatty acids on enterochromaffin cells. *FASEB J.* **2015**, *29*, 1395–1403. [CrossRef] [PubMed]
85. Ge, X.; Pan, J.; Liu, Y.; Wang, H.; Zhou, W.; Wang, X. Intestinal Crosstalk between Microbiota and Serotonin and its Impact on Gut Motility. *Curr. Pharm. Biotechnol.* **2018**, *19*, 190–195. [CrossRef]
86. Christiansen, C.B.; Gabe, M.B.N.; Svendsen, B.; Dragsted, L.O.; Rosenkilde, M.M.; Holst, J.J. The impact of short-chain fatty acids on GLP-1 and PYY secretion from the isolated perfused rat colon. *Am. J. Physiol. Gastrointest. Liver Physiol.* **2018**, *315*, G53–G65. [CrossRef]
87. Psichas, A.; Sleeth, M.L.; Murphy, K.G.; Brooks, L.; Bewick, G.A.; Hanyaloglu, A.C.; Ghatei, M.A.; Bloom, S.R.; Frost, G. The short chain fatty acid propionate stimulates GLP-1 and PYY secretion via free fatty acid receptor 2 in rodents. *Int. J. Obes. (Lond.)* **2015**, *39*, 424–429. [CrossRef]
88. Ma, Y.; Lee, E.; Yoshikawa, H.; Noda, T.; Miyamoto, J.; Kimura, I.; Hatano, R.; Miki, T. Phloretin suppresses carbohydrate-induced GLP-1 secretion via inhibiting short chain fatty acid release from gut microbiome. *Biochem. Biophys. Res. Commun.* **2022**, *621*, 176–182. [CrossRef]
89. Canfora, E.E.; Jocken, J.W.; Blaak, E.E. Short-chain fatty acids in control of body weight and insulin sensitivity. *Nat. Rev. Endocrinol.* **2015**, *11*, 577–591. [CrossRef]
90. Muller, M.; Hernandez, M.A.G.; Goossens, G.H.; Reijnders, D.; Holst, J.J.; Jocken, J.W.E.; van Eijk, H.; Canfora, E.E.; Blaak, E.E. Circulating but not faecal short-chain fatty acids are related to insulin sensitivity, lipolysis and GLP-1 concentrations in humans. *Sci. Rep.* **2019**, *9*, 12515. [CrossRef]
91. Hernandez, M.A.G.; Canfora, E.E.; Jocken, J.W.E.; Blaak, E.E. The Short-Chain Fatty Acid Acetate in Body Weight Control and Insulin Sensitivity. *Nutrients* **2019**, *11*, 1943. [CrossRef] [PubMed]

92. Wexler, H.M. Bacteroides: The good, the bad, and the nitty-gritty. *Clin. Microbiol. Rev.* **2007**, *20*, 593–621. [CrossRef] [PubMed]
93. Yang, J.Y.; Lee, Y.S.; Kim, Y.; Lee, S.H.; Ryu, S.; Fukuda, S.; Hase, K.; Yang, C.S.; Lim, H.S.; Kim, M.S.; et al. Gut commensal *Bacteroides acidifaciens* prevents obesity and improves insulin sensitivity in mice. *Mucosal Immunol.* **2017**, *10*, 104–116. [CrossRef] [PubMed]
94. Yoshida, N.; Yamashita, T.; Osone, T.; Hosooka, T.; Shinohara, M.; Kitahama, S.; Sasaki, K.; Sasaki, D.; Yoneshiro, T.; Suzuki, T.; et al. *Bacteroides* spp. promotes branched-chain amino acid catabolism in brown fat and inhibits obesity. *iScience* **2021**, *24*, 103342. [CrossRef] [PubMed]
95. Mazmanian, S.K.; Kasper, D.L. The love-hate relationship between bacterial polysaccharides and the host immune system. *Nat. Rev. Immunol.* **2006**, *6*, 849–858. [CrossRef] [PubMed]
96. Round, J.L.; Mazmanian, S.K. Inducible Foxp3+ regulatory T-cell development by a commensal bacterium of the intestinal microbiota. *Proc. Natl. Acad. Sci. USA* **2010**, *107*, 12204–12209. [CrossRef] [PubMed]
97. Ying, W.; Mahata, S.; Bandyopadhyay, G.K.; Zhou, Z.; Wollam, J.; Vu, J.; Mayoral, R.; Chi, N.W.; Webster, N.J.G.; Corti, A.; et al. Catestatin Inhibits Obesity-Induced Macrophage Infiltration and Inflammation in the Liver and Suppresses Hepatic Glucose Production, Leading to Improved Insulin Sensitivity. *Diabetes* **2018**, *67*, 841–848. [CrossRef] [PubMed]
98. Ying, W.; Tang, K.; Avolio, E.; Schilling, J.M.; Pasqua, T.; Liu, M.A.; Cheng, H.; Gao, H.; Zhang, J.; Mahata, S.; et al. Immunosuppression of Macrophages Underlies the Cardioprotective Effects of CST (Catestatin). *Hypertension* **2021**, *77*, 1670–1682. [CrossRef]
99. Muntjewerff, E.M.; Tang, K.; Lutter, L.; Christoffersson, G.; Nicolasen, M.J.T.; Gao, H.; Katkar, G.D.; Das, S.; Ter Beest, M.; Ying, W.; et al. Chromogranin A regulates gut permeability via the antagonistic actions of its proteolytic peptides. *Acta Physiol. (Oxf.)* **2021**, *232*, e13655. [CrossRef]
100. Gonzalez-Davila, P.; Schwalbe, M.; Danewalia, A.; Dalile, B.; Verbeke, K.; Mahata, S.K.; El Aidy, S. Catestatin selects for colonization of antimicrobial-resistant gut bacterial communities. *ISME J.* **2022**, *16*, 1873–1882. [CrossRef]
101. Corb Aron, R.A.; Abid, A.; Vesa, C.M.; Nechifor, A.C.; Behl, T.; Ghitea, T.C.; Munteanu, M.A.; Fratila, O.; Andronie-Cioara, F.L.; Toma, M.M.; et al. Recognizing the Benefits of Pre-/Probiotics in Metabolic Syndrome and Type 2 Diabetes Mellitus Considering the Influence of Akkermansia muciniphila as a Key Gut Bacterium. *Microorganisms* **2021**, *9*, 618. [CrossRef] [PubMed]
102. Bhutiani, N.; Schucht, J.E.; Miller, K.R.; McClave, S.A. Technical Aspects of Fecal Microbial Transplantation (FMT). *Curr. Gastroenterol. Rep.* **2018**, *20*, 30. [CrossRef] [PubMed]
103. Mattila, E.; Uusitalo-Seppala, R.; Wuorela, M.; Lehtola, L.; Nurmi, H.; Ristikankare, M.; Moilanen, V.; Salminen, K.; Seppala, M.; Mattila, P.S.; et al. Fecal transplantation, through colonoscopy, is effective therapy for recurrent Clostridium difficile infection. *Gastroenterology* **2012**, *142*, 490–496. [CrossRef] [PubMed]
104. Friedman-Korn, T.; Livovsky, D.M.; Maharshak, N.; Aviv Cohen, N.; Paz, K.; Bar-Gil Shitrit, A.; Goldin, E.; Koslowsky, B. Fecal Transplantation for Treatment of Clostridium Difficile Infection in Elderly and Debilitated Patients. *Dig. Dis. Sci.* **2018**, *63*, 198–203. [CrossRef]
105. Cohen, N.A.; Maharshak, N. Novel Indications for Fecal Microbial Transplantation: Update and Review of the Literature. *Dig. Dis. Sci.* **2017**, *62*, 1131–1145. [CrossRef]
106. Winslet, M.C.; Andrews, H.; Allan, R.N.; Keighley, M.R. Fecal diversion in the management of Crohn's disease of the colon. *Dis. Colon Rectum* **1993**, *36*, 757–762. [CrossRef]
107. Grehan, M.J.; Borody, T.J.; Leis, S.M.; Campbell, J.; Mitchell, H.; Wettstein, A. Durable alteration of the colonic microbiota by the administration of donor fecal flora. *J. Clin. Gastroenterol.* **2010**, *44*, 551–561. [CrossRef]
108. Bakken, J.S.; Borody, T.; Brandt, L.J.; Brill, J.V.; Demarco, D.C.; Franzos, M.A.; Kelly, C.; Khoruts, A.; Louie, T.; Martinelli, L.P.; et al. Treating Clostridium difficile infection with fecal microbiota transplantation. *Clin. Gastroenterol. Hepatol.* **2011**, *9*, 1044–1049. [CrossRef]
109. Gonzalez-Davila, P.; Schwalbe, M.; Danewalia, A.; Wardenaar, R.; Dalile, B.; Verbeke, K.; Mahata, S.K.; El Aidy, S. Gut microbiota transplantation drives the adoptive transfer of colonic genotype-phenotype characteristics between mice lacking catestatin and their wild type counterparts. *Gut Microbes* **2022**, *14*, 2081476. [CrossRef]
110. Schneeberger, M.; Everard, A.; Gomez-Valades, A.G.; Matamoros, S.; Ramirez, S.; Delzenne, N.M.; Gomis, R.; Claret, M.; Cani, P.D. *Akkermansia muciniphila* inversely correlates with the onset of inflammation, altered adipose tissue metabolism and metabolic disorders during obesity in mice. *Sci. Rep.* **2015**, *5*, 16643. [CrossRef]
111. Zhou, Q.; Pang, G.; Zhang, Z.; Yuan, H.; Chen, C.; Zhang, N.; Yang, Z.; Sun, L. Association Between Gut Akkermansia and Metabolic Syndrome is Dose-Dependent and Affected by Microbial Interactions: A Cross-Sectional Study. *Diabetes Metab. Syndr. Obes.* **2021**, *14*, 2177–2188. [CrossRef] [PubMed]
112. Earley, H.; Lennon, G.; Balfe, A.; Coffey, J.C.; Winter, D.C.; O'Connell, P.R. The abundance of *Akkermansia muciniphila* and its relationship with sulphated colonic mucins in health and ulcerative colitis. *Sci. Rep.* **2019**, *9*, 15683. [CrossRef]
113. Glassner, K.L.; Abraham, B.P.; Quigley, E.M.M. The microbiome and inflammatory bowel disease. *J. Allergy Clin. Immunol.* **2020**, *145*, 16–27. [CrossRef] [PubMed]
114. den Besten, G.; van Eunen, K.; Groen, A.K.; Venema, K.; Reijngoud, D.J.; Bakker, B.M. The role of short-chain fatty acids in the interplay between diet, gut microbiota, and host energy metabolism. *J. Lipid Res.* **2013**, *54*, 2325–2340. [CrossRef]
115. Bolognini, D.; Tobin, A.B.; Milligan, G.; Moss, C.E. The Pharmacology and Function of Receptors for Short-Chain Fatty Acids. *Mol. Pharmacol.* **2016**, *89*, 388–398. [CrossRef] [PubMed]

116. Morrison, D.J.; Preston, T. Formation of short chain fatty acids by the gut microbiota and their impact on human metabolism. *Gut Microbes* **2016**, *7*, 189–200. [CrossRef]
117. De Preter, V.; Geboes, K.P.; Bulteel, V.; Vandermeulen, G.; Suenaert, P.; Rutgeerts, P.; Verbeke, K. Kinetics of butyrate metabolism in the normal colon and in ulcerative colitis: The effects of substrate concentration and carnitine on the beta-oxidation pathway. *Aliment. Pharmacol. Ther.* **2011**, *34*, 526–532. [CrossRef]
118. Huda-Faujan, N.; Abdulamir, A.S.; Fatimah, A.B.; Anas, O.M.; Shuhaimi, M.; Yazid, A.M.; Loong, Y.Y. The impact of the level of the intestinal short chain Fatty acids in inflammatory bowel disease patients versus healthy subjects. *Open Biochem. J.* **2010**, *4*, 53–58. [CrossRef]
119. Wang, H.B.; Wang, P.Y.; Wang, X.; Wan, Y.L.; Liu, Y.C. Butyrate enhances intestinal epithelial barrier function via up-regulation of tight junction protein Claudin-1 transcription. *Dig. Dis. Sci.* **2012**, *57*, 3126–3135. [CrossRef]
120. Arpaia, N.; Campbell, C.; Fan, X.; Dikiy, S.; van der Veeken, J.; deRoos, P.; Liu, H.; Cross, J.R.; Pfeffer, K.; Coffer, P.J.; et al. Metabolites produced by commensal bacteria promote peripheral regulatory T-cell generation. *Nature* **2013**, *504*, 451–455. [CrossRef]
121. Furusawa, Y.; Obata, Y.; Fukuda, S.; Endo, T.A.; Nakato, G.; Takahashi, D.; Nakanishi, Y.; Uetake, C.; Kato, K.; Kato, T.; et al. Commensal microbe-derived butyrate induces the differentiation of colonic regulatory T cells. *Nature* **2013**, *504*, 446–450. [CrossRef] [PubMed]
122. Chen, L.; Sun, M.; Wu, W.; Yang, W.; Huang, X.; Xiao, Y.; Ma, C.; Xu, L.; Yao, S.; Liu, Z.; et al. Microbiota Metabolite Butyrate Differentially Regulates Th1 and Th17 Cells' Differentiation and Function in Induction of Colitis. *Inflamm. Bowel. Dis.* **2019**, *25*, 1450–1461. [CrossRef] [PubMed]
123. Fregeau, C.J.; Helgason, C.D.; Bleackley, R.C. Two cytotoxic cell proteinase genes are differentially sensitive to sodium butyrate. *Nucleic Acids Res.* **1992**, *20*, 3113–3119. [CrossRef]
124. Tsuda, H.; Ochiai, K.; Suzuki, N.; Otsuka, K. Butyrate, a bacterial metabolite, induces apoptosis and autophagic cell death in gingival epithelial cells. *J. Periodontal Res.* **2010**, *45*, 626–634. [CrossRef] [PubMed]
125. Kennedy, B.P.; Mahata, S.K.; O'Connor, D.T.; Ziegler, M.G. Mechanism of cardiovascular actions of the Chromogranin A fragment catestatin in vivo. *Peptides* **1998**, *19*, 1241–1248. [CrossRef]
126. Kruger, P.G.; Mahata, S.K.; Helle, K.B. Catestatin (CgA344–364) stimulates rat mast cell release of histamine in a manner comparable to mastoparan and other cationic charged neuropeptides. *Regul. Pept.* **2003**, *114*, 29–35. [CrossRef]
127. Kojima, M.; Ozawa, N.; Mori, Y.; Takahashi, Y.; Watanabe-Kominato, K.; Shirai, R.; Watanabe, R.; Sato, K.; Matsuyama, T.A.; Ishibashi-Ueda, H.; et al. Catestatin Prevents Macrophage-Driven Atherosclerosis but Not Arterial Injury-Induced Neointimal Hyperplasia. *Thromb. Haemost.* **2018**, *118*, 182–194. [CrossRef] [PubMed]
128. Aung, G.; Niyonsaba, F.; Ushio, H.; Kajiwara, N.; Saito, H.; Ikeda, S.; Ogawa, H.; Okumura, K. Catestatin, a neuroendocrine antimicrobial peptide, induces human mast cell migration, degranulation and production of cytokines and chemokines. *Immunology* **2011**, *132*, 527–539. [CrossRef]
129. Kljakovic-Gaspic, T.; Tokic, D.; Martinovic, D.; Kumric, M.; Supe-Domic, D.; Stojanovic Stipic, S.; Delic, N.; Vrdoljak, J.; Vilovic, M.; Ticinovic Kurir, T.; et al. Prognostic Value of Catestatin in Severe COVID-19: An ICU-Based Study. *J. Clin. Med.* **2022**, *11*, 4496. [CrossRef]
130. Jati, S.; Kundu, S.; Chakraborty, A.; Mahata, S.K.; Nizet, V.; Sen, M. Wnt5A Signaling Promotes Defense Against Bacterial Pathogens by Activating a Host Autophagy Circuit. *Front. Immunol.* **2018**, *9*, 679. [CrossRef]
131. Zhang, W.; Carravetta, V.; Plekan, O.; Feyer, V.; Richter, R.; Coreno, M.; Prince, K.C. Electronic structure of aromatic amino acids studied by soft X-ray spectroscopy. *J. Chem. Phys.* **2009**, *131*, 035103. [CrossRef] [PubMed]
132. Matta, C.F.; Hernandez-Trujillo, J.; Tang, T.H.; Bader, R.F. Hydrogen-hydrogen bonding: A stabilizing interaction in molecules and crystals. *Chemistry (Easton)* **2003**, *9*, 1940–1951. [CrossRef] [PubMed]
133. Scheiner, S.; Kar, T.; Pattanayak, J. Comparison of various types of hydrogen bonds involving aromatic amino acids. *J. Am. Chem. Soc.* **2002**, *124*, 13257–13264. [CrossRef] [PubMed]
134. Dougherty, D.A. Cation-pi interactions involving aromatic amino acids. *J. Nutr.* **2007**, *137*, 1504S–1508S. [CrossRef] [PubMed]
135. Dougherty, D.A. The cation-pi interaction. *Acc. Chem. Res.* **2013**, *46*, 885–893. [CrossRef] [PubMed]
136. Chelli, R.; Gervasio, F.L.; Procacci, P.; Schettino, V. Stacking and T-shape competition in aromatic-aromatic amino acid interactions. *J. Am. Chem. Soc.* **2002**, *124*, 6133–6143. [CrossRef]
137. Yajima, T.; Takamido, R.; Shimazaki, Y.; Odani, A.; Nakabayashi, Y.; Yamauchi, O. π-π stacking assisted binding of aromatic amino acids by copper(II)-aromatic diimine complexes. Effects of ring substituents on ternary complex stability. *Dalton Trans.* **2007**, *3*, 299–307. [CrossRef]
138. Kelkar, D.A.; Chattopadhyay, A. Membrane interfacial localization of aromatic amino acids and membrane protein function. *J. Biosci.* **2006**, *31*, 297–302. [CrossRef]
139. Brocchieri, L.; Karlin, S. Geometry of interplanar residue contacts in protein structures. *Proc. Natl. Acad. Sci. USA* **1994**, *91*, 9297–9301. [CrossRef]
140. Meyer, E.A.; Castellano, R.K.; Diederich, F. Interactions with aromatic rings in chemical and biological recognition. *Angew. Chem. Int. Ed. Engl.* **2003**, *42*, 1210–1250. [CrossRef]
141. Espinoza-Fonseca, L.M. Aromatic residues link binding and function of intrinsically disordered proteins. *Mol. Biosyst.* **2012**, *8*, 237–246. [CrossRef] [PubMed]

142. Loladze, V.V.; Ermolenko, D.N.; Makhatadze, G.I. Thermodynamic consequences of burial of polar and non-polar amino acid residues in the protein interior. *J. Mol. Biol.* **2002**, *320*, 343–357. [CrossRef] [PubMed]
143. Kellis, J.T., Jr.; Nyberg, K.; Sali, D.; Fersht, A.R. Contribution of hydrophobic interactions to protein stability. *Nature* **1988**, *333*, 784–786. [CrossRef] [PubMed]
144. Zhao, N.; Pang, B.; Shyu, C.R.; Korkin, D. Charged residues at protein interaction interfaces: Unexpected conservation and orchestrated divergence. *Protein Sci.* **2011**, *20*, 1275–1284. [CrossRef]
145. Sahu, B.S.; Obbineni, J.M.; Sahu, G.; Allu, P.K.; Subramanian, L.; Sonawane, P.J.; Singh, P.K.; Sasi, B.K.; Senapati, S.; Maji, S.K.; et al. Functional genetic variants of the catecholamine-release-inhibitory peptide catestatin in an Indian population: Allele-specific effects on metabolic traits. *J. Biol. Chem.* **2012**, *287*, 43840–43852. [CrossRef]
146. Choi, Y.; Miura, M.; Nakata, Y.; Sugasawa, T.; Nissato, S.; Otsuki, T.; Sugawara, J.; Iemitsu, M.; Kawakami, Y.; Shimano, H.; et al. A common genetic variant of the Chromogranin A-derived peptide catestatin is associated with atherogenesis and hypertension in a Japanese population. *Endocr. J.* **2015**, *62*, 797–804. [CrossRef]
147. Dhindsa, R.S.; Copeland, B.R.; Mustoe, A.M.; Goldstein, D.B. Natural Selection Shapes Codon Usage in the Human Genome. *Am. J. Hum. Genet.* **2020**, *107*, 83–95. [CrossRef]
148. McGarrah, R.W.; White, P.J. Branched-chain amino acids in cardiovascular disease. *Nat. Rev. Cardiol.* **2023**, *20*, 77–89. [CrossRef]
149. Rees, J.S.; Castellano, S.; Andres, A.M. The Genomics of Human Local Adaptation. *Trends Genet.* **2020**, *36*, 415–428. [CrossRef]
150. Brunner, J.S.; Finley, L.W.S. Metabolic determinants of tumour initiation. *Nat. Rev. Endocrinol.* **2023**, *19*, 134–150. [CrossRef]
151. Muntjewerff, E.M.; Christoffersson, G.; Mahata, S.K.; van den Bogaart, G. Putative regulation of macrophage-mediated inflammation by catestatin. *Trends Immunol.* **2022**, *43*, 41–50. [CrossRef] [PubMed]
152. Zhang, D.; Shooshtarizadeh, P.; Laventie, B.J.; Colin, D.A.; Chich, J.F.; Vidic, J.; de Barry, J.; Chasserot-Golaz, S.; Delalande, F.; Van Dorsselaer, A.; et al. Two Chromogranin A-derived peptides induce calcium entry in human neutrophils by calmodulin-regulated calcium independent phospholipase A2. *PLoS ONE* **2009**, *4*, e4501. [CrossRef] [PubMed]

pharmaceutics MDPI

Review

Neuropilin-1 and Integrins as Receptors for Chromogranin A-Derived Peptides

Angelo Corti [1,2,*], Giulia Anderluzzi [2] and Flavio Curnis [2,*]

1 Faculty of Medicine, Università Vita-Salute San Raffaele, 20132 Milan, Italy
2 Tumor Biology and Vascular Targeting Unit, Division of Experimental Oncology, IRCCS San Raffaele Scientific Institute, 20132 Milan, Italy
* Correspondence: corti.angelo@hsr.it (A.C.); curnis.flavio@hsr.it (F.C.); Tel.: +39-02-26434802 (A.C.)

Abstract: Human chromogranin A (CgA), a 439 residue-long member of the *"granin"* secretory protein family, is the precursor of several peptides and polypeptides involved in the regulation of the innate immunity, cardiovascular system, metabolism, angiogenesis, tissue repair, and tumor growth. Despite the many biological activities observed in experimental and preclinical models for CgA and its most investigated fragments (vasostatin-I and catestatin), limited information is available on the receptor mechanisms underlying these effects. The interaction of vasostatin-1 with membrane phospholipids and the binding of catestatin to nicotinic and b2-adrenergic receptors have been proposed as important mechanisms for some of their effects on the cardiovascular and sympathoadrenal systems. Recent studies have shown that neuropilin-1 and certain integrins may also work as high-affinity receptors for CgA, vasostatin-1 and other fragments. In this case, we review the results of these studies and discuss the structural requirements for the interactions of CgA-related peptides with neuropilin-1 and integrins, their biological effects, their mechanisms, and the potential exploitation of compounds that target these ligand-receptor systems for cancer diagnosis and therapy. The results obtained so far suggest that integrins (particularly the integrin avb6) and neuropilin-1 are important receptors that mediate relevant pathophysiological functions of CgA and CgA fragments in angiogenesis, wound healing, and tumor growth, and that these interactions may represent important targets for cancer imaging and therapy.

Keywords: chromogranin A; vasostatin-1; catestatin; angiogenesis; tumor diagnosis; neuropilin-1 integrin avβ6; integrin avβ8

Citation: Corti, A.; Anderluzzi, G.; Curnis, F. Neuropilin-1 and Integrins as Receptors for Chromogranin A-Derived Peptides. *Pharmaceutics* **2022**, *14*, 2555. https://doi.org/10.3390/pharmaceutics14122555

Academic Editor: Tatiana B. Tennikova

Received: 14 October 2022
Accepted: 18 November 2022
Published: 22 November 2022

1. Introduction

Human chromogranin A (CgA), a member of the *"granin"* protein family, is a 439-residues long protein present in the secretory vesicles of various normal and neoplastic neuroendocrine tissues and neurons, and exocytotically released into the blood stream upon cell stimulation [1,2].

Abnormal levels of CgA, detected by immunoassay, are present in the blood of patients with neuroendocrine tumors or with other diseases, such as cardiovascular, gastrointestinal, renal, and inflammatory diseases [3].

CgA undergoes various post-translational modifications in different cells and tissues, including phosphorylation, sulphation, glycosylation, and proteolytic cleavage [2,4]. Intra-granular and/or extra-cellular proteolytic enzymes, such as furin, cathepsin L, prohormone convertase 1 and 2, thrombin and plasmin, can cleave the full-length CgA precursor (CgA_{1-439}) at different sites to generate various biologically active fragments involved in the regulation of the innate immunity [5–8], cardiovascular system [9–12], metabolism [13–15], angiogenesis [16–18], tissue repair [19] and tumor growth [14,20,21]. These fragments include N-terminal large polypeptide fragments (e.g., CgA_{1-373}) [17], as well as shorter fragments, such as CgA_{1-76} (vasostatin-1) [9], CgA_{79-113} (vasoconstrictive-inhibitory factor) [22],

CgA_{1-113} (vasostatin-2) [23], $CgA_{250-301}$ (pancreastatin) [13], $CgA_{352-372}$ (catestatin) [24], $CgA_{411-436}$ (serpinin) [25,26], and others [3,14]. The in vitro and in vivo assays used to investigate the biological effects of all these fragments and their mechanisms are reviewed in detail elsewhere [2,6,10,12,14,18,26].

Marie Hélène Metz-Boutigue and coll. were the first to demonstrate that vasostatin-1 and catestatin, two of the most investigated fragments, are endowed of antibacterial and antifungal activities [2,6,27,28]. However, several studies have shown that these peptides can also affect the physiology of mammalian cells and exert several regulatory functions under physiological and pathological conditions. For example, catestatin and vasostatin-1 induce vasodilation [2,9,29]. In addition, catestatin inhibits nicotinic-cholinergic-stimulated catecholamine secretion [24], promotes the release of histamine from rat mast cells and stimulates monocyte chemotaxis [30]. Furthermore, vasostatin-1, catestatin, and full-length CgA_{1-439} reduce myocardial contractility and relaxation [31–33], counteract the β-adrenergic-stimulated positive inotropism, and regulate the coronary tone [12]. Additionally, CgA and vasostatin-1 can affect, in an opposite manner, the adhesion of cardiomyocytes, keratinocytes, fibroblasts, and smooth muscle cells to proteins of the extracellular matrix [2,34]. CgA and vasostatin-1 can also prevent the disassembly of vascular endothelial cadherin-dependent adherens junctions [35], inhibit vascular leakage induced by tumor necrosis factor-α [35], and exert angiogenic effects [17], whereas catestatin and CgA_{1-373} promote angiogenesis [16,17]. In human microvascular endothelial cells, vasostatin-1 inhibits the expression of tumor necrosis factor-α–induced intercellular adhesion molecule-1, the release of monocyte chemoattractant protein-1, and the relocation of high mobility group box-1 [36]. Physiological concentrations of full-length CgA_{1-439}, and vasostatin-1 may also have a regulatory role in wound healing [19] and tumor growth [21,37,38], and exert several other biological effects in the regulation of metabolism and cardiovascular system [14].

Despite the numerous activities reported for CgA, vasostatin-1, and catestatin, limited information is available on the underlying receptors. Biochemical studies have shown that vasostatin-1 can interact with phosphatidylserine and other membrane-relevant phospholipids [39]. Furthermore, a mechanism involving the binding of vasostatin-1 to heparan sulfate proteoglycans and phosphoinositide 3-kinase-dependent eNOS phosphorylation has been observed in bovine aortic endothelial cells [40]. Other studies have shown that the nicotinic acetylcholine receptor mediates the inhibitory effect of catestatin on the secretion of catecholamines from chromaffin cells [14,24]. Catestatin can also act on the β2-adrenergic receptor, as suggested by the results of a combination of experimental and computational studies [41]. Interestingly, recent studies have shown that integrins and neuropilin-1 may also act as important receptors for CgA, vasostatin-1, CgA_{1-373}, and other fragments, in endothelial cell biology, cardiovascular function, angiogenesis, wound healing, and tumor growth. Here, we review the structural requirements for the interactions of CgA and CgA-fragments with neuropilin-1 and integrins, their biological effects, their mechanisms, and the potential use of compounds targeting these ligand-receptor interactions for cancer diagnosis and therapy.

2. Neuropilin-1 as a Receptor for Chromogranin A-Derived Peptides

2.1. Biological Effects of CgA and Its Fragments in Angiogenesis and Tumor Growth

Studies in various pre-clinical models of solid tumors have shown that systemic administration of recombinant CgA_{1-439} to tumor-bearing mice enhances the endothelial barrier function, inhibits tumor neo-vascularization, and reduces tumor growth [17,21,35]. In vitro studies have shown that CgA_{1-439} is an anti-angiogenic molecule and that an anti-angiogenic site is located in the region 410–439 (i.e., the C-terminal region of the full-length protein). A latent (or less active) site is also present in the N-terminal region 1–76, this site requiring vasostatin-1 liberation by proteolytic cleavage of the $Q_{76}K_{77}$ peptide bond for full activation [17]. Physiological concentrations of CgA_{1-439} and vasostatin-1 inhibit, with U-shaped dose response curves, the pro-angiogenic activity of fibroblast growth factor-2 and vascular endothelial growth factor, two important proteins involved in angiogenesis regulation [17,42]. Studies on the mechanism of action have shown that the anti-angiogenic and anti-tumor activity of CgA_{1-439} depends on the induction of protease-nexin-1 in endothelial cells, a serine protease inhibitor endowed with potent anti-angiogenic activity [21]. Furthermore, a recent study, aimed at elucidating the pro-angiogenic mechanisms triggered by the fragment CgA_{1-373}, have shown that cleavage of the R_{373}–R_{374} dibasic site of circulating CgA in tumors and the subsequent engagement of neuropilin-1 by the fragment are crucial mechanisms for the spatio-temporal regulation of angiogenesis in cancer lesions and, consequently, for the regulation of tumor growth [37]. Opposite to CgA_{1-439} and vasostatin-1 (anti-angiogenic), the fragment CgA_{1-373} can promote angiogenesis with a bell-shaped dose-response curve and with a maximal activity at 0.2–1 nM, i.e., at concentrations found in certain cancer patients [17]; thus, the full-length CgA_{1-439} and its fragments may form a balance of anti- and pro-angiogenic factors that can be finely regulated by proteolytic cleavage at Q_{76} and R_{373}. Studies in murine models of lung carcinoma, mammary adenocarcinoma, melanoma and fibrosarcoma have shown that circulating CgA can be partially cleaved in tumors after the R_{373} residue, and that the consequent exposure of the $PGPQLR_{373}$ site is crucial for tumor progression [37]. A blockade of the exposed $PGPQLR_{373}$ site with specific polyclonal and monoclonal antibodies (unable to recognize the CgA precursor) reduced tumor vascular bed, blood flow, and tumor growth [37]. These findings suggest that cleavage of CgA by proteases and the subsequent exposure of the $PGPQLR_{373}$ site may contribute to regulate the vascular physiology in tumor tissues [37]. Given that no CgA was produced by cancer cells in the models studied, it is very likely that these fragments were generated by cleavage of bloodborne CgA in tumor lesions. It appears, therefore, that CgA molecules present in the blood can work as an "*off/on*" switch for the activation of angiogenesis in tumors after local cleavage.

2.2. Mechanisms Underlying the Biological Effects of CgA Fragments in Angiogenesis and Tumor Growth

The findings described above raise a series of questions. First, which proteases can switch on this pro-angiogenic mechanism in tumors? Second, which receptor mediates the biological activity of CgA_{1-373}? Considering that thrombin and plasmin are known to be activated in tumors, and that these enzymes can efficiently cleave the $R_{373}R_{374}$ peptide bond [17,43], both enzymes are good candidates for cleaving CgA in tumors. Although other proteases might also be involved (discussed below), it is interesting to note that full-length CgA_{1-439} can induce, in endothelial cells, the production of protease-nexin 1 (a potent inhibitor of plasmin, plasminogen activators, and thrombin [21]), and that the plasminogen activator inhibitor-1 inhibits CgA cleavage to CgA_{1-373} by cultured endothelial cells [43]. It is therefore tempting to speculate that changes in the relative levels of these protease/anti-protease molecules in tumor lesions represent a major mechanism for the regulation of the CgA-dependent angiogenic switch.

Regarding the second question, the results of biochemical studies suggest that CgA_{1-373} can bind to neuropilin-1 (NRP-1) with high affinity (Kd = 3.49 ± 0.73 nM), via its C-terminal $PGPQLR_{373}$ sequence [37]. The functional role of this ligand-receptor interaction is suggested by the fact that the pro-angiogenic effects of this fragment are blocked by anti-neuropilin-1 or anti-$PGPQLR_{373}$ antibodies. No interaction of full-length CgA and CgA_{1-372} with neuropilin-1 occurs, indicating that the $PGPQLR_{373}$ binding site is cryptic in the full-length precursor and that the C-terminal arginine residue (R_{373}, which is absent in CgA_{372}) is necessary for the binding [37].

Studies on the topology of the NRP-1-binding site showed that CgA_{1-373} and short peptides containing the PGPQLR sequence interact with a pocket of the *b1* domain of the receptor, a site that recognizes peptides and polypeptides ending with the so-called C-end Rule (CendR) motif (R/K-X-X-R/K, as in the prototypical CendR peptide RPARPAR). Remarkably, this binding pocked can also accommodate and bind the C-terminal sequence of $VEGF_{165}$, a pro-angiogenic factor that contains a CendR motif (CDKPRR) [44,45]; thus,

$CgA_{1\text{-}373}$, $VEGF_{165}$, and even the short PGPQLR and RPARPAR peptides, all with a C-terminal arginine, compete for the same binding pocket of the *b1* domain of NRP-1. Although the $PGPQLR_{373}$ sequence cannot be fully considered a CendR motif, as it lacks the first R/K residue of the consensus sequence, it is interesting to note this sequence, such as the CendR motif, ends with an arginine that is crucial for NRP-1 recognition.

The importance of the C-terminal arginine of $CgA_{1\text{-}373}$ for neuropilin-1 recognition is also supported by the results of molecular docking and molecular dynamics experiments, performed with $CgA_{352\text{-}372}$ (catestatin) and $CgA_{352\text{-}373}$ (catestatin-R). Despite these compounds differ only for the presence of an arginine (C-terminal sequence: PGPQL in catestatin; PGPQLR in catestatin-R), a clear difference in the interaction with NRP-1 was observed [46]. The interaction of catestatin-R with neuropilin-1 showed strong similarity with that of the compound EG00229, a small inhibitor of NRP-1 that contains an arginine with a free carboxyl group and whose structure in the complex with NRP-1 has been resolved by crystallography studies. In both cases, complex formation is driven by salt bridges between the guanidine moiety of the C-terminal arginine of the ligand and the carboxyl group of an aspartate residue of neuropilin-1 (D_{48}) [46]. Remarkably, despite the presence of other positively charged arginine residues and amino-groups in catestatin-R, the best binding mode was obtained with the interaction of the C-terminal R_{373} of the ligand with D_{48} of the receptor.

Another important question raised by these findings concerns the possible involvement of co-receptors. Experimental evidence showed that the pro-angiogenic activity of $CgA_{1\text{-}373}$ in assays based on endothelial cell spheroids can be inhibited by mecamylamine and α-bungarotoxin, two antagonists of nicotinic acetylcholine receptors [37]. Considering that (a) the nicotinic acetylcholine receptors are expressed on endothelial cells and are known to contribute to the regulation of angiogenesis [47–50], and (b) catestatin ($CgA_{352\text{-}372}$) is known to bind nicotinic acetylcholine receptors [51–53], this class of receptors may represent important co-receptors for $CgA_{1\text{-}373}$ signaling in endothelial cells. It cannot be excluded, however, that other receptor systems are also involved.

A final point that should be discussed concerns the issue of counterregulatory mechanisms. Experimental evidence suggests that R_{373} is rapidly removed from $CgA_{1\text{-}373}$ by plasma carboxypeptidases, when this fragment is released in circulation [37]. Given the importance of R_{373} for neuropilin-1 recognition, the cleavage of $CgA_{1\text{-}373}$ to form $CgA_{1\text{-}372}$ may represent an important mechanism to limit the $CgA_{1\text{-}373}$ activity at the site of its production (for example, in cancer lesions) and to avoid systemic effects.

Thus, the results obtained so far support a model in which cleavage of the $R_{373}R_{374}$ bond of circulating CgA, followed by neuropilin-1 engagement in tumors, and the subsequent removal of R_{373} in plasma, represent a sort of *"off/on/off"* switch for the spatiotemporal regulation of angiogenesis in tumor lesions (see Figure 1A for a schematic representation of the model).

Interestingly, it is well known that pro-hormone convertases can cleave proteins at dibasic sites (R/K-R/K), and that carboxypeptidase H/E remove the C-terminal R or K. It is possible that also these enzymes are brought into play in the regulation of the CgA-dependent angiogenic switch. Indeed, it is possible that a lower expression, or a reduced activity, of carboxypeptidase H/E in tissues in which $CgA_{1\text{-}373}$ is overproduced (see below) may contribute to activate the pro-angiogenic switch. On the other hand, the normal expression/function of carboxypeptidase H/E in other tissues might have a role in the generation of the anti-angiogenic vasostatin-1 fragment (by removal of K_{77} after cleavage of the K_{77}/K_{78} dibasic site) and $CgA_{1\text{-}372}$ (by removal of R_{373} after cleavage of the R_{373}/R_{374} dibasic site), thereby promoting an anti-angiogenic effect, a hypothesis that deserves to be tested.

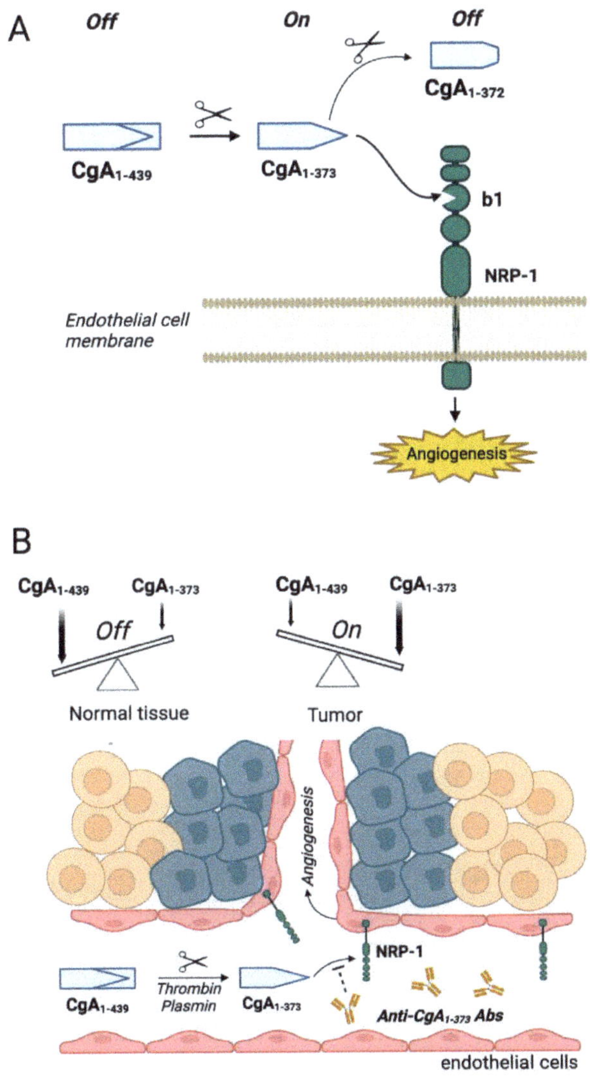

Figure 1. Hypothetical model of the CgA-dependent "*off/on/off*" switch for the regulation of angiogenesis in tumor and its inhibition by anti-PGPQLR antibodies. (**A**) Mechanisms of activation/deactivation of the NRP-1 binding site of chromogranin A. According to this model cleavage of the $R_{373}R_{374}$ peptide bond of full-length CgA ($CgA_{1\text{-}439}$) leads to exposure of the PGPQLR sequence, a site that can recognize the CendR-binding pocket of the *b1* domain of neuropilin-1 (NRP-1) on endotal cells. Removal of the R_{373} residue by carboxypeptidases causes loss of NRP-1 recognition [37]. (**B**) Mechanism of anti-tumor activity of anti-PGPQLR antibodies. Cleavage of bloodborne full-length CgA ($CgA_{1\text{-}439}$) in tumors, e.g., by plasmin or thrombin, causes loss of anti-angiogenic $CgA_{1\text{-}439}$ and generates the pro-angiogenic $CgA_{1\text{-}373}$ fragment, which may interact with NRP-1 and contribute to promote angiogenesis and tumor growth. Antibodies against the NRP-1 binding site of $CgA_{1\text{-}373}$ (anti-PGPQLR antibodies) block the $CgA_{1\text{-}373}$/NRP-1 interaction and, consequently inhibit angiogenesis and tumor growth [37]. This schematic representation has been prepared using the BioRender software.

2.3. *Potential Therapeutic Applications of Compounds That Interfere with CgA Fragment/Neuropilin-1 Interaction*

The unbalanced production of anti- and pro-angiogenic factors in tumor tissues can trigger aberrant angiogenesis and altered vascular morphology, which, in turn, may contribute to tumor cell proliferation, invasion, trafficking and formation of metastases [54–56]. The studies performed in patients with multiple myeloma have shown that CgA is cleaved into the proangiogenic form $CgA_{1\text{-}373}$ in the bone marrow and that, consequently, the ratio of pro-/anti-angiogenic forms of CgA is higher in patients compared to healthy individuals [43]. Enhanced CgA cleavage correlated with increased levels of vascular endothelial growth factor and fibroblast growth factor-2 in the bone marrow plasma, and with an increased bone marrow microvascular density [43]. Studies on the mechanism of action revealed that multiple myeloma and endothelial cells can promote CgA cleavage through the activation of the plasminogen activator/plasmin system [43].

Other studies aimed at evaluating the extent and prognostic value of CgA cleavage in patients with pancreatic ductal adenocarcinoma, an aggressive cancer arising from the exocrine component of the pancreas [57–59], have shown that cleavage of the $R_{373}R_{374}$ bond and of other sites in the C-terminal region of circulating CgA is increased in these patients. Remarkably, CgA cleavage predicts progression-free survival and overall survival in these cancer patients [38]. Experimental evidence, obtained using various pre-clinical models of pancreatic ductal adenocarcinoma, suggests that the plasminogen activator/plasmin system has a role in CgA processing in this case, and that CgA cleavage has a functional role of in the regulation of tumor vascular biology. Remarkably, anti-$PGPQLR_{373}$ antibodies capable of blocking the binding of $CgA_{1\text{-}373}$ to neuropilin-1 can reduce the growth of pancreatic ductal adenocarcinoma in mice, which implicates an important role of neuropilin-1 as mediator of these effects [38].

As cleavage of plasma CgA in tumors and the consequent interaction with neuropilin-1 may represent an important mechanism for the regulation of tumor vascular biology and growth, the assessment of the extent of CgA fragmentation in cancer patients may have a prognostic value, whereas the development of compounds that target and block this ligand-receptor interaction (e.g., anti-$PGPQLR_{373}$ monoclonal antibodies) may have a therapeutic value (see Figure 1B for a schematic representation of this concept).

2.4. *Role of CgA Fragment/Neuropilin-1 Interactions in Cardiovascular Regulation*

Global neuropilin-1 null mice develop severe cardiovascular abnormalities, indicating that neuropilin-1 has also a crucial cardiovascular function [60]. In addition, the observation that the selective knockout of neuropilin-1 in cardiomyocytes and vascular smooth muscle cells leads to cardiomyopathy, increased propensity to heart failure, and reduced survival after myocardial infarction, suggests a role for neuropilin-1 in the pathogenesis of cardiovascular diseases [61]. Based on these notions, and on the fact that CgA is the precursor of various cardio-regulatory fragments, a recent study has investigated the possibility that the fragment $CgA_{1\text{-}373}$ affects the myocardial performance by interacting with neuropilin-1 [46]. Hemodynamic assessment (performed using the Langendorff rat heart model) and studies on the mechanism of action (performed using perfused hearts and cultured cardiomyocytes) have shown that $CgA_{1\text{-}373}$ can elicit negative inotropism and vasodilation, whereas no significant effects were observed with $CgA_{1\text{-}372}$, which lacks the C-terminal arginine necessary for neuropilin-1 recognition [46]. These effects were abolished by antibodies directed against the $PGPQLR_{373}$ sequence of $CgA_{1\text{-}373}$. Furthermore, ex vivo and in vitro studies showed that these biological effects are mediated by the endothelium and involve neuropilin-1, Akt/NO/Erk1,2 activation and S-nitrosylation [46]. The effects elicited by $CgA_{1\text{-}373}$ and the lack of activity observed with $CgA_{1\text{-}372}$ suggest that $CgA_{1\text{-}373}$ is a cardio-regulatory factor and that the removal of its C-terminal arginine by carboxypeptidases may work as an important switch for "*turning off*" its cardio-regulatory activity.

3. Integrins as Receptors for CgA and CgA-Derived Peptides

3.1. The Interaction of CgA and CgA Fragments with Integrins

The first evidence for a role of integrins as receptors for full-length CgA and vasostatin-1 comes from a study on wound healing in injured mice [19]. This study has shown that CgA and vasostatin-1, at nanomolar concentrations, selectively interact with the integrin αvβ6 (see Table 1), suggesting that both polypeptides are natural ligands of this integrin. The integrin αvβ6 is an epithelial-specific cell-surface receptor of vitronectin, tenascin, fibronectin, and also of the latency associated protein of TGFβ1 [62–64]. In general, αvβ6 recognizes a site consisting of an arginine-glycine-aspartate (RGD) motif, followed by the LXXL/I motif (RGDLXXL/I) [65,66]. The latter motif folds into one α-helical turn upon binding to the receptor [65–70]. Interestingly, a short CgA-derived peptide comprising the residues 39–63 ($CgA_{39\text{-}63}$, FETLRGDERILSILRHQNLLKELQD) is sufficient for high-affinity binding and highly selective recognition of αvβ6 (Ki: 15.5 ± 3.2 nM, Table 1) [71]. This peptide exhibits a degenerate RGDLXXL/I motif, in which a glutamate residue (E_{46}) is present in place of the leucine downstream of the RGD sequence (RGDEXXL). Interestingly, in this peptide both the RGD motif ($CgA_{43\text{-}45}$) and the adjacent sequence ($CgA_{46\text{-}63}$) are crucial for αvβ6-integrin binding affinity and selectivity, as suggested by the observation that the replacement of RGD with RGE abrogates integrin recognition (Table 1), and the deletion of even a part of the C-terminal sequence markedly reduces binding affinity and selectivity [19]. The molecular determinants of αvβ6 recognition by $CgA_{39\text{-}63}$ have been elucidated by NMR, computational, and biochemical studies [71]. Homonuclear and heteronuclear multidimensional NMR analyses of this peptide in physiological conditions have shown that the region between residues E_{46} and K_{59} has an α-helical conformation, while the RGD motif is relatively flexible; the first three turns of the α-helix are amphipathic, with the hydrophilic aminoacid residues E_{46}, R_{47}, S_{50} on one side and the hydrophobic I_{48}, L_{49}, I_{51}, L_{52} on the opposite side [71]. The propensity of $CgA_{39\text{-}63}$ to form an α-helix is consistent with the results of a previous NMR study on $CgA_{47\text{-}66}$, an antifungal CgA-derived peptide, showing all-helical conformation in trifluoroethanol, an α-helix-promoting solvent [72]. Saturation transfer difference (STD) spectroscopy experiments, performed with the extracellular region of human αvβ6 and isotopically labeled ($^{13}C/^{15}N$) $CgA_{39\text{-}63}$, have shown that the hydrophobic residues I_{48}, L_{49}, I_{51}, and L_{52} of the α-helix display the strongest STD values (>75%) [71], suggesting that these aminoacids contribute to receptor binding. Molecular docking experiments led to a model of receptor-ligand interactions that is highly reminiscent of that proposed for the proTGFβ1/αvβ6 complex [71].

Table 1. Binding affinity of CgA-derived fragments for integrins.

Competitor	Competitive Binding Assay to Integrins (Ki, nM) [a]					Ref.
	αvβ6	αvβ8	αvβ3	αvβ5	α5β1	
$CgA_{1\text{-}439}$	105 ± 34	>2000	>2000	>2000	>2000	[19]
Vasostatin-1	74 ± 30	>10,000	>10,000	>10,000	>10,000	[19]
$CgA_{39\text{-}63}$	15.5 ± 3.2	7663 ± 1704	2192 ± 690	3600 ± 525	9206 ± 1810	[71]
$CgA_{39\text{-}63}$ (RGE)	>50,000	>50,000	>50,000	>50,000	>50,000	[71]
$CgA_{39\text{-}63}$ (RGDL)	1.6 ± 0.3	8.5 ± 3.7	1928 ± 226	2405 ± 592	924 ± 198	[71]
$CgA_{39\text{-}63}$ (RGDL)-Stapled	0.6 ± 0.1	3.2 ± 1.2	2453 ± 426	2741 ± 615	1310 ± 389	[71]

[a] *Ki*, equilibrium dissociation constant of the competitor (mean ± SEM). The *Ki* values were determined by competitive binding assay using an isoDGR-peroxidase conjugate as a probe for the integrin binding site [71].

No binding of $CgA_{39\text{-}63}$ has been observed to other integrins (such as α1β1, α6β4, α3β1, α9β1 α6β7, α5β1, αvβ3, αvβ5, and αvβ8) at low-nanomolar concentrations [19]. However, peptides of containing the $CgA_{39\text{-}63}$ region could recognize the integrin αvβ3 and other integrins of the RGD-family when used at high concentrations in the low-micromolar range [19]. For example, competitive binding assays performed with purified integrins showed that peptide $CgA_{39\text{-}63}$ can bind αvβ6 and αvβ3 with Ki values of 15.5 and 2192 nM, respectively (Table 1), indicating that this peptide can recognize both integrins, but with markedly different affinities [71].

3.2. *Role of CgA/Integrin Interactions in Wound Healing*

The αvβ6 integrin is barely expressed in normal adult tissues, whereas it is highly expressed during wound healing, tissue remodeling, and embryogenesis [73,74]. This integrin is involved in TGFβ1 maturation, it regulates the expression of matrix metalloproteases and modulates keratinocyte adhesion, proliferation, and migration in wound healing [19,62,64,75]. It is possible, therefore, that CgA and its fragments have also a role in the regulation of the wound healing process, by interacting with this integrin. According to this view, experimental data showed that local injection of recombinant $CgA_{1\text{-}439}$, but not of a $CgA_{1\text{-}439}$ mutant with RGE in place of RGD, can accelerate wound healing in mice [19]. Immunohistochemical analysis of skin tissue sections obtained from injured mice, showed that CgA, but not the RGE mutant, could induce keratinocyte proliferation and thickening of epidermis, suggesting that CgA can regulate the keratinocytes physiology and the process of wound healing through an RGD-dependent mechanism that likely involves the αvβ6-integrin. Interestingly, both CgA and αvβ6 are expressed in wound keratinocytes [19,76]. The fact that both ligand and receptor are expressed at injured sites lends further support to the hypothesis that the CgA/αvβ6 interaction may have a pathophysiological role in this process.

Regarding the integrin αvβ3, this cell-adhesion receptor is an important player in endothelial cell biology and angiogenesis [77,78]. Although it is unlikely that this integrin has a receptor function for the circulating CgA polypeptides, considering its micromolar affinity, significant ligand-receptor interactions can possibly occur at sites where CgA is produced and, therefore, where this protein is present at high concentrations, such as in the microenvironment of wound keratinocytes and neuroendocrine secretory cells or in the microenvironment of neuroendocrine tumors. Furthermore, this interaction might occur on αvβ3-positive endothelial cell after the interaction with other high-affinity binding sites, i.e., through a sort of ligand-passing mechanism.

3.3. *Potential Diagnostic and Therapeutic Applications of CgA-Derived Peptides That Interact with Integrins in Cancer*

The integrin αvβ6 is overexpressed by several types of cancer cells, such as head and neck squamous cell carcinoma, pancreatic ductal adenocarcinoma, breast, colon, liver, and ovarian cancers, and others [73,74,79–83]. This integrin modulates cancer cell invasion, inhibits apoptosis, and, importantly, is involved in the maturation of TGFβ1, a potent immunosuppressive cytokine. Increased expression levels of αvβ6 are prognostic indicators of poor survival in patients with various types of tumors [79,82,84–86], and various ligands of this integrin coupled to tumor imaging agents are currently being tested in cancer patients for tumor imaging purposes [87–91]; thus, the development of CgA-derived peptides capable of recognizing this integrin in tumors is of great experimental and clinical interest. Following this line of thought, experimental work has been carried out to obtain new peptides with higher affinity for αvβ6-integrin, starting from $CgA_{39\text{-}63}$ as a lead compound. The model of $CgA_{39\text{-}63}$/αvβ6 interactions, obtained by NMR and computational studies, allowed to predict that restoring the canonical RGDLXXL motif by replacing the glutamate (E) residue in the RGDERIL site of $CgA_{39\text{-}63}$ with a leucine (L) may increase its affinity for αvβ6. Intriguingly, the replacement of E_{46} with L not only increased, as expected, the binding affinity for αvβ6, but, unexpectedly, also that for the integrin αvβ8 (Table 1); thus, the $E_{46}L$ replacement converted $CgA_{39\text{-}63}$ into a bi-selective ligand of both αvβ6 and αvβ8 integrins (Ki: 1.6 ± 0.3 nM and 8.5 ± 3.7 nM, respectively) integrins [71]. Chemical "stapling" of the α-helix of the $E_{46}L$-$CgA_{39\text{-}63}$ mutant, by side-chain-to-side-chain cross linking with a triazole-bridge, further increased the affinity for both αvβ6 and αvβ8 (Ki: 0.6 ± 0.1 nM and 3.2 ± 1.2 nM, respectively) by stabilizing the α-helix [71]. Notably, the αvβ8 integrin represents another cell-surface receptor expressed by various carcinoma cells [92–94]; thus, the mutated/chemically stapled peptide (called peptide **5a**) represents a strong bi-selective ligand for these integrins, which can be potentially exploited as a tumor-homing ligand for delivering imaging and anticancer compounds to αvβ6/αvβ8 single- or double-positive tumors, such as oral and skin squamous cell carcinoma [95] (see Figure 2 for a schematic representation of this concept). This hypothesis is supported by the results of a very recent study aimed at investigating the tumor-homing properties of compounds consisting of peptide **5a** coupled with IRDye 800 CW (a near-infrared fluorescent dye) or with ^{18}F-NOTA (a label for positron emission tomography) [96]. This study showed that both conjugates can bind αvβ6 and αvβ8 with an affinity similar to that of the free peptide and that they can selectively recognize various αvβ6/αvβ8 single- or double-positive cancer cells, including cells from melanoma, pancreatic carcinoma, oral mucosa, prostate, and bladder cancer. Furthermore, biodistribution studies, performed with these conjugates in mice bearing orthotopic or subcutaneous αvβ6-positive pancreatic tumors, showed high target-specific uptake of fluorescence- and radio-labeled peptide by tumors [96]. Tumor-specific uptake of the fluorescent conjugate was also observed in mice bearing αvβ8-positive prostate tumors [96], confirming the hypothesis that peptide **5a** can home to αvβ6- and/or αvβ8-positive tumors.

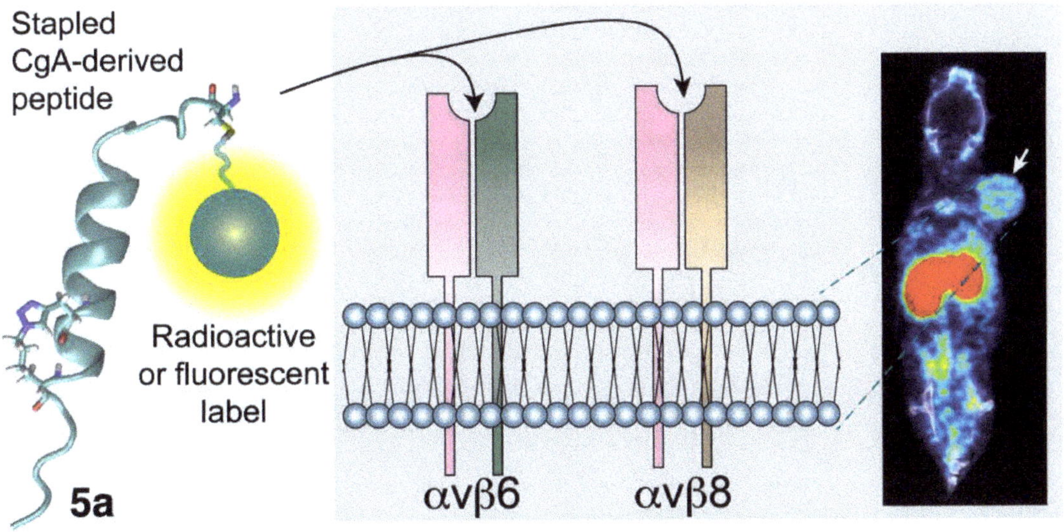

Figure 2. Use of the CgA-derived peptide **5a** (stapled) for delivering imaging or therapeutic compounds to αvβ6/αvβ8 single- or double-positive tumors. The peptide **5a**, derived from the region 38–63 of human CgA (originally published in [96]) is characterized by the sequence CFETLRGDLRILSILRX$_1$QNLX$_2$KELQ, where X$_1$ and X$_2$ are propargylglycine and azidolysine residues, respectively, which form a triazole bridge after a click chemistry reaction, thereby increasing the α-helix stability. Peptide **5a** can be exploited for delivering radioactive or fluorescent imaging compounds to αvβ6/αvβ8 single- or double-positive tumors or for developing new therapeutic tumor-homing agents. The image in the right panel shows the radiotracer uptake in a mouse bearing a pancreatic tumor implanted subcutaneously (arrow), as assessed by PET/CT scan (originally published in [96]).

Remarkably, both αvβ6 and αvβ8 integrins (which are upregulated in many tumors and, in the case of αvβ8, also in tumor infiltrating Treg cells) can activate the latency associated peptide/TGFβ complex, through interactions of integrins with the RGD sites of the complex [73,83,92]. These interactions can lead to the local activation of TGFβ in the tumor microenvironment, a potent immunosuppressive mechanism that may contribute to tumor progression. Interestingly, in vitro studies have shown that peptide **5a** can inhibit the integrin-mediated TGFβ activation [96]. Thus, in principle, the peptide **5a** can be used not only as a ligand for delivering imaging or anticancer agents to αvβ6/αvβ8 single- or double-positive tumors, but also as a tumor-homing inhibitor of these TGFβ activators.

Finally, considering the role of αvβ6/αvβ8-mediated TGFβ activation in fibrosis [97] the dual targeting capability of peptide **5a** might be also exploited in the development of anti-fibrotic drugs. This is another hypothesis that deserves to be investigated.

4. Conclusions

The results obtained so far suggest that integrins (particularly the integrin αvβ6) and neuropilin-1 are important receptors that mediate relevant pathophysiological functions of CgA and its fragments in angiogenesis, wound healing, and tumor growth. Experimental evidence indicates that these interactions may also represent important targets for cancer imaging and therapy. Although further work is necessary to clarify the receptor mechanisms of CgA and its fragments in the regulation of cardiovascular homeostasis, metabolism, and tumor growth, the results obtained so far highlight the complexity of the "CgA system", which consists of a multitude of CgA-derived peptides and various receptors. The complexity of this system is even higher if we consider that full-length CgA and some of its fragments show biphasic dose-response curves in angiogenesis assays, as well as in cardio-regulatory and tumor pre-clinical models, likely because of the activation of counterregulatory mechanisms at higher doses. These mechanisms are not clearly understood and, therefore, their full elucidation remains a challenge. A third level of complexity is related to the fact that CgA undergoes differential post-translational modifications in different cells and tissues, such as glycosylation, sulfation, and phosphorylation. As most of the studies carried out so far on the biological functions of CgA have been performed with recombinant or synthetic peptides lacking these modifications, the impact of these structural modifications on proteolytic cleavage, fragment generation, receptor recognition, and biological activity, remains to be investigated.

Author Contributions: A.C., G.A. and F.C. performed literature research, interpreted data, prepared figures, and wrote and approved the submitted manuscript. All authors have read and agreed to the published version of the manuscript.

Funding: This research was funded by Associazione Italiana per la Ricerca sul Cancro (AIRC) under IG 2019–ID. 23470 project–P.I. Angelo Corti and by Fondazione AIRC 5 per Mille 2019 (ID 22737) program, P.I. MC Bonini, Group Leader A. Corti.

Institutional Review Board Statement: Not applicable.

Informed Consent Statement: Not applicable.

Data Availability Statement: Not applicable.

Conflicts of Interest: A.C. and F.C. are inventor of a patent on CgA-derived peptides and their use in cancer imaging.

Abbreviations

Chromogranin A (CgA), neuropilin-1 (NRP-1), transforming growth factor-β1 (TGFβ1).

References

1. Taupenot, L.; Harper, K.L.; O'Connor, D.T. The chromogranin-secretogranin family. *N. Engl. J. Med.* **2003**, *348*, 1134–1149. [CrossRef]
2. Helle, K.B.; Corti, A.; Metz-Boutigue, M.H.; Tota, B. The endocrine role for chromogranin A: A prohormone for peptides with regulatory properties. *Cell Mol. Life Sci.* **2007**, *64*, 2863–2886. [CrossRef]
3. Corti, A.; Marcucci, F.; Bachetti, T. Circulating chromogranin A and its fragments as diagnostic and prognostic disease markers. *Pflugers Arch.* **2018**, *470*, 199–210. [CrossRef]

4. Gadroy, P.; Stridsberg, M.; Capon, C.; Michalski, J.C.; Strub, J.M.; Van Dorsselaer, A.; Aunis, D.; Metz-Boutigue, M.H. Phosphorylation and O-glycosylation sites of human chromogranin A (CGA79-439) from urine of patients with carcinoid tumors. *J. Biol. Chem.* **1998**, *273*, 34087–34097. [CrossRef]
5. Metz-Boutigue, M.H.; Lugardon, K.; Goumon, Y.; Raffner, R.; Strub, J.M.; Aunis, D. Antibacterial and antifungal peptides derived from chromogranins and proenkephalin-A. From structural to biological aspects. *Adv. Exp. Med. Biol.* **2000**, *482*, 299–315.
6. Shooshtarizadeh, P.; Zhang, D.; Chich, J.F.; Gasnier, C.; Schneider, F.; Haikel, Y.; Aunis, D.; Metz-Boutigue, M.H. The antimicrobial peptides derived from chromogranin/secretogranin family, new actors of innate immunity. *Regul. Pept.* **2009**, *165*, 102–110. [CrossRef]
7. Briolat, J.; Wu, S.D.; Mahata, S.K.; Gonthier, B.; Bagnard, D.; Chasserot-Golaz, S.; Helle, K.B.; Aunis, D.; Metz-Boutigue, M.H. New antimicrobial activity for the catecholamine release-inhibitory peptide from chromogranin A. *Cell. Mol. Life Sci. CMLS* **2005**, *62*, 377–385. [CrossRef]
8. Ioannidis, M.; Mahata, S.K.; van den Bogaart, G. The immunomodulatory functions of chromogranin A-derived peptide pancreastatin. *Peptides* **2022**, *158*, 170893. [CrossRef]
9. Aardal, S.; Helle, K.B. The vasoinhibitory activity of bovine chromogranin A fragment (vasostatin) and its independence of extracellular calcium in isolated segments of human blood vessels. *Regul. Pept.* **1992**, *41*, 9–18. [CrossRef]
10. Helle, K.B. The chromogranin A-derived peptides vasostatin-I and catestatin as regulatory peptides for cardiovascular functions. *Cardiovasc. Res.* **2010**, *85*, 9–16. [CrossRef]
11. Tota, B.; Mazza, R.; Angelone, T.; Nullans, G.; Metz-Boutigue, M.H.; Aunis, D.; Helle, K.B. Peptides from the N-terminal domain of chromogranin A (vasostatins) exert negative inotropic effects in the isolated frog heart. *Regul. Pept.* **2003**, *114*, 123–130. [CrossRef]
12. Tota, B.; Angelone, T.; Cerra, M.C. The surging role of Chromogranin A in cardiovascular homeostasis. *Front. Chem.* **2014**, *2*, 64. [CrossRef]
13. Tatemoto, K.; Efendic, S.; Mutt, V.; Makk, G.; Feistner, G.J.; Barchas, J.D. Pancreastatin, a novel pancreatic peptide that inhibits insulin secretion. *Nature* **1986**, *324*, 476–478. [CrossRef]
14. Mahata, S.K.; Corti, A. Chromogranin A and its fragments in cardiovascular, immunometabolic, and cancer regulation. *Ann. N. Y. Acad. Sci.* **2019**, *1455*, 34–58. [CrossRef]
15. Bandyopadhyay, G.K.; Mahata, S.K. Chromogranin A Regulation of Obesity and Peripheral Insulin Sensitivity. *Front. Endocrinol.* **2017**, *8*, 20. [CrossRef]
16. Theurl, M.; Schgoer, W.; Albrecht, K.; Jeschke, J.; Egger, M.; Beer, A.G.; Vasiljevic, D.; Rong, S.; Wolf, A.M.; Bahlmann, F.H.; et al. The neuropeptide catestatin acts as a novel angiogenic cytokine via a basic fibroblast growth factor-dependent mechanism. *Circ. Res.* **2010**, *107*, 1326–1335. [CrossRef]
17. Crippa, L.; Bianco, M.; Colombo, B.; Gasparri, A.M.; Ferrero, E.; Loh, Y.P.; Curnis, F.; Corti, A. A new chromogranin A-dependent angiogenic switch activated by thrombin. *Blood* **2013**, *121*, 392–402. [CrossRef]
18. Helle, K.B.; Corti, A. Chromogranin A: A paradoxical player in angiogenesis and vascular biology. *Cell. Mol. Life Sci.* **2015**, *78*, 339–348. [CrossRef]
19. Curnis, F.; Gasparri, A.; Longhi, R.; Colombo, B.; D'Alessio, S.; Pastorino, F.; Ponzoni, M.; Corti, A. Chromogranin A Binds to αvβ6-Integrin and Promotes Wound Healing in Mice. *Cell. Mol. Life Sci.* **2012**, *69*, 2791–2803. [CrossRef]
20. Colombo, B.; Curnis, F.; Foglieni, C.; Monno, A.; Arrigoni, G.; Corti, A. Chromogranin a expression in neoplastic cells affects tumor growth and morphogenesis in mouse models. *Cancer Res.* **2002**, *62*, 941–946.
21. Curnis, F.; Dallatomasina, A.; Bianco, M.; Gasparri, A.; Sacchi, A.; Colombo, B.; Fiocchi, M.; Perani, L.; Venturini, M.; Tacchetti, C.; et al. Regulation of tumor growth by circulating full-length chromogranin A. *Oncotarget* **2016**, *7*, 72716–72732. [CrossRef]
22. Salem, S.; Jankowski, V.; Asare, Y.; Liehn, E.; Welker, P.; Raya-Bermudez, A.; Pineda-Martos, C.; Rodriguez, M.; Munoz-Castaneda, J.R.; Bruck, H.; et al. Identification of the Vasoconstriction-Inhibiting Factor (VIF), a Potent Endogenous Cofactor of Angiotensin II Acting on the Angiotensin II Type 2 Receptor. *Circulation* **2015**, *131*, 1426–1434. [CrossRef]
23. Brekke, J.F.; Kirkeleit, J.; Lugardon, K.; Helle, K.B. Vasostatins. Dilators of bovine resistance arteries. *Adv. Exp. Med. Biol.* **2000**, *482*, 239–246.
24. Mahata, S.K.; O'Connor, D.T.; Mahata, M.; Yoo, S.H.; Taupenot, L.; Wu, H.; Gill, B.M.; Parmer, R.J. Novel autocrine feedback control of catecholamine release. A discrete chromogranin a fragment is a noncompetitive nicotinic cholinergic antagonist. *J. Clin. Investig.* **1997**, *100*, 1623–1633. [CrossRef]
25. Koshimizu, H.; Cawley, N.X.; Kim, T.; Yergey, A.L.; Loh, Y.P. Serpinin: A novel chromogranin A-derived, secreted peptide up-regulates protease nexin-1 expression and granule biogenesis in endocrine cells. *Mol. Endocrinol.* **2011**, *25*, 732–744. [CrossRef]
26. Loh, Y.P.; Koshimizu, H.; Cawley, N.X.; Tota, B. Serpinins: Role in granule biogenesis, inhibition of cell death and cardiac function. *Curr. Med. Chem.* **2012**, *19*, 4086–4092. [CrossRef]
27. Mancino, D.; Kharouf, N.; Scavello, F.; Helle, S.; Salloum-Yared, F.; Mutschler, A.; Mathieu, E.; Lavalle, P.; Metz-Boutigue, M.H.; Haikel, Y. The Catestatin-Derived Peptides Are New Actors to Fight the Development of Oral Candidosis. *Int. J. Mol. Sci.* **2022**, *23*, 2066. [CrossRef]
28. Metz-Boutigue, M.H.; Goumon, Y.; Lugardon, K.; Strub, J.M.; Aunis, D. Antibacterial peptides are present in chromaffin cell secretory granules. *Cell. Mol. Neurobiol.* **1998**, *18*, 249–266. [CrossRef]

29. Fung, M.M.; Salem, R.M.; Mehtani, P.; Thomas, B.; Lu, C.F.; Perez, B.; Rao, F.; Stridsberg, M.; Ziegler, M.G.; Mahata, S.K.; et al. Direct vasoactive effects of the chromogranin A (CHGA) peptide catestatin in humans in vivo. *Clin. Exp. Hypertens.* **2010**, *32*, 278–287. [CrossRef]
30. Kruger, P.G.; Mahata, S.K.; Helle, K.B. Catestatin (CgA344-364) stimulates rat mast cell release of histamine in a manner comparable to mastoparan and other cationic charged neuropeptides. *Regul. Pept.* **2003**, *114*, 29–35. [CrossRef]
31. Pasqua, T.; Corti, A.; Gentile, S.; Pochini, L.; Bianco, M.; Metz-Boutigue, M.H.; Cerra, M.C.; Tota, B.; Angelone, T. Full-length human Chromogranin-A cardioactivity: Myocardial, coronary and stimulus-induced processing evidence in normotensive and hypertensive male rat hearts. *Endocrinology* **2013**, *154*, 3353–3365. [CrossRef]
32. Angelone, T.; Quintieri, A.M.; Brar, B.K.; Limchaiyawat, P.T.; Tota, B.; Mahata, S.K.; Cerra, M.C. The antihypertensive chromogranin a peptide catestatin acts as a novel endocrine/paracrine modulator of cardiac inotropism and lusitropism. *Endocrinology* **2008**, *149*, 4780–4793. [CrossRef]
33. Corti, A.; Mannarino, C.; Mazza, R.; Colombo, B.; Longhi, R.; Tota, B. Vasostatins exert negative inotropism in the working heart of the frog. *Ann. N. Y. Acad. Sci.* **2002**, *971*, 362–365. [CrossRef]
34. Gasparri, A.; Sidoli, A.; Sanchez, L.P.; Longhi, R.; Siccardi, A.G.; Marchisio, P.C.; Corti, A. Chromogranin A fragments modulate cell adhesion. Identification and characterization of a pro-adhesive domain. *J. Biol. Chem.* **1997**, *272*, 20835–20843. [CrossRef]
35. Ferrero, E.; Scabini, S.; Magni, E.; Foglieni, C.; Belloni, D.; Colombo, B.; Curnis, F.; Villa, A.; Ferrero, M.E.; Corti, A. Chromogranin A protects vessels against tumor necrosis factor alpha-induced vascular leakage. *FASEB J.* **2004**, *18*, 554–556. [CrossRef]
36. Di Comite, G.; Rossi, C.M.; Marinosci, A.; Lolmede, K.; Baldissera, E.; Aiello, P.; Mueller, R.B.; Herrmann, M.; Voll, R.E.; Rovere-Querini, P.; et al. Circulating chromogranin A reveals extra-articular involvement in patients with rheumatoid arthritis and curbs TNF-alpha-elicited endothelial activation. *J. Leukoc. Biol.* **2009**, *85*, 81–87. [CrossRef]
37. Dallatomasina, A.; Gasparri, A.M.; Colombo, B.; Sacchi, A.; Bianco, M.; Daniele, T.; Esposito, A.; Pastorino, F.; Ponzoni, M.; Marcucci, F.; et al. Spatiotemporal Regulation of Tumor Angiogenesis by Circulating Chromogranin A Cleavage and Neuropilin-1 Engagement. *Cancer Res.* **2019**, *79*, 1925–1937. [CrossRef]
38. Reni, M.; Andreasi, V.; Gasparri, A.M.; Dugnani, E.; Colombo, B.; Macchini, M.; Bianco, M.; Dallatomasina, A.; Citro, A.; Assi, E.; et al. Circulating Chromogranin A Is Cleaved Into Vasoregulatory Fragments in Patients with Pancreatic Ductal Adenocarcinoma. *Front. Oncol.* **2020**, *10*, 613582. [CrossRef]
39. Blois, A.; Holmsen, H.; Martino, G.; Corti, A.; Metz-Boutigue, M.H.; Helle, K.B. Interactions of chromogranin A-derived vasostatins and monolayers of phosphatidylserine, phosphatidylcholine and phosphatidylethanolamine. *Regul. Pept.* **2006**, *134*, 30–37. [CrossRef]
40. Ramella, R.; Boero, O.; Alloatti, G.; Angelone, T.; Levi, R.; Gallo, M.P. Vasostatin 1 activates eNOS in endothelial cells through a proteoglycan-dependent mechanism. *J. Cell. Biochem.* **2010**, *110*, 70–79. [CrossRef]
41. Kiranmayi, M.; Chirasani, V.R.; Allu, P.K.; Subramanian, L.; Martelli, E.E.; Sahu, B.S.; Vishnuprabu, D.; Kumaragurubaran, R.; Sharma, S.; Bodhini, D.; et al. Catestatin Gly364Ser Variant Alters Systemic Blood Pressure and the Risk for Hypertension in Human Populations via Endothelial Nitric Oxide Pathway. *Hypertension* **2016**, *68*, 334–347. [CrossRef]
42. Belloni, D.; Scabini, S.; Foglieni, C.; Veschini, L.; Giazzon, A.; Colombo, B.; Fulgenzi, A.; Helle, K.B.; Ferrero, M.E.; Corti, A.; et al. The vasostatin-I fragment of chromogranin A inhibits VEGF-induced endothelial cell proliferation and migration. *FASEB J.* **2007**, *21*, 3052–3062. [CrossRef]
43. Bianco, M.; Gasparri, A.M.; Colombo, B.; Curnis, F.; Girlanda, S.; Ponzoni, M.; Bertilaccio, M.T.; Calcinotto, A.; Sacchi, A.; Ferrero, E.; et al. Chromogranin A is preferentially cleaved into pro-angiogenic peptides in the bone marrow of multiple myeloma patients. *Cancer Res.* **2016**, *76*, 1781–1791. [CrossRef]
44. Teesalu, T.; Sugahara, K.N.; Kotamraju, V.R.; Ruoslahti, E. C-end rule peptides mediate neuropilin-1-dependent cell, vascular, and tissue penetration. *Proc. Natl. Acad. Sci. USA* **2009**, *106*, 16157–16162. [CrossRef]
45. Teesalu, T.; Sugahara, K.N.; Ruoslahti, E. Tumor-penetrating peptides. *Front. Oncol.* **2013**, *3*, 216. [CrossRef]
46. Rocca, C.; Grande, F.; Granieri, M.C.; Colombo, B.; De Bartolo, A.; Giordano, F.; Rago, V.; Amodio, N.; Tota, B.; Cerra, M.C.; et al. The chromogranin A1-373 fragment reveals how a single change in the protein sequence exerts strong cardioregulatory effects by engaging neuropilin-1. *Acta Physiol.* **2021**, *231*, e13570. [CrossRef]
47. Pena, V.B.; Bonini, I.C.; Antollini, S.S.; Kobayashi, T.; Barrantes, F.J. Alpha 7-type acetylcholine receptor localization and its modulation by nicotine and cholesterol in vascular endothelial cells. *J. Cell. Biochem.* **2011**, *112*, 3276–3288. [CrossRef]
48. Cooke, J.P.; Ghebremariam, Y.T. Endothelial nicotinic acetylcholine receptors and angiogenesis. *Trends Cardiovasc. Med.* **2008**, *18*, 247–253. [CrossRef]
49. Arias, H.R.; Richards, V.E.; Ng, D.; Ghafoori, M.E.; Le, V.; Mousa, S.A. Role of non-neuronal nicotinic acetylcholine receptors in angiogenesis. *Int. J. Biochem. Cell. Biol.* **2009**, *41*, 1441–1451. [CrossRef]
50. Heeschen, C.; Jang, J.J.; Weis, M.; Pathak, A.; Kaji, S.; Hu, R.S.; Tsao, P.S.; Johnson, F.L.; Cooke, J.P. Nicotine stimulates angiogenesis and promotes tumor growth and atherosclerosis. *Nat. Med.* **2001**, *7*, 833–839. [CrossRef]
51. Taupenot, L.; Mahata, S.K.; Mahata, M.; Parmer, R.J.; O'Connor, D.T. Interaction of the catecholamine release-inhibitory peptide catestatin (human chromogranin A(352–372)) with the chromaffin cell surface and Torpedo electroplax: Implications for nicotinic cholinergic antagonism. *Regul. Pept.* **2000**, *95*, 9–17. [CrossRef]
52. Mahata, S.K.; Mahata, M.; Fung, M.M.; O'Connor, D.T. Catestatin: A multifunctional peptide from chromogranin A. *Regul. Pept.* **2010**, *162*, 33–43. [CrossRef]

53. Sahu, B.S.; Mohan, J.; Sahu, G.; Singh, P.K.; Sonawane, P.J.; Sasi, B.K.; Allu, P.K.; Maji, S.K.; Bera, A.K.; Senapati, S.; et al. Molecular interactions of the physiological anti-hypertensive peptide catestatin with the neuronal nicotinic acetylcholine receptor. *J. Cell. Sci.* **2012**, *125*, 2323–2337. [CrossRef]
54. Folkman, J. Angiogenesis: An organizing principle for drug discovery? *Nat. Rev. Drug Discov.* **2007**, *6*, 273–286. [CrossRef]
55. Italiano, J.E., Jr.; Richardson, J.L.; Patel-Hett, S.; Battinelli, E.; Zaslavsky, A.; Short, S.; Ryeom, S.; Folkman, J.; Klement, G.L. Angiogenesis is regulated by a novel mechanism: Pro- and antiangiogenic proteins are organized into separate platelet alpha granules and differentially released. *Blood* **2008**, *111*, 1227–1233. [CrossRef]
56. Ribatti, D. Endogenous inhibitors of angiogenesis: A historical review. *Leuk. Res.* **2009**, *33*, 638–644. [CrossRef]
57. Kleeff, J.; Korc, M.; Apte, M.; La Vecchia, C.; Johnson, C.D.; Biankin, A.V.; Neale, R.E.; Tempero, M.; Tuveson, D.A.; Hruban, R.H.; et al. Pancreatic cancer. *Nat. Rev. Dis. Prim.* **2016**, *2*, 16022. [CrossRef]
58. Rawla, P.; Sunkara, T.; Gaduputi, V. Epidemiology of Pancreatic Cancer: Global Trends, Etiology and Risk Factors. *World J. Oncol.* **2019**, *10*, 10–27. [CrossRef]
59. Conroy, T.; Desseigne, F.; Ychou, M.; Bouche, O.; Guimbaud, R.; Becouarn, Y.; Adenis, A.; Raoul, J.L.; Gourgou-Bourgade, S.; de la Fouchardiere, C.; et al. FOLFIRINOX versus gemcitabine for metastatic pancreatic cancer. *N. Engl. J. Med.* **2011**, *364*, 1817–1825. [CrossRef]
60. Kawasaki, T.; Kitsukawa, T.; Bekku, Y.; Matsuda, Y.; Sanbo, M.; Yagi, T.; Fujisawa, H. A requirement for neuropilin-1 in embryonic vessel formation. *Development* **1999**, *126*, 4895–4902. [CrossRef]
61. Wang, Y.; Cao, Y.; Yamada, S.; Thirunavukkarasu, M.; Nin, V.; Joshi, M.; Rishi, M.T.; Bhattacharya, S.; Camacho-Pereira, J.; Sharma, A.K.; et al. Cardiomyopathy and Worsened Ischemic Heart Failure in SM22-alpha Cre-Mediated Neuropilin-1 Null Mice: Dysregulation of PGC1alpha and Mitochondrial Homeostasis. *Arterioscler Thromb. Vasc. Biol.* **2015**, *35*, 1401–1412. [CrossRef]
62. Thomas, G.J.; Nystrom, M.L.; Marshall, J.F. Alphavbeta6 integrin in wound healing and cancer of the oral cavity. *J. Oral Pathol. Med.* **2006**, *35*, 1–10. [CrossRef]
63. Busk, M.; Pytela, R.; Sheppard, D. Characterization of the integrin alpha v beta 6 as a fibronectin-binding protein. *J. Biol. Chem.* **1992**, *267*, 5790–5796. [CrossRef]
64. Bandyopadhyay, A.; Raghavan, S. Defining the role of integrin alphavbeta6 in cancer. *Curr. Drug Targets* **2009**, *10*, 645–652. [CrossRef]
65. Ozawa, A.; Sato, Y.; Imabayashi, T.; Uemura, T.; Takagi, J.; Sekiguchi, K. Molecular Basis of the Ligand Binding Specificity of alphavbeta8 Integrin. *J. Biol. Chem.* **2016**, *291*, 11551–11565. [CrossRef]
66. Kraft, S.; Diefenbach, B.; Mehta, R.; Jonczyk, A.; Luckenbach, G.A.; Goodman, S.L. Definition of an unexpected ligand recognition motif for alphav beta6 integrin. *J. Biol. Chem.* **1999**, *274*, 1979–1985. [CrossRef]
67. Dong, X.; Hudson, N.E.; Lu, C.; Springer, T.A. Structural determinants of integrin beta-subunit specificity for latent TGF-beta. *Nat. Struct. Mol. Biol.* **2014**, *21*, 1091–1096. [CrossRef]
68. Dong, X.; Zhao, B.; Iacob, R.E.; Zhu, J.; Koksal, A.C.; Lu, C.; Engen, J.R.; Springer, T.A. Force interacts with macromolecular structure in activation of TGF-beta. *Nature* **2017**, *542*, 55–59. [CrossRef]
69. Kotecha, A.; Wang, Q.; Dong, X.; Ilca, S.L.; Ondiviela, M.; Zihe, R.; Seago, J.; Charleston, B.; Fry, E.E.; Abrescia, N.G.A.; et al. Rules of engagement between alphavbeta6 integrin and foot-and-mouth disease virus. *Nat. Commun.* **2017**, *8*, 15408. [CrossRef]
70. DiCara, D.; Rapisarda, C.; Sutcliffe, J.L.; Violette, S.M.; Weinreb, P.H.; Hart, I.R.; Howard, M.J.; Marshall, J.F. Structure-function analysis of Arg-Gly-Asp helix motifs in alpha v beta 6 integrin ligands. *J. Biol. Chem.* **2007**, *282*, 9657–9665. [CrossRef]
71. Nardelli, F.; Ghitti, M.; Quilici, G.; Gori, A.; Luo, Q.; Berardi, A.; Sacchi, A.; Monieri, M.; Bergamaschi, G.; Bermel, W.; et al. A stapled chromogranin A-derived peptide is a potent dual ligand for integrins alphavbeta6 and alphavbeta8. *Chem. Commun.* **2019**, *55*, 14777–14780. [CrossRef]
72. Lugardon, K.; Chasserot-Golaz, S.; Kieffer, A.E.; Maget-Dana, R.; Nullans, G.; Kieffer, B.; Aunis, D.; Metz-Boutigue, M.H. Structural and biological characterization of chromofungin, the antifungal chromogranin A-(47-66)-derived peptide. *J. Biol. Chem.* **2001**, *276*, 35875–35882. [CrossRef]
73. Koivisto, L.; Bi, J.; Hakkinen, L.; Larjava, H. Integrin alphavbeta6: Structure, function and role in health and disease. *Int. J. Biochem. Cell. Biol.* **2018**, *99*, 186–196. [CrossRef]
74. Liu, H.; Wu, Y.; Wang, F.; Liu, Z. Molecular imaging of integrin alphavbeta6 expression in living subjects. *Am. J. Nucl. Med. Mol. Imaging* **2014**, *4*, 333–345.
75. Koivisto, L.; Larjava, K.; Hakkinen, L.; Uitto, V.J.; Heino, J.; Larjava, H. Different integrins mediate cell spreading, haptotaxis and lateral migration of HaCaT keratinocytes on fibronectin. *Cell Adhes. Commun.* **1999**, *7*, 245–257. [CrossRef]
76. Radek, K.A.; Lopez-Garcia, B.; Hupe, M.; Niesman, I.R.; Elias, P.M.; Taupenot, L.; Mahata, S.K.; O'Connor, D.T.; Gallo, R.L. The neuroendocrine peptide catestatin is a cutaneous antimicrobial and induced in the skin after injury. *J. Investig. Dermatol.* **2008**, *128*, 1525–1534. [CrossRef]
77. Avraamides, C.J.; Garmy-Susini, B.; Varner, J.A. Integrins in angiogenesis and lymphangiogenesis. *Nat. Rev. Cancer* **2008**, *8*, 604–617. [CrossRef]
78. Desgrosellier, J.S.; Cheresh, D.A. Integrins in cancer: Biological implications and therapeutic opportunities. *Nat. Rev. Cancer* **2010**, *10*, 9–22. [CrossRef]

79. Elayadi, A.N.; Samli, K.N.; Prudkin, L.; Liu, Y.H.; Bian, A.; Xie, X.J.; Wistuba, I.I.; Roth, J.A.; McGuire, M.J.; Brown, K.C. A peptide selected by biopanning identifies the integrin alphavbeta6 as a prognostic biomarker for nonsmall cell lung cancer. *Cancer Res.* **2007**, *67*, 5889–5895. [CrossRef]
80. Moore, K.M.; Thomas, G.J.; Duffy, S.W.; Warwick, J.; Gabe, R.; Chou, P.; Ellis, I.O.; Green, A.R.; Haider, S.; Brouilette, K.; et al. Therapeutic targeting of integrin alphavbeta6 in breast cancer. *J. Natl. Cancer Inst.* **2014**, *106*. [CrossRef]
81. Sipos, B.; Hahn, D.; Carceller, A.; Piulats, J.; Hedderich, J.; Kalthoff, H.; Goodman, S.L.; Kosmahl, M.; Kloppel, G. Immunohistochemical screening for beta6-integrin subunit expression in adenocarcinomas using a novel monoclonal antibody reveals strong up-regulation in pancreatic ductal adenocarcinomas in vivo and in vitro. *Histopathology* **2004**, *45*, 226–236. [CrossRef]
82. Reader, C.S.; Vallath, S.; Steele, C.W.; Haider, S.; Brentnall, A.; Desai, A.; Moore, K.M.; Jamieson, N.B.; Chang, D.; Bailey, P.; et al. The integrin alphavbeta6 drives pancreatic cancer through diverse mechanisms and represents an effective target for therapy. *J. Pathol.* **2019**, *249*, 332–342. [CrossRef]
83. Niu, J.; Li, Z. The roles of integrin alphavbeta6 in cancer. *Cancer Lett.* **2017**, *403*, 128–137. [CrossRef]
84. Hazelbag, S.; Kenter, G.G.; Gorter, A.; Dreef, E.J.; Koopman, L.A.; Violette, S.M.; Weinreb, P.H.; Fleuren, G.J. Overexpression of the alpha v beta 6 integrin in cervical squamous cell carcinoma is a prognostic factor for decreased survival. *J. Pathol.* **2007**, *212*, 316–324. [CrossRef]
85. Zhang, Z.Y.; Xu, K.S.; Wang, J.S.; Yang, G.Y.; Wang, W.; Wang, J.Y.; Niu, W.B.; Liu, E.Y.; Mi, Y.T.; Niu, J. Integrin alphanvbeta6 acts as a prognostic indicator in gastric carcinoma. *Clin. Oncol.* **2008**, *20*, 61–66. [CrossRef]
86. Bates, R.C.; Bellovin, D.I.; Brown, C.; Maynard, E.; Wu, B.; Kawakatsu, H.; Sheppard, D.; Oettgen, P.; Mercurio, A.M. Transcriptional activation of integrin beta6 during the epithelial-mesenchymal transition defines a novel prognostic indicator of aggressive colon carcinoma. *J. Clin. Investig.* **2005**, *115*, 339–347. [CrossRef]
87. Altmann, A.; Sauter, M.; Roesch, S.; Mier, W.; Warta, R.; Debus, J.; Dyckhoff, G.; Herold-Mende, C.; Haberkorn, U. Identification of a Novel ITGalphavbeta6-Binding Peptide Using Protein Separation and Phage Display. *Clin. Cancer Res.* **2017**, *23*, 4170–4180. [CrossRef]
88. Roesch, S.; Lindner, T.; Sauter, M.; Loktev, A.; Flechsig, P.; Muller, M.; Mier, W.; Warta, R.; Dyckhoff, G.; Herold-Mende, C.; et al. Comparison of the RGD Motif-Containing alphavbeta6 Integrin-Binding Peptides SFLAP3 and SFITGv6 for Diagnostic Application in HNSCC. *J. Nucl. Med.* **2018**, *59*, 1679–1685. [CrossRef]
89. Quigley, N.G.; Czech, N.; Sendt, W.; Notni, J. PET/CT imaging of pancreatic carcinoma targeting the "cancer integrin" alphavbeta6. *Eur. J. Nucl. Med. Mol. Imaging* **2021**, *48*, 4107–4108. [CrossRef]
90. Kimura, R.H.; Wang, L.; Shen, B.; Huo, L.; Tummers, W.; Filipp, F.V.; Guo, H.H.; Haywood, T.; Abou-Elkacem, L.; Baratto, L.; et al. Evaluation of integrin alphavbeta6 cystine knot PET tracers to detect cancer and idiopathic pulmonary fibrosis. *Nat. Commun.* **2019**, *10*, 4673. [CrossRef]
91. Hausner, S.H.; Bold, R.J.; Cheuy, L.Y.; Chew, H.K.; Daly, M.E.; Davis, R.A.; Foster, C.C.; Kim, E.J.; Sutcliffe, J.L. Preclinical Development and First-in-Human Imaging of the Integrin αvβ6 with [18F]αvβ6-Binding Peptide in Metastatic Carcinoma. *Clin. Cancer Res.* **2019**, *25*, 1206–1215. [CrossRef]
92. McCarty, J.H. alphavbeta8 integrin adhesion and signaling pathways in development, physiology and disease. *J. Cell Sci.* **2020**, *133*. [CrossRef]
93. Takasaka, N.; Seed, R.I.; Cormier, A.; Bondesson, A.J.; Lou, J.; Elattma, A.; Ito, S.; Yanagisawa, H.; Hashimoto, M.; Ma, R.; et al. Integrin alphavbeta8-expressing tumor cells evade host immunity by regulating TGF-beta activation in immune cells. *JCI Insight* **2018**, *3*. [CrossRef]
94. Jin, S.; Lee, W.C.; Aust, D.; Pilarsky, C.; Cordes, N. beta8 Integrin Mediates Pancreatic Cancer Cell Radiochemoresistance. *Mol. Cancer Res.* **2019**, *17*, 2126–2138. [CrossRef]
95. Ahmedah, H.T.; Patterson, L.H.; Shnyder, S.D.; Sheldrake, H.M. RGD-Binding Integrins in Head and Neck Cancers. *Cancers* **2017**, *9*, 56. [CrossRef]
96. Monieri, M.; Rainone, P.; Sacchi, A.; Gori, A.; Gasparri, A.M.; Coliva, A.; Citro, A.; Ferrara, B.; Policardi, M.; Valtorta, S.; et al. A stapled chromogranin A-derived peptide homes in on tumors that express αvβ6 or αvβ8 integrins. *Int. J. Biol. Sci.* **2022**, *19*, 156–166.
97. Conroy, K.P.; Kitto, L.J.; Henderson, N.C. Alphav integrins: Key regulators of tissue fibrosis. *Cell Tissue Res.* **2016**, *365*, 511–519. [CrossRef]

pharmaceutics

Article

Antimicrobial Peptides and Biomarkers Induced by Ultraviolet Irradiation Have the Potential to Reduce Endodontic Inflammation and Facilitate Tissue Healing

Kimberly A. Morio [1], Robert H. Sternowski [2], Erliang Zeng [3] and Kim A. Brogden [4,*]

1 Apex Endodontics, Hiawatha, IA 52233, USA
2 Softronics, Ltd., Marion, IA 52302, USA
3 Division of Biostatistics and Computational Biology, College of Dentistry, The University of Iowa, Iowa City, IA 52242, USA
4 College of Dentistry, The University of Iowa, Iowa City, IA 52242, USA
* Correspondence: kim-brogden@uiowa.edu

Abstract: Background: Ultraviolet (UV) irradiation can modulate host immune responses and this approach is a novel application for treating endodontic infections and inflammation in root canals. Methods: A dataset of UV-induced molecules was compiled from a literature search. A subset of this dataset was used to calculate expression log2 ratios of endodontic tissue molecules from HEPM cells and gingival fibroblasts after 255, 405, and 255/405 nm UV irradiation. Both datasets were analyzed using ingenuity pathway analysis (IPA, Qiagen, Germantown, MD, USA). Statistical significance was calculated using Fisher's exact test and z-scores were calculated for IPA comparison analysis. Results: The dataset of 32 UV-induced molecules contained 9 antimicrobial peptides, 10 cytokines, 6 growth factors, 3 enzymes, 2 transmembrane receptors, and 2 transcription regulators. These molecules were in the IPA canonical pathway annotations for the wound healing signaling pathway (9/32, $p = 3.22 \times 10^{-11}$) and communication between immune cells (6/32, $p = 8.74 \times 10^{-11}$). In the IPA disease and function annotations, the 32 molecules were associated with an antimicrobial response, cell-to-cell signaling and interaction, cellular movement, hematological system development and function, immune cell trafficking, and inflammatory response. In IPA comparison analysis of the 13 molecules, the predicted activation or inhibition of pathways depended upon the cell type exposed, the wavelength of the UV irradiation used, and the time after exposure. Conclusions: UV irradiation activates and inhibits cellular pathways and immune functions. These results suggested that UV irradiation can activate innate and adaptive immune responses, which may supplement endodontic procedures to reduce infection, inflammation, and pain and assist tissues to heal.

Keywords: ultraviolet irradiation; UV; UVC; UVB; UVA; antimicrobial peptides; chemokines; cytokines; endodontic; inflammation; pain; tissue healing

Citation: Morio, K.A.; Sternowski, R.H.; Zeng, E.; Brogden, K.A. Antimicrobial Peptides and Biomarkers Induced by Ultraviolet Irradiation Have the Potential to Reduce Endodontic Inflammation and Facilitate Tissue Healing. *Pharmaceutics* **2022**, *14*, 1979. https://doi.org/10.3390/pharmaceutics14091979

Academic Editors: Scavello Francesco, Jean-Eric Ghia and Amiche Mohamed

Received: 18 August 2022
Accepted: 13 September 2022
Published: 19 September 2022

1. Introduction

Physical traumas, factures, erosions, and local infections, including caries and periodontal disease, on human teeth are among the conditions that lead to endodontic disease [1]. These conditions often create 'barrier defects' that allow entry of opportunistic oral microbiota into the underlying dental pulp tissue. Invading microorganisms can then develop into polymicrobial biofilms containing *Archaebacteria*, *Eubacteria*, yeast, and fungal species [2–4]. The resulting infections can induce inflammation and pain in the root canal systems. The associated craniofacial pain can be severe and significantly impact the patient's quality of life and day-to-day comfort.

Current standard-of-care treatment for endodontic emergencies and treatment include opening the tooth; exposing underlying inflammatory tissue or infection; and creating an open instrumented root canal to the tooth apex, removing the inflammatory tissue, canal exudate, necrotic tissue, and tissue debris [3]. The root canal is then irrigated with sodium hypochlorite (NaOCl) to cleanse and dissolve the canal of the remaining tissue and debris, followed by the irrigation of ethylenediaminetetraacetic acid (EDTA) to remove the smear layer and open the dentinal tubules. The root canal is completed by sealing the canal space with endodontic sealer and gutta percha. The access opening is filled with amalgam or a composite material, and in some cases, a crown is advocated to enhance tooth integrity and adequate seal.

Unfortunately, the standard-of-care treatment is unable to completely clean the canal space and residual pulp tissue debris, missed areas of infection, and remnant microorganisms in the dentin tubules along the sides of the root canal can be left; resulting in potential persist reinfections and apical chronic inflammation [3]. Irrigation with NaOCl is the gold standard, but extrusion of this irrigant apically can be extremely detrimental to the apical tissues. The damaged apical tissues by the NaOCl have little chance to regenerate and heal with no opportunities for activation of local innate or adaptive immune responses in the area.

An emerging and novel concept is to treat instrumented root canals with ultraviolet (UV) irradiation during the standard-of-care treatment procedure [5,6]. UVC irradiation kills microorganisms isolated from endodontic infections [6–8]. Brief treatment with UV irradiation would also modulate host immune responses [9,10]. UV irradiation induces an influx of cells and the production of antimicrobial peptides (AMPs); chemokines, cytokines, and biomarkers (CCBMs); and other molecules that alter lesion pathogenesis and facilitates local tissue healing [9,10]. UVC induces the secretion of CCBMs in HEPM cells and gingival fibroblasts related to endodontic tissue regeneration [6].

In this study, we were interested in identifying molecules expressed or secreted from cells and tissues after UV irradiation. We first searched the PubMed literature to identify AMPs or CCBMs reported to have up- or down-levels of mRNA expression or secretion in response to UV irradiation treatment. We used a bioinformatics approach with ingenuity pathway analysis (IPA, Qiagen, Germantown, MD, USA) to associate their expression with specific innate and adaptive immune responses. We then used the concentrations of 13 CCBMs in tissue culture media of HEPM cells and gingival fibroblasts after treatment with 255, 405, or 255/405 UV irradiation [6] as (i) wet lab experimental data to validate claims from the bioinformatics data, and (ii) to assess the ability of UV irradiation to activate or inhibit cellular pathways related to innate and adaptive immune responses. We hypothesized that UV irradiation would induce host cells and tissues to express AMPs and CCBMs (Figure 1) and these molecules would be important in future treatments designed to reduce endodontic infection and inflammation, modulate endodontic pain, and assist in endodontic tissue healing.

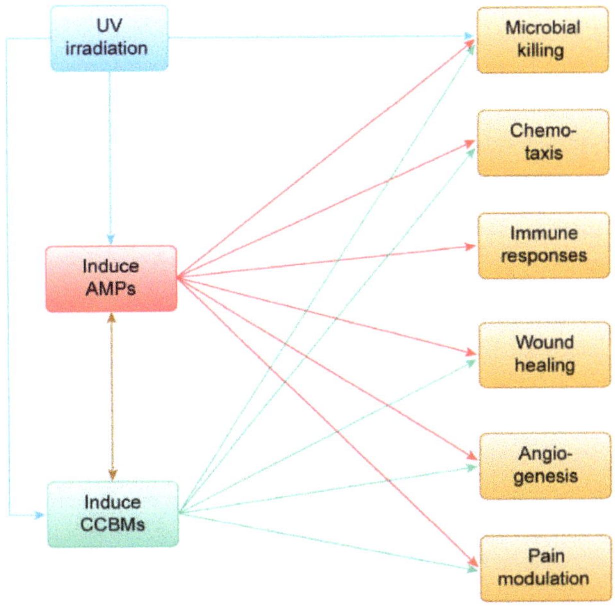

Figure 1. A schematic diagram of the proposed effects of UV irradiation on endodontic infection and inflammation, pain, and tissue healing. UV irradiation can kill microorganisms directly (blue line). UV irradiation can also induce host cells to express antimicrobial peptides (AMPs) and chemokines, cytokines, and biomarkers (CCBMs) (blue lines). AMPs can kill microorganisms (red line); induce the production of CCBMs (brown line); and induce chemotaxis, modulate immune responses, assist in wound healing, play a role in angiogenesis, and reduce pain (red lines). CCBMs can kill microorganisms (green line); induce the production of AMPs (brown line); and induce chemotaxis, modulate immune responses, assist in wound healing, play a role in angiogenesis, and reduce pain (green lines).

2. Materials and Methods

2.1. Dataset of UV-Induced Molecules

We searched the PubMed literature using UVC, UVB, UVA, chemokines, cytokines, and antimicrobial peptides as search terms linked in various combinations, using Boolean operators to identify AMPs or CCBMs reported to have up- or down-levels of mRNA expression or secretion in response to UV irradiation. AMPs and CCBMs were combined into a single dataset (Table 1).

Table 1. Antimicrobial peptides (AMPs); chemokines, cytokines, and biomarkers (CCBMs); and other mediators were identified from ingenuity pathway analysis (IPA, Qiagen, Germantown, MD, USA), combined into a single dataset below, annotated for their IPA symbol, Entrez Gene name, Entrez Gene ID (human), cellular location, and function type.

Symbol	Entrez Gene Name	Entrez Gene ID (Human)	Location	Function Type
BMP10	Bone morphogenetic protein 10	27302	Extracellular space	Growth factor
CAMP	Cathelicidin antimicrobial peptide	820	Cytoplasm	Other
CCL2	C-C motif chemokine ligand 2	6347	Extracellular space	Cytokine
CCL20	C-C motif chemokine ligand 20	6364	Extracellular space	Cytokine

Table 1. *Cont.*

Symbol	Entrez Gene Name	Entrez Gene ID (Human)	Location	Function Type
CSF2	Colony stimulating factor 2	1437	Extracellular space	Cytokine
CXCL1	C-X-C motif chemokine ligand 1	2919	Extracellular space	Cytokine
CXCL2	C-X-C motif chemokine ligand 2	2920	Extracellular space	Cytokine
CXCL3	C-X-C motif chemokine ligand 3	2921	Extracellular space	Cytokine
CXCL8	C-X-C motif chemokine ligand 8	3576	Extracellular space	Cytokine
DEFB1	Defensin beta 1	1672	Extracellular space	Other
DEFB103B	Defensin beta 103B	55894	Extracellular space	Other
DEFB4A	Defensin beta 4A	1673	Extracellular space	Other
FGF1	Fibroblast growth factor 1	2246	Extracellular space	Growth factor
FGF2	Fibroblast growth factor 2	2247	Extracellular space	Growth factor
FN1	Fibronectin 1	2335	Extracellular space	Enzyme
ICAM1	Intercellular adhesion molecule 1	3383	Plasma membrane	Transmembrane receptor
IL6	Interleukin 6	3569	Extracellular space	Cytokine
IL10	Interleukin 10	3586	Extracellular space	Cytokine
PI3	Peptidase inhibitor 3	5266	Extracellular space	Other
PIGF	Phosphatidylinositol glycan anchor biosynthesis class F	5281	Cytoplasm	Enzyme
RNASE7	Ribonuclease A family member 7	84659	Extracellular space	Enzyme
S100A7	S100 calcium binding protein A7	6278	Cytoplasm	Other
S100A8	S100 calcium binding protein A8	6279	Cytoplasm	Other
S100A9	S100 calcium binding protein A9	6280	Cytoplasm	Other
S100A12	S100 calcium binding protein A12	6283	Cytoplasm	Other
SELE	Selectin E	6401	Plasma membrane	Transmembrane receptor
SMAD3	SMAD family member 3	4088	Nucleus	Transcription regulator
SMAD4	SMAD family member 4	4089	Nucleus	Transcription regulator
TGFA	Transforming growth factor alpha	7039	Extracellular space	Growth factor
TGFB1	Transforming growth factor beta 1	7040	Extracellular space	Growth factor
TNF	Tumor necrosis factor	7124	Extracellular space	Cytokine
VEGFA	Vascular endothelial growth factor A	7422	Extracellular space	Growth factor

2.2. Subset of Endodontic Tissue Molecules

Morio et al. reported the concentrations of 13 CCBMs in tissue culture media of HEPM cells and gingival fibroblasts at 0, 24, and 48 h after treatment with 255, 405, or 255/405 UV irradiation [6]. We used these CCBMs as (i) wet lab experimental data to validate claims from the bioinformatics data, and (ii) to assess the ability of UV irradiation to activate or inhibit cellular pathways related to immune functions. For this, levels of CCBM expression were calculated as the expression log2 ratios of the mean of each treatment after UV irradiation, divided by the mean of the respective control for the same treatment (Figure 2).

	Fibroblasts 0 hours Group 2 255 nm	Fibroblasts 0 hours Group 3 405 nm	Fibroblasts 0 hours Group 4 255/405 nm	Fibroblasts 24 hours Group 6 255 nm	Fibroblasts 24 hours Group 7 405 nm	Fibroblasts 24 hours Group 8 255/405 nm	Fibroblasts 48 hours Group 10 255 nm	Fibroblasts 48 hours Group 11 405 nm	Fibroblasts 48 hours Group 12 255/405 nm
TNFA	−2.03953	−2.52449	−1.75002	0.18137	0.08246	0.22474	0.13245	0.08092	0.00000
IL6	−0.58863	−0.62770	−0.53459	0.86355	−0.00450	1.09492	1.25599	0.16628	1.44830
CXCL3	−0.70020	−0.69271	−0.65765	0.09560	−0.18049	0.06643	0.30141	0.34028	0.13151
EGF	−0.27009	−0.31316	−0.29146	−0.15605	−0.12167	−0.02609	0.09818	0.07907	0.07907
VEGFA	−0.64134	−0.68109	−0.79277	−0.08053	0.03464	−0.23631	−0.12635	0.19443	−0.09860
PDGFA	−1.45799	−1.49005	−0.85775	−0.06325	−0.08019	−0.10554	0.06529	0.06529	0.06529
TGFA	−0.35257	−0.53766	−0.68450	0.24644	0.19794	0.31057	−0.03562	−0.07215	−0.18763
BMP2	−0.40166	−0.44745	−0.43487	0.07528	0.32285	0.26028	0.14420	−0.01558	0.06091
BMP4	−0.14202	−0.25853	−0.25853	0.05697	−0.07717	0.13029	0.03045	0.06028	−0.13289
GDF2	−0.63346	−0.45670	−0.59994	0.13290	−0.02708	0.07266	−0.02438	0.27107	0.85603
BMP10	−0.61778	−0.61095	−0.60719	−0.01875	−0.03683	−0.09946	0.11520	0.09723	0.15328
FGF1	0.16163	0.11746	0.75870	−0.21728	−0.25538	−0.17722	−0.02804	0.00000	0.23772
P1GF	0.05247	0.02647	0.47393	0.03526	0.10357	0.07764	−0.18772	−0.24391	−0.60547

	HEPM cells 0 hours Group 2 255 nm	HEPM cells 0 hours Group 3 405 nm	HEPM cells 0 hours Group 4 255/405 nm	HEPM cells 24 hours Group 6 255 nm	HEPM cells 24 hours Group 7 405 nm	HEPM cells 24 hours Group 8 255/405 nm	HEPM cells 48 hours Group 10 255 nm	HEPM cells 48 hours Group 11 405 nm	HEPM cells 48 hours Group 12 255/405 nm
TNFA	0.95936	1.81143	0.95936	−0.18209	−0.47520	0.50055	0.43610	−0.04307	0.16046
IL6	−0.08489	−0.08489	−0.86985	0.58486	−0.01004	0.68498	1.23808	0.15370	0.75529
CXCL3	−0.05219	−0.03458	−0.34589	−0.00065	0.02748	0.22280	0.06307	0.03771	0.20408
EGF	−0.02106	−0.02106	−0.04244	0.00000	−0.06875	−0.12063	0.00000	0.00000	0.00000
VEGFA	0.39652	−0.24548	−0.81597	0.14570	−0.28323	−1.07768	−0.46651	−0.53225	0.11240
PDGFA	0.88679	−0.33560	0.38060	−0.07514	−0.12825	0.14978	0.19686	0.33703	0.12131
TGFA	−0.25376	−0.04731	−0.09622	0.21573	0.25526	−0.07973	−0.19265	−0.29956	0.12928
BMP2	0.08549	0.03317	0.01668	0.07733	0.30806	0.00587	0.26732	0.05311	0.22607
BMP4	0.12873	−0.92504	−0.73697	−0.13347	−0.17113	0.02476	0.17945	−0.00959	0.15391
GDF2	−0.01934	0.01391	−0.08823	−0.02580	−0.18741	−0.47923	−0.72477	−0.77892	0.16481
BMP10	0.11792	0.08450	0.04181	0.10998	0.03329	−0.21360	−0.38145	−0.18510	−0.08476
FGF1	−0.12218	−0.19465	−0.11251	−0.29223	−0.24737	−0.29679	−0.05221	−0.10547	0.14309
P1GF	−0.31194	−0.58496	−0.64638	0.03743	−0.10706	−0.59471	−0.38878	−0.23553	−0.52704

Figure 2. Expression Log2 ratios of 13 chemokine, cytokine, and biomarker (CCBMs) concentrations reported in tissue culture media of HEPM cells and gingival fibroblasts at 0, 24, and 48 h after treatment with 255 nm, 405 nm, or 255/405 nm UV irradiation. Expression was calculated as the log2 ratio of the mean of each treatment after UV irradiation over the mean of the untreated control for that same cell type, UV irradiation wavelength, and time period. Groups are shown as a heatmap, where blue represents inhibition (negative values), white represents midpoint, and orange represents activation (positive values).

2.3. Analysis

We used ingenuity pathway analysis (IPA, Qiagen, Germantown, MD, USA) to assess whether the molecules in this study were related to the activation of innate and immune mechanisms. Two types of analysis were performed.

IPA core analysis was run on the list of 32 molecules from the literature dataset in Table 1 and used to assess whether the IPA canonical pathway and IPA diseases and function annotations were predicted to be associated with relevant diseases, immune pathways, and immune functions. Statistical significance was calculated using Fisher's exact test and significant p values ($p < 0.05$) were reported.

IPA comparison analysis was run on the expression log2 ratios of the 13 molecules and used to assess whether the IPA canonical pathway and IPA diseases and function annotations were predicted to be activated or inhibited after UV irradiation. Statistical differences were determined using activation z-scores calculated from the mean of each treatment expression log2 ratio. The activation z-score was determined by IPA as reported by Kramer et al. [11] and makes predictions based on the direction of gene activation or inhibition.

IPA comparison analysis was also used to assess whether the IPA canonical pathway predicted any effects of the activated or inhibited 13 molecules on downstream regulation of gene expression for other innate or adaptive immune functions.

3. Results

3.1. Dataset of 32 UV-Induced Molecules

We identified 32 unique molecules reported to be expressed after UV irradiation of cells and tissues (Table 1). There were 9 antimicrobial peptides (Supplementary Table S1), 10 cytokines, 6 growth factors, 3 enzymes, 2 transmembrane receptors, and 2 transcription regulators (Supplementary Table S2).

IPA analysis predicted that UV irradiation can induce molecules that are involved in innate and adaptive immune responses. IPA canonical pathway annotations of these 32 molecules (Table 1) were predicted to be associated with cellular stress and injury. This category included the wound healing signaling pathway (9/32, $p = 3.22 \times 10^{-11}$). Annotations were also associated with cellular immune responses, and this category included the role of cytokines in mediating communication between immune cells (6/32, $p = 8.74 \times 10^{-11}$), communication between innate and adaptive immune cells (5/32, $p = 7.44 \times 10^{-3}$), the Th1 and Th2 activation pathway (4/32, $p = 8.01 \times 10^{-5}$), the Th1 pathway (3/32, $p = 5.81 \times 10^{-4}$), and the Th2 pathway (3/32, $p = 8.13 \times 10^{-4}$). The top relevant pathway annotations are listed in Table 2 and all the relevant annotations ranked by their $-\log(p$ value) are listed in Supplementary Table S3. This list contains relevant annotations within categories on cytokine signaling; growth factor signaling; intracellular and second messenger signaling; cellular growth, proliferation, and development; cellular immune response; and organismal growth and development.

Table 2. Ingenuity pathway analysis (IPA, Qiagen, Germantown, MD, USA) was used to assess whether the biomarkers in the literature dataset (n = 32) would participate in the activation of innate and immune mechanisms applicable to reducing endodontic infection, reducing inflammation, and assisting in endodontic tissue healing. Representative IPA canonical pathways annotations were associated with cellular stress and injury; cytokine signaling; cellular immune response. Representative IPA diseases or functions annotations were associated with antimicrobial response, cell-to-cell signaling and interaction, cellular movement, hematological system development and function, immune cell trafficking, and inflammatory response.

IPA Function	p-Value	No.	Identification of Molecules
Canonical Pathway Annotations			
Cellular Stress and Injury			
Wound healing signaling pathway	3.22×10^{-11}	9	CSF2, CXCL8, FGF2, FN1, IL6, TGFA, TGFB1, TNF, VEGFA
Cytokine Signaling			
IL17 signaling	8.43×10^{-20}	13	CCL2, CCL20, CSF2, CXCL1, CXCL3, CXCL8, DEFB1, DEFB103A/DEFB103B, DEFB4A/DEFB4B, IL6, TGFB1, TNF, VEGFA
IL6 signaling	2.53×10^{-5}	4	CXCL8, IL6, TNF, VEGFA
IL10 signaling	1.23×10^{-4}	3	IL10, IL6, TNF
IL8 signaling	1.73×10^{-4}	4	CXCL1, CXCL8, ICAM1, VEGFA
Cellular Immune Response			
Role of cytokines in mediating communication between immune cells	8.74×10^{-11}	6	CSF2, CXCL8, IL10, IL6, TGFB1, TNF
Th1 and Th2 activation pathway	8.01×10^{-5}	4	ICAM1, IL10, IL6, TGFB1
Th1 pathway	5.81×10^{-4}	3	ICAM1, IL10, IL6
Th2 pathway	8.13×10^{-4}	3	ICAM1, IL10, TGFB1
Communication between innate and adaptive immune cells	7.44×10^{-3}	5	CSF2, CXCL8, IL10, IL6, TNF
Diseases or Functions Annotations			
Antimicrobial Response, Inflammatory Response			
Antibacterial response	1.66×10^{-22}	13	CAMP, CCL20, DEFB1, DEFB103A/DEFB103B, DEFB4A/DEFB4B, IL10, IL6, RNASE7, S100A12, S100A7, S100A8, S100A9, TNF
Antimicrobial Response, Inflammatory Response			
Chemoattraction	5.09×10^{-23}	12	CAMP, CCL2, CCL20, CSF2, CXCL1, CXCL3, CXCL8, DEFB4A/DEFB4B, FN1, TGFB1, TNF, VEGFA
Cellular Movement, Hematological System Development and Function, Immune Cell Trafficking, Inflammatory Response			
Chemotaxis	4.28×10^{-37}	27	CAMP, CCL2, CCL20, CSF2, CXCL1, CXCL2, CXCL3, CXCL8, DEFB1, DEFB103A/DEFB103B, DEFB4A/DEFB4B, FGF2, FN1, ICAM1, IL10, IL6, S100A12, S100A7, S100A8, S100A9, SELE, SMAD3, SMAD4, TGFA, TGFB1, TNF, VEGFA

Table 2. *Cont.*

IPA Function	*p*-Value	No.	Identification of Molecules
Chemotaxis of leukocytes	1.43×10^{-35}	24	CAMP, CCL2, CCL20, CSF2, CXCL1, CXCL2, CXCL3, CXCL8, DEFB1, DEFB103A/DEFB103B, DEFB4A/DEFB4B, FN1, ICAM1, IL10, IL6, S100A12, S100A7, S100A8, S100A9, SELE, SMAD3, TGFB1, TNF, VEGFA
Inflammatory Response			
Inflammatory response	9.34×10^{-29}	26	CAMP, CCL2, CCL20, CSF2, CXCL1, CXCL2, CXCL3, CXCL8, DEFB1, DEFB103A/DEFB103B, DEFB4A/DEFB4B, FGF1, FGF2, FN1, ICAM1, IL10, IL6, S100A12, S100A7, S100A8, S100A9, SELE, SMAD3, TGFB1, TNF, VEGFA
Proinflammatory response	4.92×10^{-15}	7	CCL2, CXCL3, CXCL8, IL10, IL6, TNF, VEGFA
Innate immune response	5.94×10^{-14}	10	CAMP, CXCL1, CXCL8, FN1, IL10, IL6, RNASE7, S100A12, SMAD3, TNF
Tissue Development			
Healing of wound	1.53×10^{-20}	13	CSF2, FGF1, FGF2, FN1, ICAM1, IL10, IL6, SMAD3, SMAD4, TGFA, TGFB1, TNF, VEGFA
Cell-To-Cell Signaling and Interaction, Cellular Movement, Hematological System Development and Function, Immune Cell Trafficking, Inflammatory Response			
Cell movement of monocytes	1.37×10^{-31}	19	CAMP, CCL2, CCL20, CSF2, CXCL3, CXCL8, DEFB1, DEFB103A/DEFB103B, FN1, ICAM1, IL10, IL6, S100A12, S100A7, SELE, SMAD3, TGFB1, TNF, VEGFA
Cell movement of neutrophils	2.01×10^{-28}	21	CAMP, CCL2, CSF2, CXCL1, CXCL2, CXCL3, CXCL8, DEFB1, DEFB103A/DEFB103B, DEFB4A/DEFB4B, FN1, ICAM1, IL10, IL6, S100A12, S100A8, S100A9, SELE, SMAD3, TGFB1, TNF
Cell-To-Cell Signaling and Interaction, Cellular Movement, Hematological System Development and Function, Immune Cell Trafficking, Inflammatory Response			
Recruitment of cells	7.85×10^{-29}	21	BMP10, CAMP, CCL2, CCL20, CSF2, CXCL1, CXCL2, CXCL3, CXCL8, DEFB4A/DEFB4B, FGF2, FN1, ICAM1, IL10, IL6, S100A8, SELE, SMAD3, TGFB1, TNF, VEGFA
Recruitment of leukocytes	1.05×10^{-25}	19	BMP10, CAMP, CCL2, CCL20, CSF2, CXCL1, CXCL2, CXCL3, CXCL8, DEFB4A/DEFB4B, FN1, ICAM1, IL10, IL6, S100A8, SELE, SMAD3, TGFB1, TNF
Cell-To-Cell Signaling and Interaction, Inflammatory Response			
Immune response of cells	2.70×10^{-18}	18	CAMP, CCL2, CCL20, CSF2, CXCL1, CXCL3, CXCL8, FN1, ICAM1, IL10, IL6, S100A12, S100A8, S100A9, SMAD3, TGFB1, TNF, VEGFA
Immune response of myeloid cells	2.85×10^{-17}	13	CAMP, CCL2, CSF2, CXCL1, CXCL3, CXCL8, FN1, ICAM1, IL10, IL6, S100A9, TGFB1, TNF
Cellular Growth and Proliferation			
Angiogenesis	1.74×10^{-23}	24	BMP10, CAMP, CCL2, CSF2, CXCL1, CXCL2, CXCL8, FGF1, FGF2, FN1, ICAM1, IL10, IL6, PIGF, S100A12, S100A8, S100A9, SELE, SMAD3, SMAD4, TGFA, TGFB1, TNF, VEGFA
Proliferation of vascular cells	3.32×10^{-20}	16	CAMP, CCL2, CXCL1, CXCL8, FGF1, FGF2, FN1, IL10, IL6, S100A8, S100A9, SMAD3, SMAD4, TGFB1, TNF, VEGFA

Table 2. *Cont.*

IPA Function	p-Value	No.	Identification of Molecules
Cellular Movement, Hematological System Development and Function, Immune Cell Trafficking			
Cell survival	1.73×10^{-17}	24	CAMP, CCL2, CSF2, CXCL1, CXCL2, CXCL3, CXCL8, DEFB103A/DEFB103B, DEFB4A/DEFB4B, FGF1, FGF2, FN1, ICAM1, IL10, IL6, S100A8, S100A9, SELE, SMAD3, SMAD4, TGFA, TGFB1, TNF, VEGFA
Cell viability	2.74×10^{-15}	22	CAMP, CCL2, CSF2, CXCL1, CXCL2, CXCL3, CXCL8, FGF1, FGF2, FN1, ICAM1, IL10, IL6, S100A8, S100A9, SELE, SMAD3, SMAD4, TGFA, TGFB1, TNF, VEGFA

IPA disease and function annotations of these 32 molecules (Table 1) were predicted to be associated with innate and adaptive immune responses, applicable to reducing infection and inflammation and assisting in tissue healing. These included antimicrobial response, inflammatory response, cell-to-cell signaling and interaction, cellular movement, immune cell trafficking, cell death and survival, cellular growth and proliferation, cellular development, hematological system development and function, hematopoiesis, lymphoid tissue structure and development, tissue development, and tissue morphology. Examples of these functions are listed in Table 2 and all the relevant IPA disease and function annotations are listed in Supplementary Table S4.

3.2. Subset of 13 Endodontic Tissue Molecules

IPA analysis also predicted that different wavelengths of UV irradiation might selectively regulate gene expression for innate or adaptive immune functions. IPA comparison analysis of the log2 ratios of the concentrations of 13 CCBMs in tissue culture media of HEPM cells and gingival fibroblasts at 0, 24, and 48 h after treatment with 255, 405, or 255/405 UV irradiation identified differences in expression across 18 observations, representing 2 cell lines, 3 UV treatments, and 3 time periods (Figure 3). The expression log2 ratios of the 13 CCBMs varied from −2.6245 to 1.6114. Relevant IPA canonical pathways in Figure 3 were related to cellular stress and injury (wound healing signaling pathway, the CLEAR signaling pathway, HIF1α signaling pathway, and autophagy), cytokine signaling (IL17 signaling, IL6 signaling, and NF-κB signaling), growth factor signaling (regulation of the epithelial mesenchymal transition by growth factors pathway), and organismal growth and development (ID1 signaling pathway). IPA comparison analysis predicted that UV irradiation activated or inhibited pathways depending upon the cell type, wavelength of treatment, and time after treatment. Numerous IPA canonical pathway annotations were inhibited shortly after irradiation (0 h) but activated at 24 and 48 h. Fibroblasts and HEPM cells both were strongly activated by 405 nm and 255/405 nm UV irradiation treatments (Figure 3).

Within the wound healing signaling pathway, for example, after 0 h of 255, 405, and 255/405 nm irradiation, fibroblasts and HEPM cells (observations 1–3, 10–12) had negative z-scores, predicting inhibition of pathway signaling, whereas after 24 and 48 h after 255, 405, and 255/405 nm irradiation, fibroblasts (observations 4–9) and after 48 h of 255, 405, and 255/405 nm irradiation, HEPM cells (observations 16–18) had positive z-scores, predicting activation of pathway signaling. There were also conditions predicted to activate some pathways yet inhibit others (observations 3, 9, 12, 15, and 18). Here, treatment with cells with 255/405 nm irradiation contained both inhibited and activated pathway signaling. After 48 h of 255/405 nm irradiation, HEPM cells (observation 18) had a positive z-score but contained both inhibited and activated pathway signaling (Figure 4A–C). Signaling was predicted to occur through binding of molecules to the TNF receptor, EGFR, and TGFBR and signaling through JNK and ERK1/2 to transcription factors NF-κB, CEBPB, and AP-1. This was predicted to activate additional CCBMs, leading to proinflammatory responses, disruption of desmosomes, chemoattraction of leukocytes, migration and proliferation of fibroblasts and cells, collagen matrix remodeling, and wound healing pathways.

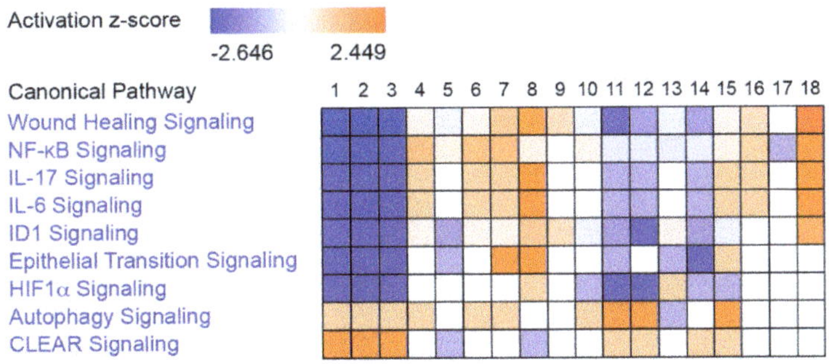

Figure 3. IPA comparison analysis of 18 observations from fibroblasts (observations 1–9) and HEPM cells (observations 10–18) at 0 h (observations 1–3, 10–12), 24 h (observations 4–6, 13–15), and 48 h (observations 7–9, 16–18) after treatment with 255 nm (observations 1, 4, 7, 10, 13, and 16), 405 nm (observations 2, 5, 8, 11, 14, 17), or 255/405 nm (observations 3, 6, 9, 12, 15, and 18) irradiation. Groups are shown as a heatmap, where blue represents inhibition (negative values), white represents midpoint, and orange represents activation (positive values). Numerous IPA canonical pathways were inhibited shortly after irradiation (0 h) but activated at 24 and 48 h. Fibroblasts and HEPM cells both were strongly activated by 405 nm and 255/405 nm UV irradiation treatments.

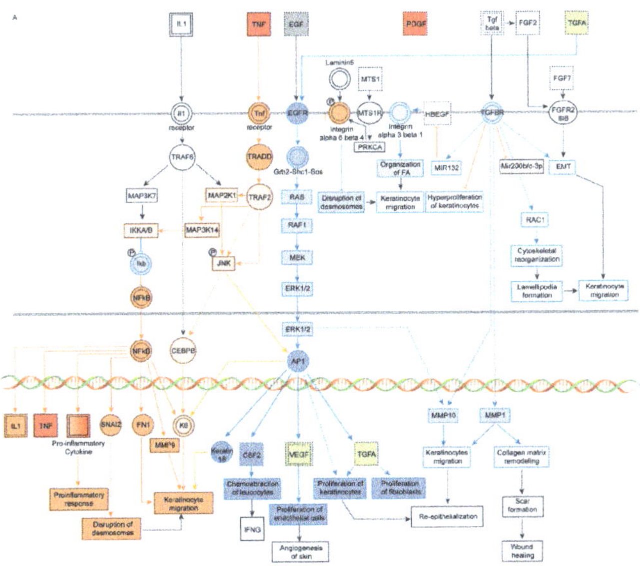

Figure 4. *Cont.*

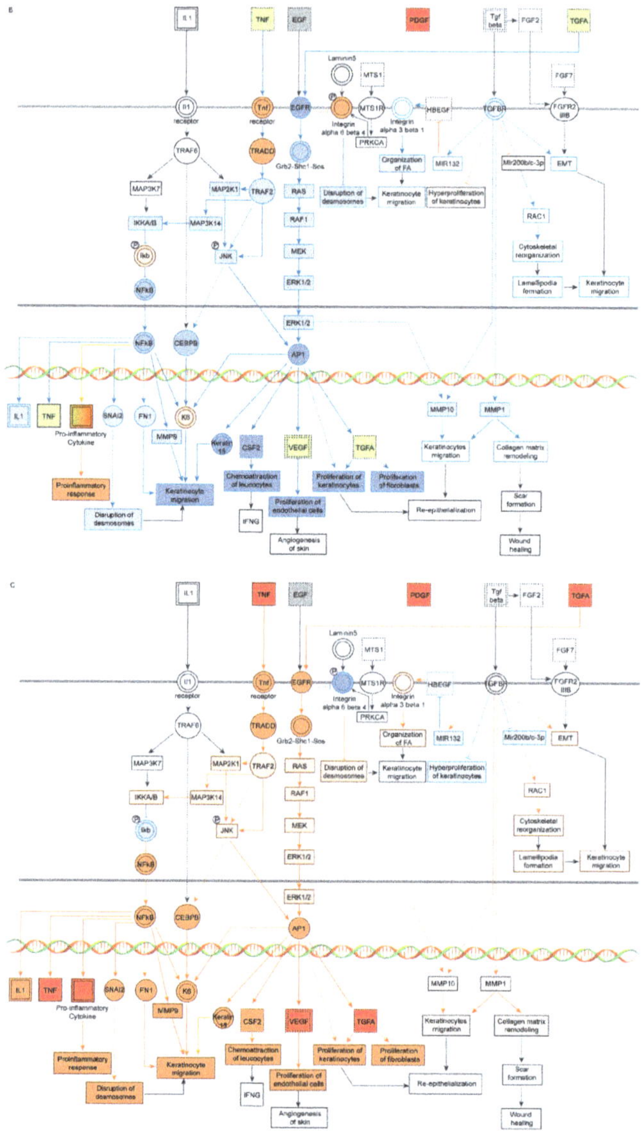

Figure 4. Schematic diagrams of the wound healing signaling pathway, prepared using Ingenuity Pathway Analysis software (IPA, Qiagen, Germantown, MD), showing both inhibited and activated pathway signaling in HEPM cells 48 h after (**A**) 255 nm irradiation, (**B**) 405 nm irradiation, and (**C**) 255/405 nm irradiation. Signaling starts via TNF binding to the TNF receptor; EGF and TGFA binding to the EGFR; and TGFB binding to the TGFBR. These pathways signal through TRADD and TRAF2 to JNK and through RAS, RAF, and MEK to ERK1/2. Signaling continues to NF-κB, CEBPB, and AP-1 to activate additional CCBMs, leading to proinflammatory responses, disruption of desmosomes, chemoattraction of leukocytes, migration and proliferation of fibroblasts and cells, collagen matrix remodeling, and wound healing pathways. Pathway molecules in red indicate activation and molecules in green indicated inhibition. Signaling connections in orange indicate pathway activation and signaling connections in blue indicate pathway inhibition.

3.3. Modulating Endodontic Pain

While UV irradiation is not known to be directly anti-nociceptive, it does act on ERK, p38 MAPK, JNK, and NF-κB signaling pathways (Supplementary Figure S1) that produce AMPs and CCBMs that are involved in pain reduction (Supplementary Figure S2). Many of the molecules induced by UV irradiation (Table 1) can be pro-nociceptive, nociceptive, and anti-nociceptive stimuli. AMPs can be anti-nociceptive stimuli or pro-nociceptive stimuli, chemokines can be pro-nociceptive or nociceptive stimuli, and growth factors can be anti-nociceptive stimuli.

4. Discussion

We identified a list of 32 molecules expressed or secreted from cells and tissues after UV irradiation and used a bioinformatics approach to show that they were related to wound healing and innate and adaptive immune functions, including chemotaxis, movement, growth, and proliferation of cells. We then used a subset of 13 osteoinductive, angiogenic, proliferative, and proinflammatory molecules to show that HEPM cells and gingival fibroblasts treated with 255, 405, and 255/405 nm UV irradiation had different expression profiles. These results suggested that UV irradiation can activate innate and adaptive immune responses, which may supplement endodontic procedures to reduce infection, inflammation, and pain and assist tissues to heal.

Wound healing is a complex process, whereby secreted molecules from infected, injured, or damaged cells and tissues attract a variety of inflammatory cells to the injured site (inflammatory phase). Once present, inflammatory cells release additional CCBMs. These include growth factors that attract and transform fibroblasts and molecules that stimulate attracted cells to proliferate and stimulate other cells to begin forming new capillaries and blood vessels (proliferative phase). Tissue development, angiogenesis, and vasculogenesis occurs (maturation and remodeling phase). IPA identified 13/32 molecules ($p = 1.53 \times 10^{-20}$) in the IPA canonical pathway annotations associated with these phases in the wound healing signaling pathway (Table 2, Supplementary Table S3). Of the 13 molecules, TNFA, a proinflammatory cytokine, is secreted by activated monocytes, macrophages, B-cells, T-cells, and fibroblasts; IL6 regulates immune and inflammatory responses, including B-cell differentiation and antibody production; and IL10 inhibits the expression of pro-inflammatory cytokines, but enhances humoral immune responses and attenuates cell mediated immune reactions. CSF2 stimulates the development of neutrophils and macrophages. Growth factors including FGF1, FGF2, TGFA, TGFB1, and VEGFA are involved in cell motility; cell proliferation; cell growth and differentiation; redistribution of tissue; angiogenesis and vascular permeability of endothelial cells; and synthesis and deposition of the extracellular matrix. FN1 and collagen from fibroblasts allow tissues to contract [12], SMAD3 and SMAD4 play roles in the signaling of TGFB1 [13], and ICAM1 provides adhesion between endothelial cells and leukocytes after stress or injury.

Innate immunity is a type of nonspecific host resistance without memory, involving soluble molecules and cells [14]. Stimulation of receptors activates several cellular pathways, resulting in the production of AMPs and inflammatory cytokines. Stimulation also leads to changes in cellular metabolism, upregulation of numerous genes involved in cell defense and pathogen restriction, and the induction of regulated cell death [14].

AMPs are a large component of innate immune responses and UV irradiation induces their transcription and secretion (Supplementary Table S1). Their ability to modulate both innate immune responses, cellular immunity, and angiogenesis are very well known [15–18]. IPA identified 14/32 molecules involved in antimicrobial and antibacterial response annotations (Supplementary Table S4). These included CAMP (LL37), the defensin family (DEFB1, DEFB103A/DEFB103B, and DEFB4A/DEFB4B), the S100 family of calcium binding proteins (S100A7, S100A8, S100A9, and S100A12), and RNASE7. It also included the chemokine CCL20, which has antimicrobial activity [19] and cytokines IL6, IL10, TNFA, and TGFB1, which contribute to the antibacterial response. So far, AMP expression and secretion has been reported to occur in a narrow range from 280 to 400 nm (Supplementary

Table S1). AMP expression is increased after 280–313 nm irradiation [20,21], but not after 340 to 400 nm irradiation [22,23]. Thus, UV irradiation of an infected root canal would not have to kill 100% of microorganisms, but simply reduce the microbial infection burden to a level that could be managed by UV-induced innate immune mechanisms.

CCBMs are large components of innate immune responses and UV irradiation induces their transcription and secretion (Supplementary Table S2), CCBM expression and secretion has been reported to occur in a wider range from 254 to 404 nm (Supplementary Table S2). In irradiated cells, CCBM expression is increased for many CCBMs after 200 to 320 nm irradiation but decreased for others, such as BMP10 and FGF1 in HEPM cells and VEGF in fibroblasts [6]. At 340 to 405 nm irradiation, cells had increased levels of IL6, CXCL8, and CSF mRNA expression and secretion in keratinocytes [24,25] but decreased levels of secreted FGF1 in fibroblasts [6]. In irradiated human and murine skin at 2–3 MED (minimal erythemal dose), there were increased levels of immunostaining for CCL2, CCL20, CXCL1, CXCL8, ICAM1, IL1, IL10, SELE, SMAD3, SMAD4, TGFA, TGFB, TNFA, and VEGF 24–48 h after exposure [26–28].

UV-induced AMPs and CCBMs would have a variety of common functions (Figure 1). In addition to their potent antimicrobial activity mentioned above, these UV-induced molecules can chemoattract a variety of cells important to both immune protection and wound healing. Defensins attract keratinocytes, dendritic cells, and T-cells [29,30] and CAMP (LL37) attracts fibroblasts, microvascular endothelial cells, and human umbilical vein endothelial cells [31]. AMPs can regulate proinflammatory CCBM production [32]. At lower concentrations, defensins do not induce TNFA or IL1B expression in monocytes or macrophages [33,34]. However, at higher concentrations, defensins induce CCBM production in epithelial cells, keratinocytes, monocytes, and macrophages [30,35,36] and CAMP (LL37) induces CXCL8 in epithelial cells and macrophages [37]. Finally, AMPs and CCBMs play a direct role in wound healing, angiogenesis, and vasculogenesis [18]. DEFB4A (HBD2) increases keratinocyte proliferation [30,38] and CAMP (LL37) increases fibroblast proliferation, induces human microvascular endothelial cell and human umbilical vein endothelial cell proliferation and stimulates re-epithelialization [31,39,40].

UV irradiation also suppresses cellular immunity and acts primarily on T-cell-mediated immune reactions [41–43]. This application has been used to treat several T-cell-mediated diseases, including graft-versus-host disease and systemic scleroderma [44]. UV irradiation alters antigen specificity, alters antigen-presenting cell function, acts on effector and regulatory T-cells [41] and induces the production of CCBMs [42]. For example, UV irradiated dendritic cells do not present antigens effectively, and thus induce regulatory T-cells ($CD4^+CD25^+$), but not effector T-cells [41]. UV irradiation can lead to T-cell tolerance and prevents the priming of antigen-specific $CD8^+$ T-cells (in models of contact hypersensitivity) independent of conventional $CD4^+$ regulatory T-cells [44]. Tolerant $CD8^+$ T-cells prevented migration of dendritic cells and prevented priming of other $CD8^+$ T-cells. TGFB and immunosuppressive IL10 are regulatory T-cell-associated cytokines [10,45].

There are differences based on the specific wavelength. UVB induces the infiltration of immature inflammatory myeloid $CD11c^+$ $bDCA1^-$ dendritic cells, which may have a suppressive function [10]. UVA1 does not induce IL10, but does suppress the production of TNFA and IL12, and contributes to cis-UCA isomerization [10,46]. Immune suppression may be dependent upon the extent of UV irradiation-induced damage to DNA [41]. However, to what extent the secondary immunostimulatory effects of UV-induced AMPs and CCBMs offset the immunosuppressive effects of immune cells is not yet known.

Many oral related infections, inflammation, and tissue injury/peripheral nerve injury can be stimuli that activate receptors on the surface of cells, initiating signal transduction through MAPK and NF-κB signaling pathways (Supplementary Figure S2) [47]. The MAPK pathway regulates proinflammatory and pronociceptive molecules involved in inflammation and pain [48]. Nociceptive activity or nerve injury stimuli signal through raf and MEK1/2 to ERK in the cytoplasm and transcription factor CREB in the nucleus [48]. Chemokines (FKN), cytokines (TNFA), and nerve injury stimuli signal through TAK1 and MKK3,6 to p38 MAPK in the cytoplasm and on to transcription factor ATF-2 in the nucleus [48]. Cytokines (TNFA), growth factors BFGF (FGF2), and nerve injury stimuli signal through MLK3/MEKK1 and MKK4,7 to JNK in the cytoplasm and on to transcription factor c-Jun (AP-1) in the nucleus [48].

UV irradiation also alters the expression of ERK, p38 MAPK, JNK, and NF-κB signaling pathways [49] (Supplementary Figure S1) and it is possible that treatment of endodontic infections, inflammation, and tissue injury/peripheral nerve injury can modulate downstream production of CCBMs and be a potential intervention at these nodes to mitigate acute or chronic pain. Phosphorylation of ERK in nerve injury is induced early, is long lasting, and is involved in the induction of pain (Supplementary Figure S2). Suppressing this step in ERK is thought to be a promising strategy for treatment of neuropathic pain [50]. Likewise, targeting the p38 MAPK pathway and its signaling is also thought to be a potential therapeutic strategy for pain management [51].

Our analysis and results support those of Ou and Peterson [9] and Vieyra-Garcia et al. [10] and also suggest that UV irradiation can induce the production of AMPs and CCBMs. Our results also suggest that the production of these molecules can induce the innate and adaptive immune responses involved in attenuating infection, inflammation, and pain and enhancing healing and regeneration of tissue. However, these results are based on bioinformatics analysis of molecules induced by UV irradiation reported in the literature and produced in culture from cells treated with UV irradiation. These concepts and results form a strong hypothesis for future studies and should be examined in detail.

5. Conclusions

In summary, UV irradiation has the ability to kill microorganisms, but could also be used to activate innate and adaptive immune mechanisms in endodontic root canals directly or through UV-induced molecules. UV irradiation-induced effects appear to be wavelength specific and could supplement procedures to reduce infection, to reduce inflammation, and to facilitate local tissue healing.

Supplementary Materials: The following supporting information can be downloaded at: https://www.mdpi.com/article/10.3390/pharmaceutics14091979/s1, Supplementary Figure S1: A schematic diagram of UV-induced MAPK signaling; Supplementary Figure S2: A schematic diagram of neuropathic pain signaling; Supplementary Table S1: Antimicrobial peptides (AMPs) expressed or secreted in cells, tissues, and tissue explants after irradiation with UVC, UVB, or UVA; Supplementary Table S2: Chemokines, cytokines, and biomarkers (CCBMs) expressed or secreted in cells, tissues, and tissue explants after irradiation with UVC, UVB, or UVA; Supplementary Table S3: Antimicrobial peptides and biomarkers in the literature dataset (n = 32) participating in the activation of innate and immune mechanisms, as determined using IPA canonical pathways annotations; and Supplementary Table S4: Antimicrobial peptides and biomarkers in the literature dataset (n = 32) participating in the activation of innate and immune mechanisms as determined using IPA diseases and function annotations. Refs. [52–71] are cited in Supplementary Materials.

Author Contributions: The authors contributed to the study in the following aspects: Conceptualization, K.A.M., R.H.S. and K.A.B.; Methodology, K.A.M., E.Z. and K.A.B.; Validation, E.Z.; Formal Analysis, K.A.M., E.Z. and K.A.B.; Data Curation, K.A.M. and K.A.B.; Writing—Original Draft Preparation, K.A.M., R.H.S. and K.A.B.; Writing—Review and Editing, K.A.M., R.H.S., E.Z. and K.A.B.; and Project Administration, K.A.M. and K.A.B. All authors have read and agreed to the published version of the manuscript.

Funding: This research received no external funding.

Institutional Review Board Statement: Not applicable.

Informed Consent Statement: Not applicable.

Data Availability Statement: Morio et al. reported the concentrations of 13 CCBMs in tissue culture media of HEPM cells and gingival fibroblasts at 0, 24, and 48 h after treatment with 255, 405, or

255/405 UV irradiation [6]. In the current study, we used these CCBM values as a subset of data to assess the ability of UV irradiation to activate or inhibit cellular pathways related to immune functions.

Conflicts of Interest: The authors declare no conflict of interest. We have no financial affiliation (e.g., employment, direct payment, stock holdings, retainers, consultantships, patent licensing arrangements, or honoraria), or involvement with any commercial organization with direct financial interest in the subject or materials discussed in this manuscript, nor have any such arrangements existed in the past 3 years. Kimberly A. Morio is an Endodontist at Apex Endodontics and an Adjunct Instructor at the University of Iowa; Robert H. Sternowski is the President of Softronics, Ltd.; Erliang Zeng is a Biostatistician and a Computational Biologist at the University of Iowa; and Kim A. Brogden is an Emeritus Professor at the University of Iowa. The funders had no role in the design of the study; in the collection, analyses, or interpretation of data; in the writing of the manuscript; or in the decision to publish the results. The company had no role in the design of the study; in the collection, analyses, or interpretation of data; in the writing of the manuscript, and in the decision to publish the results.

References

1. Fouad, A.F. Microbiological Aspects of Traumatic Injuries. *J. Endod.* **2019**, *45*, S39–S48. [CrossRef] [PubMed]
2. Delikan, E.; Caliskan, S.; Cankilic, M.Y.; Aksu, S.; Kesim, B.; Ulger, S.T. Microbiota of endodontically infected primary and permanent teeth. *Pediatr. Dent.* **2021**, *43*, 102–110. [PubMed]
3. Wong, J.; Manoil, D.; Näsman, P.; Belibasakis, G.N.; Neelakantan, P. Microbiological aspects of root canal infections and disinfection strategies: An update review on the current knowledge and challenges. *Front. Oral Health* **2021**, *2*, 672887. [CrossRef]
4. Siqueira, J.F., Jr.; Rocas, I.N. Present status and future directions: Microbiology of endodontic infections. *Int. Endod. J.* **2022**, *55*, 512–530. [CrossRef] [PubMed]
5. Metzger, Z.; Better, H.; Abramovitz, I. Immediate root canal disinfection with ultraviolet light: An ex vivo feasibility study. *Oral Surg. Oral Med. Oral Pathol. Oral Radiol. Endod.* **2007**, *104*, 425–433. [CrossRef] [PubMed]
6. Morio, K.A.; Thayer, E.L.; Bates, A.M.; Brogden, K.A. 255-nm light-emitting diode kills *Enterococcus faecalis* and induces the production of cellular biomarkers in human embryonic palatal mesenchyme cells and gingival fibroblasts. *J. Endod.* **2019**, *45*, 774–783.e6. [CrossRef]
7. Metzger, Z.; Dotan, M.; Better, H.; Abramovitz, I. Sensitivity of oral bacteria to 254 nm ultraviolet light. *Int. Endod. J.* **2007**, *40*, 120–127. [CrossRef]
8. Morio, K.A.; Sternowski, R.H.; Brogden, K.A. Dataset of endodontic microorganisms killed at 265 nm wavelength by an ultraviolet C light emitting diode in root canals of extracted, instrumented teeth. *Data Brief* **2022**, *40*, 107750. [CrossRef]
9. Ou, Y.; Petersen, P.M. Application of ultraviolet light sources for in vivo disinfection. *Jpn. J. Appl. Phys.* **2021**, *60*, 100501. [CrossRef]
10. Vieyra-Garcia, P.A.; Wolf, P. A deep dive into UV-based phototherapy: Mechanisms of action and emerging molecular targets in inflammation and cancer. *Pharmacol. Ther.* **2021**, *222*, 107784. [CrossRef]
11. Kramer, A.; Green, J.; Pollard, J., Jr.; Tugendreich, S. Causal analysis approaches in Ingenuity Pathway Analysis. *Bioinformatics* **2014**, *30*, 523–530. [CrossRef] [PubMed]
12. Morykwas, M.J.; Mark, M.W. Effects of ultraviolet light on fibroblast fibronectin production and lattice contraction. *Wounds* **1998**, *10*, 111–117.
13. Dennler, S.; Itoh, S.; Vivien, D.; Dijke, P.; Huet, S.; Gauthier, J.M. Direct binding of Smad3 and Smad4 to critical TGF beta-inducible elements in the promoter of human plasminogen activator inhibitor-type 1 gene. *EMBO J.* **1998**, *17*, 3091–3100. [CrossRef] [PubMed]
14. Hagai, T.; Chen, X.; Miragaia, R.J.; Rostom, R.; Gomes, T.; Kunowska, N.; Henriksson, J.; Park, J.E.; Proserpio, V.; Donati, G.; et al. Gene expression variability across cells and species shapes innate immunity. *Nature* **2018**, *563*, 197–202. [CrossRef]
15. Lai, Y.; Gallo, R.L. AMPed up immunity: How antimicrobial peptides have multiple roles in immune defense. *Trends Immunol.* **2009**, *30*, 131–141. [CrossRef] [PubMed]
16. Brogden, K.A.; Bates, A.M.; Fischer, C.L. Antimicrobial peptides in host defense: Functions beyond antimicrobial activity. In *Antimicrobial Peptides—Role in Human Health and Disease*; Harder, J., Schroeder, J.M., Kaufmann, S.H., Mercer, A.A., Weber, B., Eds.; Birkhauser Advances in Infectious Diseases; Springer International Publishing: Cham, Switzerland, 2016; pp. 129–146. [CrossRef]
17. Drayton, M.; Deisinger, J.P.; Ludwig, K.C.; Raheem, N.; Muller, A.; Schneider, T.; Straus, S.K. Host defense peptides: Dual antimicrobial and immunomodulatory action. *Int. J. Mol. Sci.* **2021**, *22*, 11172. [CrossRef]
18. Takahashi, M.; Umehara, Y.; Yue, H.; Trujillo-Paez, J.V.; Peng, G.; Nguyen, H.L.T.; Ikutama, R.; Okumura, K.; Ogawa, H.; Ikeda, S.; et al. The antimicrobial peptide human beta-defensin-3 accelerates wound healing by promoting angiogenesis, cell migration, and proliferation through the FGFR/JAK2/STAT3 signaling pathway. *Front. Immunol.* **2021**, *12*, 712781. [CrossRef] [PubMed]
19. Wang, G. Human antimicrobial peptides and proteins. *Pharmaceuticals* **2014**, *7*, 545–594. [CrossRef]
20. Mallbris, L.; Edstrom, D.W.; Sundblad, L.; Granath, F.; Stahle, M. UVB upregulates the antimicrobial protein hCAP18 mRNA in human skin. *J. Investig. Dermatol.* **2005**, *125*, 1072–1074. [CrossRef]

21. Glaser, R.; Navid, F.; Schuller, W.; Jantschitsch, C.; Harder, J.; Schroder, J.M.; Schwarz, A.; Schwarz, T. UV-B radiation induces the expression of antimicrobial peptides in human keratinocytes in vitro and in vivo. *J. Allergy Clin. Immunol.* **2009**, *123*, 1117–1123. [CrossRef]
22. Kreuter, A.; Hyun, J.; Skrygan, M.; Sommer, A.; Bastian, A.; Altmeyer, P.; Gambichler, T. Ultraviolet A1-induced downregulation of human beta-defensins and interleukin-6 and interleukin-8 correlates with clinical improvement in localized scleroderma. *Br. J. Dermatol.* **2006**, *155*, 600–607. [CrossRef]
23. Gambichler, T.; Skrygan, M.; Tomi, N.S.; Altmeyer, P.; Kreuter, A. Changes of antimicrobial peptide mRNA expression in atopic eczema following phototherapy. *Br. J. Dermatol.* **2006**, *155*, 1275–1278. [CrossRef]
24. Park, K.C.; Jung, H.C.; Hwang, J.H.; Youn, S.W.; Ahn, J.S.; Park, S.B.; Kim, K.H.; Chung, J.H.; Youn, J.I. GM-CSF production by epithelial cell line: Upregulation by ultraviolet A. *Photodermatol. Photoimmunol. Photomed.* **1997**, *13*, 133–138. [CrossRef] [PubMed]
25. Imokawa, G.; Yada, Y.; Kimura, M.; Morisaki, N. Granulocyte/macrophage colony-stimulating factor is an intrinsic keratinocyte-derived growth factor for human melanocytes in UVA-induced melanosis. *Biochem. J.* **1996**, *313*, 625–631. [CrossRef] [PubMed]
26. Blaudschun, R.; Sunderkotter, C.; Brenneisen, P.; Hinrichs, R.; Peters, T.; Schneider, L.; Razi-Wolf, Z.; Hunzelmann, N.; Scharffetter-Kochanek, K. Vascular endothelial growth factor causally contributes to the angiogenic response upon ultraviolet B irradiation in vivo. *Br. J. Dermatol.* **2002**, *146*, 581–587. [CrossRef] [PubMed]
27. Oxholm, A.; Oxholm, P.; Staberg, B.; Bendtzen, K. Immunohistological detection of interleukin I-like molecules and tumour necrosis factor in human epidermis before and after UVB-irradiation in vivo. *Br. J. Dermatol.* **1988**, *118*, 369–376. [CrossRef]
28. Crispin, M.K.; Fuentes-Duculan, J.; Gulati, N.; Johnson-Huang, L.M.; Lentini, T.; Sullivan-Whalen, M.; Gilleaudeau, P.; Cueto, I.; Suárez-Fariñas, M.; Lowes, M.A.; et al. Gene profiling of narrowband UVB-induced skin injury defines cellular and molecular innate immune responses. *J. Investig. Dermatol.* **2013**, *133*, 692–701. [CrossRef]
29. Yang, D.; Chertov, O.; Bykovskaia, S.N.; Chen, Q.; Buffo, M.J.; Shogan, J.; Anderson, M.; Schroder, J.M.; Wang, J.M.; Howard, O.M.; et al. B-defensins: Linking innate and adaptive immunity through dendritic and T cell CCR6. *Science* **1999**, *286*, 525–528. [CrossRef]
30. Niyonsaba, F.; Ushio, H.; Nakano, N.; Ng, W.; Sayama, K.; Hashimoto, K.; Nagaoka, I.; Okumura, K.; Ogawa, H. Antimicrobial peptides human beta-defensins stimulate epidermal keratinocyte migration, proliferation and production of proinflammatory cytokines and chemokines. *J. Investig. Dermatol.* **2007**, *127*, 594–604. [CrossRef]
31. Ramos, R.; Silva, J.P.; Rodrigues, A.C.; Costa, R.; Guardao, L.; Schmitt, F.; Soares, R.; Vilanova, M.; Domingues, L.; Gama, M. Wound healing activity of the human antimicrobial peptide LL37. *Peptides* **2011**, *32*, 1469–1476. [CrossRef]
32. Yang, D.; Biragyn, A.; Kwak, L.W.; Oppenheim, J.J. Mammalian defensins in immunity: More than just microbicidal. *Trends Immunol.* **2002**, *23*, 291–296. [CrossRef]
33. Chaly, Y.V.; Paleolog, E.M.; Kolesnikova, T.S.; Tikhonov, I.I.; Petratchenko, E.V.; Voitenok, N.N. Neutrophil alpha-defensin human neutrophil peptide modulates cytokine production in human monocytes and adhesion molecule expression in endothelial cells. *Eur. Cytokine Netw.* **2000**, *11*, 257–266. [PubMed]
34. Barabas, N.; Rohrl, J.; Holler, E.; Hehlgans, T. Beta-defensins activate macrophages and synergize in pro-inflammatory cytokine expression induced by TLR ligands. *Immunobiology* **2013**, *218*, 1005–1011. [CrossRef] [PubMed]
35. Van Wetering, S.; MannesseLazeroms, S.P.G.; Dijkman, J.H.; Hiemstra, P.S. Effect of neutrophil serine proteinases and defensins on lung epithelial cells: Modulation of cytotoxicity and IL-8 production. *J. Leukoc. Biol.* **1997**, *62*, 217–226. [CrossRef] [PubMed]
36. Petrov, V.; Funderburg, N.; Weinberg, A.; Sieg, S. Human beta defensin-3 induces chemokines from monocytes and macrophages: Diminished activity in cells from HIV-infected persons. *Immunology* **2013**, *140*, 413–420. [CrossRef] [PubMed]
37. Scott, M.G.; Davidson, D.J.; Gold, M.R.; Bowdish, D.; Hancock, R.E. The human antimicrobial peptide LL-37 is a multifunctional modulator of innate immune responses. *J. Immunol.* **2002**, *169*, 3883–3891. [CrossRef]
38. Warnke, P.H.; Voss, E.; Russo, P.A.; Stephens, S.; Kleine, M.; Terheyden, H.; Liu, Q. Antimicrobial peptide coating of dental implants: Biocompatibility assessment of recombinant human beta defensin-2 for human cells. *Int. J. Oral Maxillofac. Implant.* **2013**, *28*, 982–988. [CrossRef]
39. Heilborn, J.D.; Nilsson, M.F.; Kratz, G.; Weber, G.; Sorensen, O.; Borregaard, N.; Stahle-Backdahl, M. The cathelicidin anti-microbial peptide LL-37 is involved in re-epithelialization of human skin wounds and is lacking in chronic ulcer epithelium. *J. Investig. Dermatol.* **2003**, *120*, 379–389. [CrossRef]
40. Nakatsuji, T.; Gallo, R.L. Antimicrobial peptides: Old molecules with new ideas. *J. Investig. Dermatol.* **2012**, *132*, 887–895. [CrossRef]
41. Schwarz, T.; Beissert, S. Milestones in photoimmunology. *J. Investig. Dermatol.* **2013**, *133*, E7–E10. [CrossRef]
42. Matos, T.R.; Sheth, V. The symbiosis of phototherapy and photoimmunology. *Clin. Dermatol.* **2016**, *34*, 538–547. [CrossRef] [PubMed]
43. Bruhs, A.; Schwarz, T. Ultraviolet radiation-induced immunosuppression: Induction of regulatory T cells. *Methods Mol. Biol.* **2017**, *1559*, 63–73. [CrossRef] [PubMed]
44. Hequet, O.; Nosbaum, A.; Guironnet-Paquet, A.; Blasco, E.; Nicolas-Virelizier, E.; Griffith, T.S.; Rigal, D.; Cognasse, F.; Nicolas, J.F.; Vocanson, M. CD8(+) T cells mediate ultraviolet A-induced immunomodulation in a model of extracorporeal photochemotherapy. *Eur. J. Immunol.* **2020**, *50*, 725–735. [CrossRef] [PubMed]
45. Schwarz, T. Regulatory T cells induced by ultraviolet radiation. *Int. Arch. Allergy Immunol.* **2005**, *137*, 187–193. [CrossRef]

46. Bernard, J.J.; Gallo, R.L.; Krutmann, J. Photoimmunology: How ultraviolet radiation affects the immune system. *Nat. Rev. Immunol.* **2019**, *19*, 688–701. [CrossRef]
47. Patil, C.S.; Kirkwood, K.L. p38 MAPK signaling in oral-related diseases. *J. Dent. Res.* **2007**, *86*, 812–825. [CrossRef]
48. Ji, R.R.; Gereau, R.W.; Malcangio, M.; Strichartz, G.R. MAP kinase and pain. *Brain Res. Rev.* **2009**, *60*, 135–148. [CrossRef]
49. Muthusamy, V.; Piva, T.J. A comparative study of UV-induced cell signalling pathways in human keratinocyte-derived cell lines. *Arch. Dermatol. Res.* **2013**, *305*, 817–833. [CrossRef]
50. Ma, W.; Quirion, R. The ERK/MAPK pathway, as a target for the treatment of neuropathic pain. *Expert Opin. Ther. Targets* **2005**, *9*, 699–713. [CrossRef]
51. Lin, X.; Wang, M.; Zhang, J.; Xu, R. p38 MAPK: A potential target of chronic pain. *Curr. Med. Chem.* **2014**, *21*, 4405–4418. [CrossRef]
52. Bode, A.M.; Dong, Z. Mitogen-activated protein kinase activation in UV-induced signal transduction. *Sci. STKE* **2003**, *2003*, 167. [CrossRef] [PubMed]
53. Bender, K.; Blattner, C.; Knebel, A.; Iordanov, M.; Herrlich, P.; Rahmsdorf, H.J. UV-induced signal transduction. *J. Photochem. Photobiol. B.* **1997**, *37*, 1–17. [CrossRef]
54. Adachi, M.; Gazel, A.; Pintucci, G.; Shuck, A.; Shifteh, S.; Ginsburg, D.; Rao, L.S.; Kaneko, T.; Freedberg, I.M.; Tamaki, K.; et al. Specificity in stress response: Epidermal keratinocytes exhibit specialized UV-responsive signal transduction pathways. *DNA Cell Biol.* **2003**, *22*, 665–677. [CrossRef] [PubMed]
55. Seo, S.J.; Ahn, S.W.; Hong, C.K.; Ro, B.I. Expressions of beta-defensins in human keratinocyte cell lines. *J. Dermatol. Sci.* **2001**, *27*, 183–191. [CrossRef]
56. Suo, W.; Guo, H.; Wang, X.; Wang, D. Effect of ultraviolet C light on the expression of basic fibroblast growth factor in rat wounds. *Chinese J. Phys. Med. Rehabil.* **2003**, *25*, 651–654.
57. Suo, W.; Wang, X.; Wang, D. Effect of ultraviolet C irradiation on expression of transforming growth factor-β in wound. *Chin. J. Rehabil. Theory Pract.* **2002**, *8*, 5–7.
58. Gallo, R.L.; Staszewski, R.; Sauder, D.N.; Knisely, T.L.; Granstein, R.D. Regulation of GM-CSF and IL-3 production from the murine keratinocyte cell line PAM 212 following exposure to ultraviolet radiation. *J. Investig. Dermatol.* **1991**, *97*, 203–209. [CrossRef]
59. Kirnbauer, R.; Koch, A.; Kurtmann, J.; Schwarz, T.; Urbanski, A.; Luger, T.A. Different effects of UVA and UVB irradiation on epidermal cell-IL6 expression and release. In Proceedings of the ESDR-JSID-SID Tricontinental Meeting, Washington, DC, USA, 26 April 1989; pp. 393–548.
60. Gallo, R.L.; Brownstein, E.; Granstein, R.D. Secretion of interleukin 3 activity from a transformed murine keratinocyte line after exposure to ultraviolet radiation: Role of membrane signal transduction mechanisms. In Proceedings of the ESDR-JSID-SID Tricontinental Meeting, Washington, DC, USA, 26 April 1989; pp. 393–548.
61. Clingen, P.H.; Berneburg, M.; Petit-Frere, C.; Woollons, A.; Lowe, J.E.; Arlett, C.F.; Green, M.H. Contrasting effects of an ultraviolet B and an ultraviolet A tanning lamp on interleukin-6, tumour necrosis factor-alpha and intercellular adhesion molecule-1 expression. *Br. J. Dermatol.* **2001**, *145*, 54–62. [CrossRef]
62. Chung, K.Y.; Chang, N.S.; Park, Y.K.; Lee, K.H. Effect of ultraviolet light on the expression of adhesion molecules and T lymphocyte adhesion to human dermal microvascular endothelial cells. *Yonsei Med. J.* **2002**, *43*, 165–174. [CrossRef]
63. Ansel, J.C.; Luger, T.A.; Green, I. The effect of in vitro and in vivo UV irradiation on the production of ETAF activity by human and murine keratinocytes. *J. Investig. Dermatol.* **1983**, *81*, 519–523. [CrossRef]
64. Kupper, T.S.; Chua, A.O.; Flood, P.; McGuire, J.; Gubler, U. Interleukin 1 gene expression in cultured human keratinocytes is augmented by ultraviolet irradiation. *J. Clin. Investig.* **1987**, *80*, 430–436. [CrossRef] [PubMed]
65. Schwarz, T.; Luger, T.A. New trends in photobiology: Effect of UV irradiation on epidermal cell cytokine production. *J. Photochem. Photobiol. B: Biol.* **1989**, *4*, 1–13. [CrossRef]
66. Rhodes, L.E.; Joyce, M.; West, D.C.; Strickland, I.; Friedmann, P.S. Comparison of changes in endothelial adhesion molecule expression following UVB irradiation of skin and a human dermal microvascular cell line (HMEC-1). *Photodermatol. Photoimmunol. Photomed.* **1996**, *12*, 114–121. [CrossRef] [PubMed]
67. Kennedy-Crispin, M.; Billick, E.; Mitsui, H.; Gulati, N.; Fujita, H.; Gilleaudeau, P.; Sullivan-Whalen, M.; Johnson-Huang, L.M.; Suarez-Farinas, M.; Krueger, J.G. Human keratinocytes' response to injury upregulates CCL20 and other genes linking innate and adaptive immunity. *J. Investig. Dermatol.* **2012**, *132*, 105–113. [CrossRef] [PubMed]
68. Kirnbauer, R.; Kock, A.; Neuner, P.; Forster, E.; Krutmann, J.; Urbanski, A.; Schauer, E.; Ansel, J.C.; Schwarz, T.; Luger, T.A. Regulation of epidermal cell interleukin-6 production by UV light and corticosteroids. *J. Investig. Dermatol.* **1991**, *96*, 484–489. [CrossRef]
69. Yarosh, D.; Both, D.; Kibitel, J.; Anderson, C.; Elmets, C.; Brash, D.; Brown, D. Regulation of TNFalpha production and release in human and mouse keratinocytes and mouse skin after UV-B irradiation. *Photodermatol. Photoimmunol. Photomed.* **2000**, *16*, 263–270. [CrossRef]
70. Gambichler, T.; Tomi, N.S.; Skrygan, M.; Altmeyer, P.; Kreuter, A. Alterations of TGF-beta/Smad mRNA expression in atopic dermatitis following narrow-band ultraviolet B phototherapy: Results of a pilot study. *J. Dermatol. Sci.* **2006**, *44*, 56–58. [CrossRef]
71. James, L.C.; Moore, A.M.; Wheeler, L.A.; Murphy, G.M.; Dowd, P.M.; Greaves, M.W. Transforming growth factor alpha: In vivo release by normal human skin following UV irradiation and abrasion. *Skin Pharmacol.* **1991**, *4*, 61–64. [CrossRef]

pharmaceutics

Article

In Vivo Evaluation of the Efficacy of a Nisin–Biogel as a New Approach for Canine Periodontal Disease Control

Eva Cunha [1,2,*], Luís Miguel Carreira [1,2], Telmo Nunes [1,2], Marta Videira [3], Luís Tavares [1,2], Ana Salomé Veiga [4] and Manuela Oliveira [1,2]

1 CIISA—Centro de Investigação Interdisciplinar em Sanidade Animal, Faculdade de Medicina Veterinária, Universidade de Lisboa, Av. da Universidade Técnica, 1300-477 Lisbon, Portugal
2 Laboratório Associado para Ciência Animal e Veterinária (AL4AnimalS), 1300-477 Lisbon, Portugal
3 Casa dos Animais de Lisboa, Estrada da Pimenteira, 1300-459 Lisbon, Portugal
4 Instituto de Medicina Molecular, Faculdade de Medicina, Universidade de Lisboa, Avenida Professor Egas Moniz, 1649-028 Lisbon, Portugal
* Correspondence: evacunha@fmv.ulisboa.pt

Abstract: Periodontal disease (PD) is a common oral disease in dogs. Recent in vitro research revealed that nisin–biogel is a promising compound for canine PD control. In this work, a clinical trial was developed to assess the in vivo efficacy of nisin–biogel in dogs by determining the dental plaque index (DPI), gingivitis index (GI), and periodontal pocket depth (PPD) after dental administration. The biogel's influence on aerobic bacteria counts was also evaluated, as well as its acceptance/adverse effects in dogs. Twenty animals were allocated to one of two groups: a treatment group (TG) subjected to a dental topical application of nisin–biogel for 90 days and a control group (CG) with no treatment. Besides daily monitoring, on day 1 (T0) and at the end of the assay (T90), animals were subjected to blood analysis, periodontal evaluation, dental plaque sampling, scaling, and polishing. Statistical analysis with mixed models showed a significant reduction in mean PPD (estimate = −0.371, p-value < 0.001) and DPI (estimate = −0.146, p-value < 0.05) in the TG animals at T90. A reduction in the GI (estimate = −0.056, p-value > 0.05) was also observed but with no statistical significance. No influence on total bacterial counts was observed, and no adverse effects were detected. The nisin–biogel was revealed to be a promising compound for canine PD control.

Keywords: nisin–biogel; periodontal disease; dogs; clinical trial

Citation: Cunha, E.; Carreira, L.M.; Nunes, T.; Videira, M.; Tavares, L.; Veiga, A.S.; Oliveira, M. In Vivo Evaluation of the Efficacy of a Nisin–Biogel as a New Approach for Canine Periodontal Disease Control. *Pharmaceutics* **2022**, *14*, 2716. https://doi.org/10.3390/pharmaceutics14122716

Academic Editors: Scavello Francesco, Jean-Eric Ghia and Amiche Mohamed

Received: 28 October 2022
Accepted: 1 December 2022
Published: 4 December 2022

1. Introduction

Periodontal disease (PD) is one of the most prevalent inflammatory diseases in dogs [1]. Affecting over 80% of animals over two years old, PD is initiated by the formation of a polymicrobial biofilm on the tooth surface; this is also known as dental plaque and is responsible for a subsequent local host inflammatory reaction [1,2]. PD can evolve from a reversible stage of gingivitis to an irreversible stage of periodontitis [3]. The periodontal damage can be assessed by evaluating gingivitis, furcation, and mobility indices or stages and by measuring the periodontal pocket depth and the clinical attachment level [4]. According to the severity of the periodontium damage, four stages can be considered in PD classification. Stage 1 includes animals with only gingivitis; stage 2 includes animals with early periodontitis with less than 25% of attachment loss and/or stage 1 furcation; stage 3 comprises animals with moderate periodontitis, revealing an attachment loss of 25 to 50% and/or stage 2 furcation involvement; and the final stage, stage 4, includes dogs with advanced periodontitis, more than 50% of attachment loss and/or stage 3 furcation [4,5]. Besides damaging the periodontium, the persistent contact of dental plaque bacteria with periodontal structures facilitates their migration to the bloodstream and, consequently, the appearance of PD-related systemic consequences in distant organs [6–9].

Several strategies can be used for PD control. Professional removal of the dental plaque along with the application of measures that inhibit dental plaque formation, such as daily toothbrushing (the gold standard method), application of chemical or natural anti-plaque compounds, administration of dental treats and biscuits, or a specific dental diet, are useful for PD control [4]. However, when periodontitis is present, apart from the previously mentioned measures and according to the case severity, PD treatment may include specific surgical approaches, the use of host modulation drugs, and even antimicrobial therapy [5].

Recently, an innovative approach based on the application of the antimicrobial peptide nisin incorporated in a delivery system formed by guar gum gel (nisin–biogel) has shown inhibitory and eradication abilities against pre-formed biofilms composed by PD enterococci, as well as towards canine dental plaque polymicrobial biofilms, in vitro [10,11]. Previous studies have also revealed that nisin–biogel activity is maintained in the presence of canine saliva and over a long-term storage period at distinct temperatures. Furthermore, nisin–biogel has shown an absence of cytotoxicity up to 200 µg/mL towards several cell lines, reinforcing the potential of nisin–biogel as a promising compound for canine PD control [12,13].

The main goal of this in vivo study was to evaluate the influence of the long-term dental application of the nisin–biogel on the dental plaque and gingivitis indices, the periodontal pocket depth, and total oral bacterial counts in dogs through a randomized controlled clinical trial. In addition, hematological and biochemical parameters were monitored during the trial, as well as general side effects.

2. Materials and Methods

2.1. Nisin–Biogel Preparation

Nisin–biogel was prepared as described elsewhere [10,11,13,14]. According to previous studies, a final concentration of 200 µg/mL was selected to be used in the clinical trial [10–13]. After preparation, the nisin–biogel was stored at 4 °C until further use.

2.2. Dog Selection

Animals were selected according to the Veterinary Oral Health Council (VOHC) guidelines for trials testing compounds for PD prevention. The dogs were from an official animal rescue institution ("Casa dos Animais de Lisboa"), and all experimental procedures were approved by the Ethical Committee for Research and Teaching (CEIE) of the Faculty of Veterinary Medicine, University of Lisbon, Portugal (N/Ref 014/2020).

The inclusion criteria were as follows: healthy dogs over 2 years old without severe PD and with no history of antimicrobial therapy in the last month. All animals were submitted to a clinical examination, oral handling, and complete blood analysis (hemogram and measurement of urea, creatinine, alanine aminotransferase, alkaline phosphatase, glucose, albumin, and total blood proteins) to detect any deviations that would prevent their inclusion in the study.

2.3. Clinical Trial

A total of twenty dogs were selected and submitted to a complete periodontal evaluation, dental plaque sampling (for bacterial total counts), scaling, and dental polishing (timepoint 0). The complete removal of dental plaque and calculus was checked using a disclosure solution (GC Tri Plaque ID Gel®, Tokyo, Japan). All procedures were performed under general anesthesia using acepromazine (0.01 mg/kg, IM), propofol (2 mg/kg, IV), and isoflurane. Intraoperative meloxicam (0.2 mg/Kg, SC) and amoxicillin/clavulanate (8.75 mg/kg, SC) were administered to all animals [15]. After that, each dog was randomly allocated to one of two groups: a treatment group (TG, N = 10) or a control group (CG, N = 10). Animals in the treatment group were submitted to a topical dental application of the nisin–biogel (200 µg/mL) every 48 h, as mentioned in Table 1. The control group was composed of dogs that were not submitted to any treatment. Animals were kept in the trial for 90 days. At the end of the clinical trial (T90), all animals were subjected to a new complete blood analysis, complete periodontal evaluation, dental plaque sampling, scaling, and dental polishing.

Table 1. Posology of the nisin–biogel (200 μg/mL) according to the animal's weight.

Animal's Weight	Volume of Nisin–Biogel
<20 kg	2 mL
20–40 kg	3 mL
>40 kg	4 mL

In addition, two intermediate dental plaque evaluations of all animals were performed on days 30 and 60. Intermediate evaluations included an awake dental plaque evaluation using a disclosure solution (GC Tri Plaque ID Gel®, Tokyo, Japan) that dyed the dental plaque according to its accumulation. Then, a photographic register of the vestibular margins was performed on all animals. After that, the dental plaque coverage of 4 teeth (2 canines and 2 premolars) of each dog was evaluated using the IMAGEJ® program. The percentage of dental plaque coverage was determined after measuring the total tooth vestibular margin and the area of the dyed dental plaque.

Animals in both groups were fed the same dry food and housed in the same building. All animals were observed daily to detect any general side effects, such as changes in feeding habits and behavior, prostration, vomiting, or diarrhea.

2.4. Periodontal Evaluation

The periodontal examination was blinded and performed by a trained veterinarian. The clinician determined the dental plaque and gingivitis scoring and performed six measures of the distance between the gingival margin and the bottom of the periodontal pocket (periodontal pocket depth—PPD) of each tooth (three measures were performed in the vestibular face and the other three in the palatine face) [16,17]. Every tooth in each animal was evaluated. The indexes used are presented in Tables 2 and 3.

Table 2. Dental plaque index (DPI) based on Holmstrom et al. [16].

Score	Description
1	No plaque on the dental surface
2	Thin film of plaque at gingival margin detectable with probing
3	Moderate amount of plaque at gingival margin, plaque is directly visible
4	High abundance of dental plaque accumulation in the gingival margin and/or dental surface, including interdental space

Table 3. Gingivitis index (GI) according to the modified Talbott method [17].

Score	Description
0	Normal gingiva; no inflammation, discoloration, or bleeding
1	Mild inflammation, slight color change, mild alteration of gingival surface, no bleeding upon probing
2	Moderate inflammation, erythema, swelling, or bleeding upon probing or when pressure applied
3	Severe inflammation, severe erythema and swelling, tendency toward spontaneous hemorrhage, some ulceration

2.5. Dental Plaque Sample Processing

A dental plaque sample was collected from all dogs on day 0 and day 90 using a swab (AMIES, VWR, Amadora, Portugal), which was applied to the entire dental surface. Swabs were transported to the Laboratory of Microbiology and Immunology, Faculty of Veterinary Medicine, University of Lisbon, and processed for total bacterial quantification, according to Belo et al. (2018) [18]. Briefly, the collected swabs were placed in test tubes with 1 mL of sterile saline and vortexed, and the resulting suspension was diluted (10^{-1} to 10^{-8}). From each dilution, 100 µL were collected and inoculated on Brain Heart Infusion agar (VWR, Amadora, Portugal) and incubated at 37 °C for 48 h. Afterward, bacterial quantification was performed by determining the colony-forming units [18].

2.6. Statistical Analysis

Statistical analysis of the data was carried out using RStudio® software version 1.1.383 (Boston, MA, USA) and Microsoft Excel 2016® (Redmond, WA, USA). Variables were evaluated by plotting the data into a histogram to confirm that they followed a normal distribution. Linear mixed models were used to evaluate differences in the dependent variables: mean PPD, GI, and DPI. In the mixed model, the variables group, timepoint, and tooth type (superior or inferior, incisive, canine, pre-molar, and molar) were defined as fixed effects, and the variables animal, weight, and tooth number were considered random effects. The interaction between fixed effects was investigated, and the Akaike Information Criterion was used to select the model.

Quantitative variables were expressed as mean values ± standard deviation. A confidence interval of 95% was considered in this study, with a p-value ≤ 0.05 indicating statistical significance.

3. Results

A total of 20 dogs were selected to participate in this clinical trial. The animals' gender distribution and mean age and weight are presented in Table 4. The results of the hemogram and biochemical parameters of all animals included in the trial agreed with the reference values established for healthy dogs at both timepoints (0 and 90) (Supplementary File S1).

Table 4. Gender, age, and weight distribution by group of animals in this study.

Global Gender Distribution (Female/Male)	8 F/12 M
Treatment group	4 F/6 M
Control group	4 F/6 M
Global age distribution (years) (mean values ± SD)	5.25 ± 1.69
Treatment group (TG)	6.1 ± 1.52
Control group (CG)	4.4 ± 1.58
Global weight distribution (Kg) (mean values ± SD)	23.34 ± 6.30
Treatment group (TG)	25.4 ± 6.59
Control group (CG)	21.35 ± 6.04

F—Female; M—Male; SD—Standard deviation.

3.1. Periodontal Evaluation

Complete periodontal evaluations of every tooth of each dog were performed at timepoints 0 and 90. The mean results for each dog are presented in Table 5. At timepoint 0, after periodontal assessment, all dogs were submitted to scaling and dental polishing before proceeding to the clinical trial.

Statistical analysis with mixed models showed a significant reduction in mean PPD (estimate = −0.371, p-value < 0.001) for all tooth types evaluated (incisive, canine, molar, and pre-molar) in the animals in the treatment group at T90. For the DPI, a statistically significant reduction (estimate = −0.145, p-value < 0.05) was observed in the animals in the treatment group at T90, with a marked reduction (p-value < 0.001) in incisive teeth. In addition, a reduction in the GI was observed in the animals in the treatment group at T90, but it was not statistically significant (estimate = −0.056, p-value > 0.05). For the GI, it was possible to observe a significant reduction (p-value < 0.05) in the mandibular incisive teeth and the maxillary molars and pre-molars.

Table 5. Mean gingivitis, dental plaque, and periodontal pocket depth of each dog by timepoint.

Animal ID	Group	Mean GI		Mean DPI		Mean Palatine PPD		Mean Vestibular PPD		Total Mean PPD	
		T0	**T90**	**T0**	**T90**	**T0**	**T90**	**T0**	**T90**	**T0**	**T90**
1	TG	1.93	0.45	3.71	2.21	1.33	1.06	1.63	1.08	1.48	1.07
3	TG	2.32	0.14	2.89	1.57	1.68	1.56	2.40	1.62	2.04	1.58
9	TG	2.33	0.64	2.76	1.83	1.25	1.22	1.35	1.41	1.30	1.32
10	TG	1.83	1.59	3.40	2.10	1.99	1.36	2.25	2.10	2.12	1.67
11	TG	2.19	1.64	3.86	1.79	1.68	1.42	1.98	1.65	1.83	1.54
12	TG	1.95	0.49	2.98	2.02	1.27	1.17	1.36	1.31	1.31	1.24
13	TG	2.17	0.12	2.67	1.98	1.28	1.15	1.29	1.14	1.29	1.14
14	TG	2.63	0.00	2.83	1.19	1.75	1.08	2.12	1.05	1.93	1.06
18	CG	2.31	1.29	2.98	1.83	1.41	1.33	1.41	1.29	1.41	1.31
19	CG	2.79	0.71	3.38	2.14	1.15	1.10	1.15	1.17	1.15	1.14
20	CG	2.33	1.19	3.19	2.21	1.31	1.27	1.30	1.18	1.31	1.23
22	TG	2.21	0.71	3.24	1.26	1.45	1.17	1.57	1.29	1.51	1.23
23	TG	2.41	1.50	3.02	2.14	1.76	1.21	2.44	1.34	2.10	1.27
27	CG	1.38	0.10	2.14	1.21	1.13	1.14	1.06	1.07	1.09	1.11
28	CG	2.19	0.43	3.14	1.85	1.06	1.12	1.10	1.14	1.08	1.13
31	CG	2.05	0.10	2.88	1.36	1.21	1.15	1.20	1.07	1.20	1.11
33	CG	1.95	0.21	2.31	1.57	1.28	1.35	1.25	1.29	1.26	1.29
34	CG	2.33	1.18	3.36	2.10	2.17	2.20	2.44	1.87	2.31	2.03
35	CG	1.26	1.40	2.93	1.90	1.29	1.24	1.42	2.01	1.35	1.62
37	CG	2.24	0.12	2.68	1.57	1.04	1.22	1.20	1.25	1.12	1.23
Mean ± SD		**2.14** ± 0.61	**0.7** ± 0.85	**3.02** ± 0.97	**1.82** ± 0.83	**1.43** ± 0.72	**1.27** ± 0.52	**1.59** ± 0.97	**1.36** ± 0.66	**1.51** ± 0.86	**1.32** ± 0.59
					1.84 ± 0.83						**1.33** ± 0.61
Mean T ± SD		**2.16** ± 0.53	**0.63** ± 0.81	**3.14** ± 1	*	**1.53** ± 0.79	**1.25** ± 0.51	**1.79** ±1.08	**1.41** ± 0.69	**1.66** ± 0.96	*
Mean C ± SD		**2.12** ± 0.65	**0.74** ± 0.86	**2.94** ± 0.94	**1.8** ± 0.82	**1.36** ± 0.67	**1.29** ± 0.53	**1.46** ± 0.86	**1.32** ± 0.63	**1.41** ± 0.77	**1.31** ± 0.58

ID—identification; SD—standard deviation; TG—treatment group; CG—control group; T0—timepoint 0; T90—timepoint 90; *—statistical significance (p-value < 0.05).

Intermediate evaluation of dental plaque coverage performed on days 30 and 60 using a disclosure solution revealed that the animals in the treatment group presented a mean dental plaque coverage of 24.58% and 33.08%, respectively. On the other hand, the dogs in the control group presented a mean dental plaque coverage of 42.37% on day 30 and 51.61% on day 60.

The results of the intermediate dental plaque coverage obtained by IMAGEJ are presented in Table 6.

Table 6. Mean dental plaque coverage (%) of each dog, determined after application of a disclosure solution and IMAGEJ analysis, on days 30 and 60.

Animal ID	Group	Dental Plaque Coverage (%)	
		30 Days	60 Days
1	TG	16.73	17.73
3	TG	16.82	22.52
9	TG	27.83	28.31
10	TG	26.87	32.91
11	TG	22.74	32.31
12	TG	24.54	36.99
13	TG	32.92	45.44
14	TG	13.65	28.17
18	CG	53.25	44.83
19	CG	62.16	77.03
20	CG	61.36	78.15
22	TG	38.74	39.68
23	TG	18.74	46.76
27	CG	16.88	22.65
28	CG	44.11	43.94
31	CG	40.29	36.79
33	CG	33.95	57.85
34	CG	48.11	64.12
35	CG	29.02	35.22
37	CG	34.63	55.53
Mean ± SD		33.38 ± 14.73	42.35 ± 17.02
Mean T ± SD		24.58 ± 7.89	33.08 ± 9.39
Mean C ± SD		42.37 ± 14.41	51.61 ± 18.24

ID—identification; SD—standard deviation; TG—treatment group; CG—control group.

3.2. *Dental Plaque Sample Processing*

Total oral aerobic bacteria counts were performed on days 0 and 90 using the dental plaque swab samples obtained from each animal. At T0, the samples revealed a mean total count of 8.6×10^7 CFU/mL, with those from the treatment group presenting 6.9×10^7 CFU/mL and those from the control group presenting 1×10^8 CFU/mL. At T90, the mean total counts were 7.4×10^7 CFU/mL, with the samples from the treatment group presenting 8.3×10^7 CFU/mL and those from the control group presenting 6.6×10^7 CFU/mL.

4. Discussion

Periodontal disease (PD) is well established as one of the most common oral inflammatory diseases in dogs, with 80% of these animals presenting some degree of PD by two years of age [2,19]. Its high prevalence, along with its potential local and systemic consequences, reinforces the need to improve PD control measures [6–9]. Nisin–biogel has been shown to be a promising compound against canine oral biofilms in vitro [10,11], with the ability to act on periodontopathogens present in the canine oral microbiome [20]. In this study, an in vivo clinical trial was performed to assess the efficacy of nisin–biogel dental topical application in dogs. A total of twenty animals were included in the trial after health parameter assessment through blood analysis to confirm the animals' health.

During the study, all animals were observed daily to identify any adverse clinical signs or behavior deviations. Previous reports have described an absence of cytotoxicity of nisin at 200 μg/mL [12,13,21,22], which was confirmed by our study, in which none of the animals showed side effects during the 3-month trial. Moreover, a blood analysis was performed in all dogs at the end of the study, and it was observed that none presented deviations in the evaluated parameters.

In addition, oral samples were collected from all animals at T0 and T90 to evaluate the influence of the nisin–biogel on oral aerobic bacteria through total bacterial counts. No reduction in total bacteria counts was observed in the samples collected after the application of the nisin–biogel for 3 months, which is in agreement with previous reports that have suggested resilient behavior of dental plaque and oral microbiota after professional dental cleaning [15,23]. In addition, Cunha and collaborators (2021), who studied the influence of the dental application of the nisin–biogel in the dynamics of the oral microbiome of dogs, detected an increase in bacterial diversity after one week of application of this compound [20]. Considering these studies, the oral bacteria population seems to suffer a rearrangement after antimicrobial compound administration or physical aggression, maintaining microbiota dynamics and relative concentrations [20,23].

One of the main goals of this study was to evaluate the in vivo efficacy of nisin–biogel for PD control using periodontal measures, such as DPI, GI, and PPD, and it was observed that the nisin–biogel had the ability to reduce dental plaque accumulation. This reduction was particularly evident after 1 month of dental topical application, which resulted in 24.58% dental plaque coverage in the treatment group and 42.37% in the control group. Similar results were obtained by Howell et al. (1993), who applied a nisin mouth rinse in dogs for 88 days. In that study, researchers observed a 34–38% reduction in the dental plaque index in the animals submitted to nisin application [24]. Dental plaque is one of the key factors in PD onset, being responsible for the initial aggression to the periodontium [1,4]. Considering this, dental plaque reduction is an essential step for PD control [5]. At the end of our trial, a significant reduction in DPI was observed in the animals in the treatment group in comparison with those in the control group, reinforcing the potential of the nisin–biogel for PD management. This reduction may be caused by the direct action of nisin–biogel on the inner layers of the biofilm. Previous reports have shown that this compound can not only inhibit biofilm formation but also eradicate mono and polymicrobial oral biofilms [10,11]. It has been suggested that nisin's biochemical structure and mechanism of action contribute to its high activity against biofilms [25]. Specifically, nisin can penetrate these structures without being neutralized or bound by the biofilm cells or matrix and is able to target the extracellular polymeric substances of the surrounding matrix [26,27].

The other periodontal indicator evaluated was the gingivitis index. PD can be classified into four different stages, the first of which is characterized by gingivitis [4]. This is a reversible stage in which most periodontal structures are undamaged, with animals showing inflammation and swelling of the gingiva with or without bleeding upon probing [1,4]. At the end of our clinical trial, it was possible to observe that the application of the nisin–biogel induced a slight reduction in GI. Howell and collaborators (1993) also observed a reduction in GI after nisin oral application for 88 days, which may be related to the potential immunomodulatory effect of this antimicrobial peptide [24]. In fact, besides having antimicrobial activity against several periodontopathogens, nisin presents immunomodulatory and hound healing abilities [12,28]. It acts by binding to Lipid II, present in the bacterial cell membrane, and by interfering with cell wall biosynthesis, leading to bacterial death [10]. In addition, it has been shown that nisin promotes a reduction in pro-inflammatory cytokines, which contributes to PD progression, and presents some effects on immune cells [28–30]. In our trial, a significant PPD decrease was observed in the animals from the treatment group at T90, which may be directly related to dental plaque and gingivitis reduction. It is known that gingiva swelling can increase pocket depth, so the reduction of gingivitis after nisin–biogel therapy may have decreased local inflammation, leading to gingival shrinkage and reducing PPD [4].

It is important to note that the animals in the control group also exhibited a slight reduction in GI and PPD at T90, with no statistical significance, which was potentially caused by the scaling and dental polishing performed at the beginning of the study. In fact, dental scaling, polishing, irrigation, and home dental care are the standard procedures for the treatment of PD stage one cases (gingivitis), so a reduction in this periodontal parameter was expected. Nevertheless, the use of nisin–biogel promoted a higher and statistically significant decrease of the DPI and PPD and a reduction in the GI compared with the results from the control group. Moreover, nisin–biogel was well accepted by the animals, showing no adverse effects over a 3-month period of application.

This study has some limitations, such as the number of animals included, and a larger in vivo study would help to validate the nisin–biogel as a potential future compound to be applied by clinicians. Yet, the fact that it was necessary to use dogs with similar housing conditions, which needed to be maintained over a long period of time, prevented the inclusion of more animals in the trial. In addition, although the oral cavity is a complex and diverse environment [11,20], in our study we focused on supragingival dental plaque and, consequently, on the aerobic microbiota. However, anaerobic bacteria are also present in dental plaque and should be evaluated in a future trial [11,20]. Finally, as PD establishment is a multifactorial process [31,32] influenced by the oral microbiome, the host immune system and the local oral environment [1,4,33], in upcoming studies the influence of the nisin–biogel in the oral immune-inflammatory response could be a relevant issue to explore [33].

Considering the high prevalence and impact of canine PD, several other natural compounds are being studied to evaluate their potential for the control of this disease. Promising strategies include essential oils, alcoholic herbal products, and algae [34,35]. Most of these compounds have already been evaluated in vitro regarding their antimicrobial abilities. Moreover, in vivo studies have already shown the efficacy of oral products, including *Calendula officinalis* or the brown alga *Ascophyllum nodosum*, for PD control [34,35].

The nisin–biogel seems to be a promising compound for veterinary dentistry and human medicine [31,32]. Reports have described the potential application of this compound in human odontology and other areas, including for the treatment of diabetic foot infections [36], reinforcing the versatility of the nisin–biogel use in the biomedical field.

5. Conclusions

Periodontal disease is an inflammatory disease that is highly prevalent in dogs. The development of effective measures to control this disease is essential. Previous in vitro studies have revealed that the nisin–biogel is a safe antibiofilm agent with the potential to be used in canine PD control. Our study aimed to evaluate the in vivo efficacy of the nisin–biogel in dogs. Nisin–biogel showed the ability to reduce the DPI, GI, and PPD without any adverse effects on the animals in the study. These results suggest that the topical dental application of nisin–biogel may be used as an adjuvant measure for the prevention and control of canine PD. In the future, a large in vivo trial would be useful to fully validate this compound for commercial use. In addition, considering the high similarity between dog and human PD [31,32], nisin–biogel could also be a valuable compound to use in human dentistry.

Supplementary Materials: The following supporting information can be downloaded at: https://www.mdpi.com/article/10.3390/pharmaceutics14122716/s1, Supplementary File S1: Data blood analysis.

Author Contributions: Conceptualization, E.C, L.M.C. and M.O.; methodology, E.C., L.M.C. and M.V.; software, T.N. and E.C.; validation, E.C. and M.O.; formal analysis, E.C.; investigation, E.C., L.M.C., M.V. and A.S.V.; resources, L.T. and M.O.; data curation, M.O.; writing—original draft preparation E.C.; writing—review and editing, L.T. and M.O.; visualization, L.M.C. and M.O.; supervision, A.S.V. and M.O.; project administration, M.O.; funding acquisition, L.T. and M.O. All authors have read and agreed to the published version of the manuscript.

Funding: The authors would like to acknowledge the Foundation for Science and Technology (Eva Cunha Ph.D. fellowship SFRH/BD/131384/2017); CIISA—Centro de Investigação Interdisciplinar em Sanidade Animal, Faculdade de Medicina Veterinária, Universidade de Lisboa, Project UIDB/00276/2020 (funded by FCT); and Laboratório Associado para Ciência Animal e Veterinária (LA/P/0059/2020—AL4AnimalS) for the financial support of this work. The funding bodies did not participate in the design, analysis, or report of this study.

Institutional Review Board Statement: All experimental procedures were approved by the Ethical Committee for Research and Teaching (CEIE) of the Faculty of Veterinary Medicine, University of Lisbon, Portugal (N/Ref 014/2020). This study complied with relevant institutional, national, and ARRIVE guidelines for the care and use of animals. All experiments were performed in accordance with relevant institutional, national, and ARRIVE guidelines for the care and use of animals. Written informed consent was obtained from the institution responsible for the animals.

Informed Consent Statement: Not applicable.

Data Availability Statement: The datasets used and/or analyzed in the current study are available from the corresponding author upon reasonable request.

Acknowledgments: The authors would like to acknowledge all technical workers at the Casa dos Animais de Lisboa and the Centro de Medicina Veterinária Anjos de Assis for their support in this study.

Conflicts of Interest: The authors declare no conflict of interest.

References

1. Niemiec, B.A. Periodontal disease. *Top. Companion Anim. Med.* **2008**, *23*, 72–80. [CrossRef] [PubMed]
2. Stella, J.L.; Bauer, A.E.; Croney, C.C. A cross-sectional study to estimate prevalence of periodontal disease in a population of dogs (*Canis familiaris*) in commercial breeding facilities in Indiana and Illinois. *PLoS ONE* **2018**, *13*, e0191395. [CrossRef] [PubMed]
3. Wallis, C.; Holcombe, L.J. A review of the frequency and impact of periodontal disease in dogs. *J. Small Anim. Pract.* **2020**, *61*, 529–540. [CrossRef] [PubMed]
4. Stepaniuk, K. Periodontology. In *Wiggs's Veterinary Dentistry*; Lobprise, H.B., Dodd, J.R., Eds.; John Wiley & Sons Inc.: Hoboken, NJ, USA, 2019; pp. 81–108.
5. Bellows, J.; Berg, M.L.; Dennis, S.; Harvey, R.; Lobprise, H.B.; Snyder, C.J.; Stone, A.E.S.; Van de Wetering, A.G. 2019 AAHA Dental care guidelines for dogs and cats. *J. Am. Anim. Hosp. Assoc.* **2019**, *55*, 49–69. [CrossRef] [PubMed]
6. Pavlica, Z.; Petelin, M.; Juntes, P.; Erzen, D.; Crossley, D.; Skaleric, U. Periodontal disease burden and pathologic changes in organs of dogs. *J. Vet. Dent.* **2008**, *25*, 97–105. [CrossRef] [PubMed]
7. Glickman, L.T.; Glickman, N.W.; Moore, G.E.; Goldstein, G.S.; Lewis, H.B. Evaluation of the risk of endocarditis and other cardiovascular events on the basis of the severity of periodontal disease in dogs. *J. Am. Vet. Med. Assoc.* **2009**, *234*, 486–494. [CrossRef]
8. Glickman, L.T.; Glickman, N.W.; Moore, G.E.; Lund, E.M.; Lantz, G.C.; Pressler, B.M. Association between chronic azotemic kidney disease and the severity of periodontal disease in dogs. *Prev. Vet. Med.* **2011**, *99*, 193–200. [CrossRef]
9. Pereira Dos Santos, J.D.; Cunha, E.; Nunes, T.; Tavares, L.; Oliveira, M. Relation between periodontal disease and systemic diseases in dogs. *Res. Vet. Sci.* **2019**, *125*, 136–140. [CrossRef]
10. Cunha, E.; Trovão, T.; Pinheiro, A.; Nunes, T.; Santos, R.; Moreira Da Silva, J.; São Braz, B.; Tavares, L.; Veiga, A.S.; Oliveira, M. Potential of two delivery systems for nisin topical application to dental plaque biofilms in dogs. *BMC Vet. Res.* **2018**, *14*, 1–10. [CrossRef]
11. Cunha, E.; Rebelo, S.; Carneiro, C.; Tavares, L.; Carreira, L.M.; Oliveira, M. A polymicrobial biofilm model for testing the antimicrobial potential of a nisin-biogel for canine periodontal disease control. *BMC Vet. Res.* **2020**, *16*, 469. [CrossRef]
12. Shin, J.M.; Ateia, I.; Paulus, J.R.; Liu, H.; Fenno, J.C.; Rickard, A.H.; Kapila, Y.L. Antimicrobial nisin acts against saliva derived multi-species biofilms without cytotoxicity to human oral cells. *Front. Microbiol.* **2015**, *6*, 617. [CrossRef] [PubMed]
13. Cunha, E.; Freitas, F.B.; São Braz, B.; Moreira Da Silva, J.; Tavares, L.; Veiga, A.S.; Oliveira, M. Polyphasic validation of a nisin-biogel to control canine periodontal disease. *Antibiotics* **2020**, *9*, 180. [CrossRef] [PubMed]
14. Santos, R.; Gomes, D.; Macedo, H.; Barros, D.; Tibério, C.; Veiga, A.S.; Tavares, L.; Castanho, M.; Oliveira, M. Guar gum as a new antimicrobial peptide delivery system against diabetic foot ulcers *Staphylococcus aureus* isolates. *J. Med. Microbiol.* **2016**, *65*, 1092–1099. [CrossRef] [PubMed]
15. Ramsey, I. *BSAVA Small Animal Formulary*, 8th ed.; BSAVA Publications: Gloucester, UK, 2014; ISBN 9781905319657.
16. Holmstrom, S.E.; Frost, P.; Eisner, E.R. Dental Records. In *Veterinary Dental Techniques for the Small Animal Practitioner*, 3rd ed.; Holmstrom, S.E., Frost, P., Eisner, E.R., Eds.; WB Saunders: Philadelphia, PA, USA, 2004; p. 20.
17. Sitzman, C. Evaluation of a hydrophilic gingival dental sealant in beagle dogs. *J. Vet. Dent.* **2013**, *30*, 150–155. [CrossRef] [PubMed]

18. Belo, L.; Serrano, I.; Cunha, E.; Carneiro, C.; Tavares, L.; Carreira, L.M.; Oliveira, M. Skin asepsis protocols as a preventive measure of surgical site infections in dogs: Chlorhexidine-alcohol versus povidone-iodine. *BMC Vet. Res.* **2018**, *14*, 95. [CrossRef]
19. Marshall, M.; Wallis, C.; Milella, L.; Colyer, A.; Tweedie, A.; Harris, S. A longitudinal assessment of periodontal disease in 52 miniature schnauzers. *BMC Vet. Res.* **2014**, *10*, 1–13. [CrossRef]
20. Cunha, E.; Valente, S.; Nascimento, M.; Pereira, M.; Tavares, L.; Dias, R.; Oliveira, M. Influence of the dental topical application of a nisin-biogel in the oral microbiome of dogs: A pilot study. *PeerJ* **2021**, *9*, e11626. [CrossRef]
21. Murinda, S.E.; Rashid, K.A.; Roberts, R.F. In vitro assessment of the cytotoxicity of nisin, pediocin, and selected colicins on simian virus 40–transfected human colon and Vero monkey kidney cells with trypan blue staining viability assays. *J. Food Prot.* **2003**, *66*, 847–853. [CrossRef]
22. Dreyer, L.; Smith, C.; Deane, S.M.; Dicks, L.M.T.; van Staden, A.D. Migration of bacteriocins across gastrointestinal epithelial and vascular endothelial cells, as determined using in vitro simulations. *Sci. Rep.* **2019**, *9*, 11481. [CrossRef]
23. Flancman, R.; Singh, A.; Weese, J.S. Evaluation of the impact of dental prophylaxis on the oral microbiota of dogs. *PLoS ONE* **2018**, *13*, e0199676. [CrossRef]
24. Howell, T.H.; Fiorellini, J.P.; Blackburn, P.; Projan, S.J.; de la Harpe, J.; Williams, R.C. The effect of a mouthrinse based on nisin, a bacteriocin, on developing plaque and gingivitis in beagle dogs. *J. Clin. Periodontol.* **1993**, *20*, 335–339. [CrossRef] [PubMed]
25. Okuda, K.; Zendo, T.; Sugimoto, S.; Iwase, T.; Tajima, A.; Yamada, S.; Sonomoto, K.; Mizunoe, Y. Effects of bacteriocins on methicillin-resistant *Staphylococcus aureus* biofilm. *Antimicrob. Agents Chemother.* **2013**, *57*, 5572–5579. [CrossRef] [PubMed]
26. Davison, W.M.; Pitts, B.; Stewart, P.S. Spatial and temporal patterns of biocide action against *Staphylococcus epidermidis* biofilms. *Antimicrob. Agents Chemother.* **2010**, *54*, 2920–2927. [CrossRef] [PubMed]
27. Zhao, M.; Qu, Y.; Liu, J.; Mai, S.; Gu, L. A universal adhesive incorporating antimicrobial peptide nisin: Effects on *Streptococcus mutans* and saliva-derived multispecies biofilms. *Odontology* **2020**, *108*, 376–385. [CrossRef] [PubMed]
28. Mouritzen, M.V.; Andrea, A.; Qvist, K.; Poulsen, S.S.; Jenssen, H. Immunomodulatory potential of Nisin A with application in wound healing. *Wound Repair Regen.* **2019**, *27*, 650–660. [CrossRef]
29. Begde, D.; Bundale, S.; Mashitha, P.; Rudra, J.; Nashikkar, N.; Upadhyay, A. Immunomodulatory efficacy of nisin—A bacterial lantibiotic peptide. *J. Pept. Sci.* **2011**, *17*, 438–444. [CrossRef]
30. Małaczewska, J.; Kaczorek-Łukowska, E.; Wójcik, R.; Rękawek, W.; Siwicki, A.K. In vitro immunomodulatory effect of nisin on porcine leucocytes. *J. Anim. Physiol. Anim. Nutr. (Berl.)* **2019**, *103*, 882–893. [CrossRef]
31. Albuquerque, C.; Morinha, F.; Requicha, J.; Martins, T.; Dias, I.; Guedes-Pinto, H.; Bastos, E.; Viegas, C. Canine periodontitis: The dog as an important model for periodontal studies. *Vet. J.* **2012**, *191*, 299–305. [CrossRef]
32. Cunha, E.; Trovão, T.; Santos, R.; Santos, J.D.; Moreira da Silva, J.; São Braz, B.; Veiga, A.S.; Tavares, L.; Oliveira, M. Canine periodontal disease and its systemic implications—A review. *RPCV* **2017**, *112*, 12–22.
33. Teles, R.; Teles, F.; Frias-Lopez, J.; Paster, B.; Haffajee, A. Lessons learned and unlearned in periodontal microbiology. *Periodontology 2000* **2013**, *62*, 95–162. [CrossRef]
34. Diaz, M.A.N.; Carvalho, I.; Diaz, G. Herbal Dentifrices for Children. In *Emerging Trends in Oral Health Sciences and Dentistry*; Virdi, M.S., Ed.; IntechOpen: London, UK, 2015; Available online: https://www.intechopen.com/chapters/48160 (accessed on 20 October 2022). [CrossRef]
35. Gawor, J.; Jank, M.; Jodkowska, K.; Klim, E.; Svensson, U.K. Effects of Edible Treats Containing Ascophyllum nodosum on the Oral Health of Dogs: A Double-Blind, Randomized, Placebo-Controlled Single-Center Study. *Front. Vet. Sci.* **2018**, *5*, 168. [CrossRef] [PubMed]
36. Santos, R.; Ruza, D.; Cunha, E.; Tavares, L.; Oliveira, M. Diabetic foot infections: Application of a nisin-biogel to complement the activity of conventional antibiotics and antiseptics against *Staphylococcus aureus* biofilms. *PLoS ONE* **2019**, *14*, e0220000. [CrossRef] [PubMed]

pharmaceutics

MDPI

Article

Validation of a Dendritic Cell and CD4+ T Cell Restimulation Assay Contributing to the Immunogenicity Risk Evaluation of Biotherapeutics

Michel Siegel [1], Guido Steiner [1], Linnea C. Franssen [1], Francesca Carratu [2], James Herron [2], Katharina Hartman [1], Cary M. Looney [1], Axel Ducret [1], Katharine Bray-French [1], Olivier Rohr [3], Timothy P. Hickling [1], Noel Smith [2] and Céline Marban-Doran [1,*]

[1] Roche Pharmaceutical Research and Early Development, Pharmaceutical Sciences, Roche Innovation Center Basel, 4070 Basel, Switzerland
[2] Lonza Biologics, Chesterford Research Park, Saffron Walden CB10 1XL, UK
[3] UR 7292, IUT Louis Pasteur, Université de Strasbourg, 67300 Schiltigheim, France
[*] Correspondence: celine.marban-doran@roche.com

Citation: Siegel, M.; Steiner, G.; Franssen, L.C.; Carratu, F.; Herron, J.; Hartman, K.; Looney, C.M.; Ducret, A.; Bray-French, K.; Rohr, O.; et al. Validation of a Dendritic Cell and CD4+ T Cell Restimulation Assay Contributing to the Immunogenicity Risk Evaluation of Biotherapeutics. *Pharmaceutics* **2022**, *14*, 2672. https://doi.org/10.3390/pharmaceutics14122672

Academic Editors: Scavello Francesco, Jean-Eric Ghia and Amiche Mohamed

Received: 31 October 2022
Accepted: 25 November 2022
Published: 1 December 2022

Abstract: Immunogenicity, defined as the ability to provoke an immune response, can be either wanted (i.e., vaccines) or unwanted. The latter refers to an immune response to protein or peptide therapeutics, characterized by the production of anti-drug antibodies, which may affect the efficacy and/or the safety profiles of these drugs. Consequently, evaluation of the risk of immunogenicity early in the development of biotherapeutics is of critical importance for defining their efficacy and safety profiles. Here, we describe and validate a fit-for-purpose FluoroSpot-based in vitro assay for the evaluation of drug-specific T cell responses. A panel of 24 biotherapeutics with a wide range of clinical anti-drug antibody response rates were tested in this assay. We demonstrated that using suitable cutoffs and donor cohort sizes, this assay could identify most of the compounds with high clinical immunogenicity rates (71% and 78% for sensitivity and specificity, respectively) while we characterized the main sources of assay variability. Overall, these data indicate that the dendritic cell and CD4+ T cell restimulation assay published herein could be a valuable tool to assess the risk of drug-specific T cell responses and contribute to the selection of clinical candidates in early development.

Keywords: immunogenicity; immunomodulation; biotherapeutics; in vitro T cell assay; assay validation

1. Introduction

Despite success in the clinic, a substantial number of biotherapeutics elicit unwanted immune or immunogenic responses—termed immunogenicity. One of the hallmarks of immunogenicity is the onset of anti-drug antibodies (ADAs). Due to ADAs exhibiting major consequences for both patient's safety and treatment efficacy, it is of utmost importance to assess this risk as early as possible during drug development [1,2].

Partially or fully humanized biotherapeutics (i.e., antibodies with minimal non-germline amino acid sequences) are usually at lesser risk of an unwanted immunogenicity response; however, this measure may not completely abrogate ADA formation. It is now established that a compound immunogenicity risk assessment must include multiple complex factors ranging from product-related risks, such as protein structure, formulation, or impurities [3]; patient and disease-related factors, including genetic factors, age, concomitant treatment; and route of administration [4]. In the case of immunomodulatory drugs, adverse events may also be caused by target binding in healthy tissues, or enhanced pharmacology attenuating the activity of target molecules on cells.

Consequently, an integrated preclinical risk assessment should be considered as a key element in biotherapeutics development. Regulatory bodies, such as the European

Medicines Agency (EMA) and the Food and Drug Administration (FDA), are now encouraging drug developers to consider risk factors related to the product and to the patient, mentioned above, as early as possible in the development process. An integrated approach relies on the use of specific tools and methods to identify relevant immunogenicity factors and to develop corresponding risk mitigation strategies [5]. Currently, these tools include in silico screening algorithms to scan for sequence liabilities, in vitro cell-based assays to measure various readouts from the immune response (dendritic cell internalization, activation and presentation, T cell activation), and the use of transgenic animal models designed to study the intimate mechanisms of an immune response from a mechanistic viewpoint [6]. However, most of these tools have not undergone a formal qualification process, and factors contributing to assay variability are not always understood. For example, T cell-dependent responses are the major drivers for immunogenicity, and in vitro T cell assays are frequently used to identify and measure CD4+ T cell responses to biotherapeutics. These assays have been derived in different formats and reviewed elsewhere [6–8]. However, the sensitivity of these assays is usually quite low, as the size of the pre-existing CD4+ T cell repertoire reactive to the drug is very small, ranging from 1 to 10 cells out of 10^8 T cells [9].

Here, we describe and characterize a dendritic cell and CD4+ T cell restimulation assay and discuss the potential of such an assay to assess a CD4+ T cell-driven immunogenicity risk. This assay consists of a co-culture between monocyte-derived dendritic cells (moDCs) and autologous CD4+ T cells, including a re-stimulation step to increase assay sensitivity. The main goals of this study were to establish an assay threshold to distinguish between positive and negative responses, to determine the optimal cohort size for the assay, and to identify factors affecting assay variability. We are currently using this assay as part of an integrated approach to rank candidate biotherapeutics during the initial selection process, enabling the selection of lower-risk clinical leads for subsequent large-scale production and clinical trials.

2. Materials and Methods

2.1. Compounds

Stock solutions of keyhole limpet hemocyanin (KLH-Imject Maleimide-Activated mcKLH, Thermo Fisher Scientific, Basel, Switzerland, #77600) were reconstituted and stored at −80 °C in single-use aliquots according to the manufacturer's recommendations under sterile conditions. All biotherapeutics were bought from Runge Pharma GmbH & Co (Lörrach, Germany) in their respective formulation and stored according to the manufacturer's recommendations. Peptides were synthesized by Cambridge Research Biochemicals and reconstituted in sterile ultra-pure water (Invitrogen, Basel, Switzerland, #10977015) and 50% Acetonitrile (≥99.95%, VWR, #83639.320). Biotherapeutics were used at a final concentration of 0.3 µM (peptides were used at a final concentration of 10 µg/mL) for both the DC stimulation stage and re-stimulation stage.

2.2. Healthy Donor Cohort

Healthy donors were recruited at Phase I clinical trial units in the UK. All samples were collected under an ethical protocol approved by a local Research Ethics Committee (reference number: 21/LO/0474), and written informed consent was obtained from each donor prior to sample donation. All samples were stored according to the terms of Lonza's Human Tissue Authority license for the use of samples in research. Peripheral blood mononuclear cells (PBMC) from healthy donors were prepared from whole blood or leukopaks using Lymphorep density gradient medium (Cedarlane, # CL5120) within six hours of blood withdrawal. PBMC were controlled-rate frozen and stored in vapor-phase nitrogen at −196 °C until used in the assays. The quality and functionality of each PBMC preparation were analyzed after seven days of activation, with positive controls such as KLH to assess naïve T cell responses. For each screen, the donor cohorts consisted of typically 30 donors selected to represent the world population in terms of their HLA-DRB1 allele frequency distribution [5] (Supplementary Figure S1).

2.3. DC:CD4+ Re-Stimulation Assay (Epibase® IV, Lonza)

Monocytes were isolated from frozen PBMC samples by magnetic bead selection using CD14 microbeads (Miltenyi Biotec # 130-050-201 on an AutoMACS Pro system) and differentiated into immature DC (iDC) using 1000 IU/mL of granulocyte-macrophage colony-stimulating factor (GM-CSF) and 1000 IU/mL of IL-4 in a serum-free mediun (CellGenix # 20805-0500, supplemented with 0.05 mg/mL Gentamicin Lonza # 17-518L) for 5 days at 37 °C, 5% CO_2. iDC were then harvested, washed and loaded with each test protein/peptide individually for 4 h at 37 °C, 5% CO_2. A DC maturation cocktail containing TNFα (800 IU/mL) and IL-1β (100 IU/mL) was then added for a further 40–42 h to activate/mature the DC (mDC). The expression of key DC surface markers (CD11c-3.9, CD14-63D3, CD40-5C3, CD80-2D10, CD83-HB15E, CD86-BU63, CD209-9E9A8 and HLA-DR-L243) at both the immature and mature stage were assessed by flow cytometry (Bio-Rad ZE5 Cell Analyzer) to ensure the DC were activated prior to T cell interaction. After a thorough washing procedure, 100,000 mDCs were then co-cultured with 1 million autologous CD4+ T cells (isolated by magnetic bead selection, Miltenyi Biotec # 130-045-101 on an AutoMACS Pro system) in a deep-well plate (final volume of 1.2 mL, Greiner # 780271). The DC:CD4+ T cells ratio is 1:10 and the co-culture is incubated for 6 days at 37 °C, 5% CO_2 in a humidified atmosphere. On day 6, autologous monocytes were isolated from PBMC using magnetic bead selection (Miltenyi Biotec # 130-050-201 on an AutoMACS Pro system) and loaded with the selected protein or peptide that were initially used to load the DC. After incubation at 37 °C, 5% CO_2 in a humidified atmosphere for 4 h, the monocytes were washed and then added to anti-IFN-γ/anti-IL-5 pre-coated FluoroSpot plates (Mabtech # FSP-0108-10) along with the corresponding DC:CD4 co-culture in quadruplicate (25,000 monocytes: 250,000 CD4+ T cells in a final volume of 200 μL). The FluoroSpot plates were incubated for 40–42 h at 37 °C, 5% CO_2 in a humidified atmosphere. After incubation, the FluoroSpot plates were developed according to the manufacturer's procedure (IRIS FluoroSpot reader, Mabtech) and the number of spot-forming cells (SFC) per well were assessed for each test condition in an automated and unbiased manner.

2.4. Data Analysis

Data management and statistical analysis were performed in the R programming language (https://www.R-project.org/, accessed on 28 October 2022, versions 3.6.1 up to 4.1.2), including essential packages for handling generalized linear models (nlme, emmeans) and carrying out variance component analyses (VCA, version 1.4.3).

The calculation of Stimulation Indices (SI) was performed as follows. Spot forming cells (SFC) from the FluoroSpot assay were transformed to a log2 scale, and a generalized linear model (GLM) was applied to estimate the SI (i.e., the ratio between a treatment condition and the donor-matched blank on a linear signal scale) and associated confidence intervals. Quadruplicate SFC measurements were implicitly aggregated by the GLM to yield one SI value for each combination of a specific test compound, donor, and screen. The screens were analyzed sequentially and independently from each other, with the linear model considering a specific cytokine readout of an entire screen as input. The processing workflow was tailored to address a few peculiarities of the given data. Specifically, we used an exponential type of heteroscedasticity adjustment in the GLM to achieve scale-invariance of residuals and injected some Gaussian noise at the low end of the SFC scale to support model convergence with the frequent presence of ties of discrete values around zero. (The standard deviation of this normally distributed, zero-centered noise was chosen to correspond to the replicate variability inferred by the GLM in the limit of zero SFC counts at the low end of the SFC scale and drops down exponentially by a factor of $\exp(-2) = 0.14$ for every unit increase of the log2 SFC). Furthermore, we observed a consistent trend in the data to the effect that higher blank values of a donor corresponded to systematically lower SI values for that donor. The relation between 'pre-stimulation' of the blank and observed stimulation indices could be well captured by linear regressions performed for each treatment within a screen. We corrected the raw SI values then for every donor-treatment pair with the respective linear model, basically extrapolating to the value which would have been observed with a common blank value of 0.

Standard quality control plots were generated for every data set, including the visualization of DC differentiation markers, the reproducibility of reference compound data across studies, and (if possible) the variability of repeated compound testing with the same donor. We also looked at the individual stimulation profile of each donor within a study, as the overall inducibility of T-cell response could vary from person to person; simultaneously, this enabled us to rule out the presence of generally inert sample material. A donor response was recorded as "positive" if a SI fold-change of 2 or above (compared to its blank control) was measured at a statistical significance of $p < 0.05$ (using non-adjusted p-values from the GLM). The fraction of positive donor responses (within a cohort of typically 30 healthy donors per screen) provided the response rate for the treatment in a specific screen.

3. Results

3.1. DC:CD4+ T Cell Restimulation Assay Workflow

The general workflow of the assay is illustrated on Figure 1a. Test items were investigated in independent screens of the DC:CD4+ T cell restimulation assay over a time span of several years. Therefore, various controls were employed to ensure a consistent and comprehensive analysis of the data.

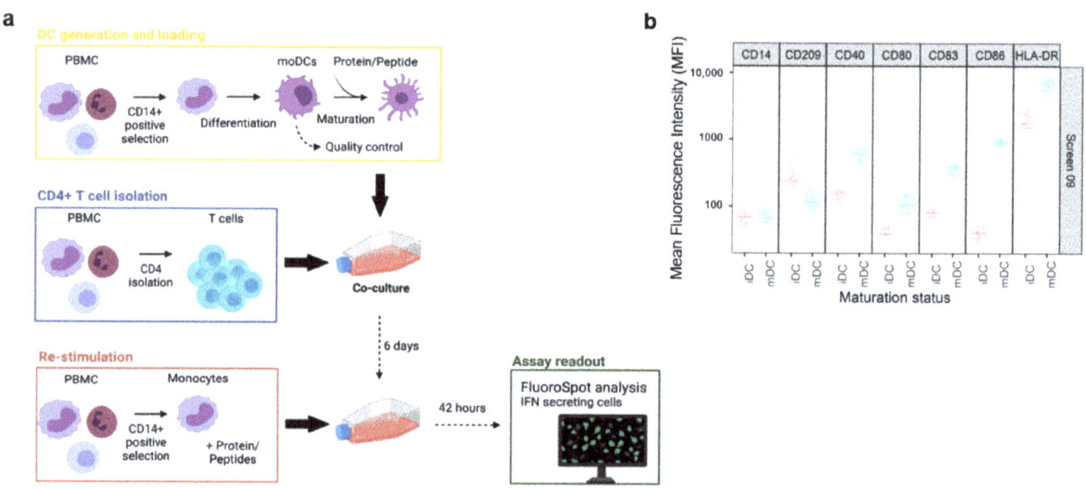

Figure 1. The DC:CD4+ T cell restimulation assay. (**a**) Experimental setup of the DC:CD4+ restimulation assay. The assay starts with the isolation of monocytes from healthy donor PBMCs, followed by the loading of the protein of interest and maturation of the monocyte derived Dendritic Cells (moDC). Autologous CD4+ T cells are isolated and co-cultured with the loaded moDCs. After an incubation of 6 days, freshly isolated monocytes are challenged with the same protein and added to the co-culture for an additional 42 h before analyzing the production of IFN-γ by FluoroSpot. (**b**) DC were characterized by the expression of the following cell surface markers: CD11c, CD14, CD80, CD83, CD86, CD209, and HLA-DR before and after DC activation by addition of TNF-α and IL-1β to ensure good cell fitness. Created with BioRender.com, accessed on 27 November 2022.

For each screen, 30 healthy donors were selected based on their HLA-DRB1 alleles to reflect the world population [5] (Supplementary Figure S1). In addition, a characterization of the dendritic cells (DCs) was included in every screen to assess the phenotype of these cells before and after maturation by flow cytometry. Activation of the DCs was determined by upregulation of key maturation markers on the cell surface that are known to be correlated with T-cell priming capacity: CD40, CD80, CD83, CD86, and HLA-DR [10]. Moreover, CD209, a pathogen-recognition receptor expressed on the surface of immature DCs, is internalized together with other markers, thus resulting in efficient presentation [11]. Accordingly, the downregulation of CD209 is the consequence of a shift from an immature to a mature DC phenotype. A representative distribution of cell surface marker expression at both the immature (iDC) and mature stage (mDC) is shown in Figure 1b. The addition of the DC maturation cocktail, composed of TNF-α and IL-1β, led to a slightly higher expression of CD40, CD80, CD83, and HLA-DR, but also a substantial increase in CD86 expression, resulting in a more than ten-fold increase in the average MFI for this surface marker. In addition, we also observed a moderate decrease in CD209 expression. Altogether, this analysis confirmed that DCs from all donors of the cohort have the potential to be activated prior to their interaction with autologous CD4+ T cells. Moreover, the assay is qualified for a given immunomodulatory protein by treating the DCs together with KLH to assess what impact the protein has on the KLH-induced T cell response. This enables us to highlight proteins that may influence the DC-induced activation of T cells.

3.2. DC:CD4+ T Cell Restimulation Assay Precision Assessment and Comparators

We investigated first the repeatability of the assay by testing the IFN-γ response of donors to KLH and Avastin (same production batch) in multiple assay screens. To this aim, we plotted the SI for KLH and Avastin for all donors, grouped by batches. All the donors analyzed over 24 screens consistently showed high SIs (with a geometric mean of 225 across all screens) upon treatment with KLH (a widely accepted positive control), while SIs obtained with treatment with bevacizumab were distributed around 1 (Figure 2a), suggesting that there was no substantial change in IFN-γ release compared to the blank. Moreover, very few donors (40/607, 6.6%) in this treatment group showed a two-fold SI change or above (our criterion for calling a positive response, see Section 2). Based on these findings, we recommend the use of bevacizumab as a negative comparator in this assay. We used KLH as the technical positive control in our analyses, as highly immunogenic biopharmaceuticals tend not to reach marketing authorization [12].

We used the DC:CD4+ T-cell restimulation assay to investigate 24 biotherapeutics developed by a range of pharmaceutical companies, comprising a broad range of drug formats and targets. Details about the molecules were extracted from the corresponding FDA label [13] and are summarized in Table 1.

However, for most of the labels, important data about the trial were missing, ultimately limiting the interpretability of the reported ADA rates. Moreover, for a number of trials, drugs were administered in combination with radiotherapy, which is known to impact the immune system and the subsequent production of ADA [14]. In other cases, biopharmaceuticals were administered with corticoid pre-treatment to dampen the immune response, which also influences the production of ADA. In this manuscript, data from combination trials were omitted, except for Alemtuzumab, Cetuximab, Daratumumab, Elotuzumab, Sarilumab, and Tocilizumab, which are always co-administered with other drugs.

Results are summarized in Figure 2b,c. Most of the tested biopharmaceuticals elicited low levels of IFN-γ release (alirocumab, avelumab, benralizumab, bevacizumab, brentuximab, certolizumab, cetuximab, durvalumab, evolocumab, galcanezumab, necitumumab, nivolumab, sarilumab, tocilizumab, ustekinumab, vedolizumab). However, we saw stronger T cell responses with alemtuzumab, elotuzumab, pembrolizumab, infliximab, and daratumumab, for which more than 10% of the donors showed a SI statistically significant above 2. Interestingly, antibodies with identical modes of action (i.e., infliximab, adalimumab, and certolizumab all target TNF-α) triggered different T cell responses with regards to IFN-γ production. In addition, when compounds were tested several times in different screens, we observed that SIs and the derived response rates showed a significant variability (Figure 2c). We observed that for adalimumab, for which screens 02 and 06 resulted in 23.3% and 26.7% of positive donors, respectively, whereas it dropped to 0% in screen 07. These discrepancies are seen for pembrolizumab, atezolizumab, and elotuzumab, as well. One explanation for this observation could be a compound batch effect, as illustrated for adalimumab in Figure 2d.

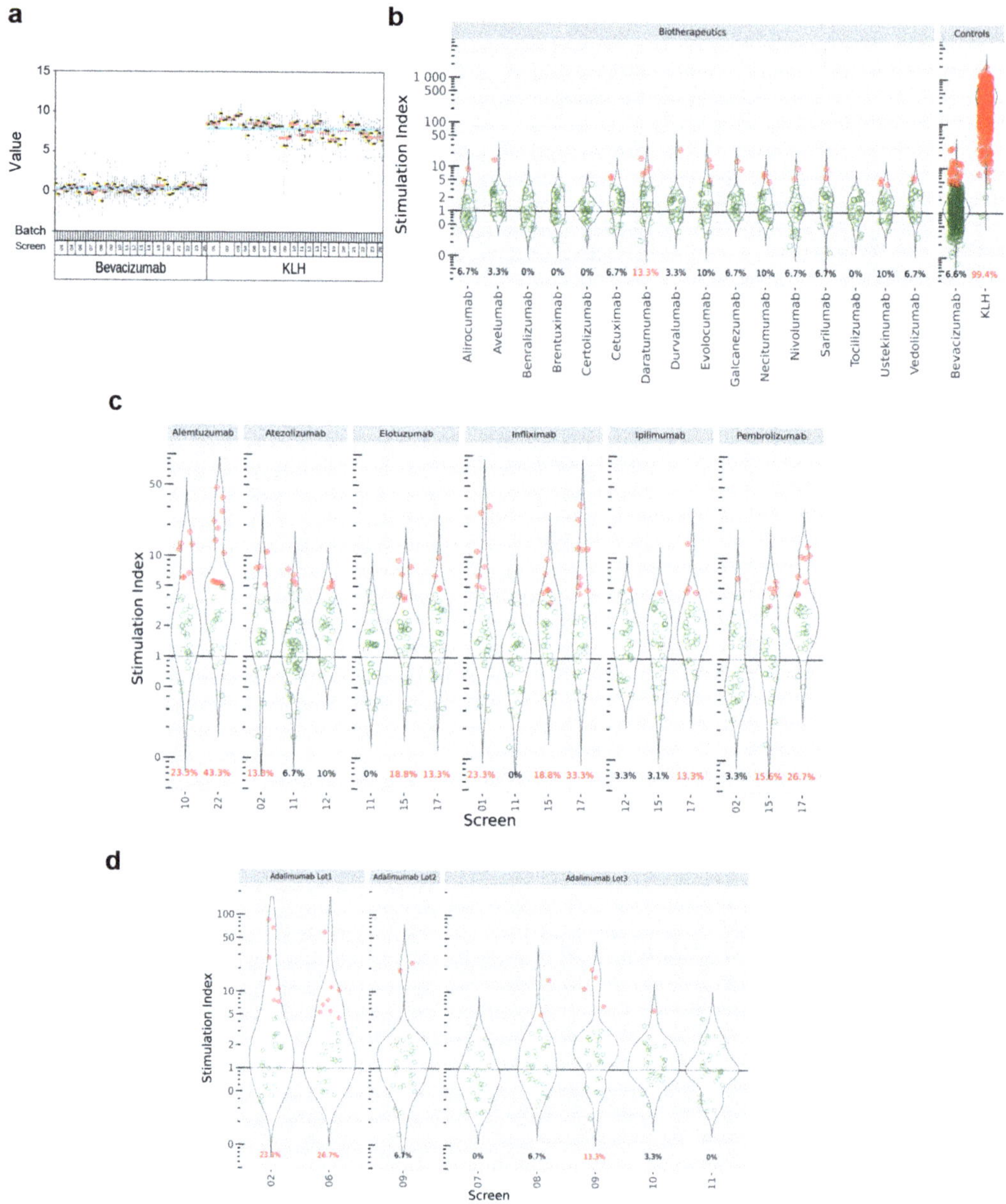

Figure 2. Overview of the stimulation indexes (SI) obtained in the DC:CD4+ T cell restimulation assay. (**a**) Stability of the controls over different assay screens and donor batches. Data were generated for the set of benchmark compounds in single (**b**) or multiple screens (**c**,**d**). SI represents the number of IFN-γ positive cells over baseline. If a datapoint is significantly above the SI threshold of 2, the donor is considered as positive for this condition and appears in red.

Table 1. Overview of the test items and their respective clinical ADA rates. Alemtuzumab, cetuximab, daratumumab, elotuzumab, sarilumab and tocilizumab are part of a co-treatment. Therefore, consideration should be taken when looking at the reported ADA rates. The information was extracted from FDA labels [13]. If several clinical ADA rates were reported, studies mentioning a co-treatment were excluded and the mean value for the remaining study outcomes was taken. In many cases, larger deviations may be due to systematic differences in the treated patient populations, as well as different analytical methods.

Antibody Name	Trade Name	Format	Target	Main Target Patient Population	Clinical ADA Rate	Screens
Adalimumab	Humira	Human IgG1	TNF-α	Rheumatoid Arthritis	23	02; 06; 07; 08; 09; 10; 11
Alemtuzumab	Lemtrada	Humanized IgG1	CD-52	Multiple Sclerosis	35	10; 22
Alirocumab	Praluent	Human IgG1	PCSK9	Cardiovascular disease	5	10
Atezolizumab	Tecentriq	Human IgG1 no-Glyco	PD-L1	Non-Small-Cell Lung Carcinoma (NSCLC)	44	02; 11; 12
Avelumab	Bavencio	Human IgG1	PD-L1	Urothelial Carcinoma	17	12
Benralizumab	Fasenra	Humanized IgG1	CD-125	Asthma	13	11
Bevacizumab	Avastin	Humanized IgG1	VEGF	Solid Tumor	0.6	ALL
Brentuximab	Adcetris	Chimeric IgG1-ADC	CD-30	Classical Hodgkin Lymphoma (late stage)	30	11
Certolizumab	Cimzia	FabPEG	TNF-α	Crohn Disease and Rheumatoid Arthritis	8	10
Cetuximab	Erbitux	Chimeric IgG1	EGFR	Head, Colorectal and Neck Cancer	5	12
Daratumumab	Darzalex	Human IgG1	CD-38	Multiple myeloma	0	12
Durvalumab	Imfinzi	Human IgG1	PD-L1	Locally advanced or Metastatic Urothelial Carcinoma, NSCLC	3	10
Elotuzumab	Empliciti	Human IgG1	SLAMF7	Multiple Myeloma	27	11; 15; 17
Evolocumab	Repatha	Human IgG2	PCSK9	Cardiovascular Disease	0.3	10
Galcanezumab	Emgality	Humanized IgG4	Calcitonin	Migraine	5	10
Infliximab	Remicade	Chimeric IgG1	TNF-α	Psoriatic Arthritis	27	01; 11; 15; 17
Ipilimumab	Yervoy	Human IgG1	CTLA-4	Metastatic melanoma, advanced renal cell carcinoma, metastatic colorectal cancer	8	12; 15; 17
Necitumumab	Portrazza	Human IgG1	EGFR	NSCLC	4	12
Nivolumab	Opdivo	Human IgG4-CPPC	PD-1	NSCLC	11	02
Pembrolizumab	Keytruda	Humanized IgG4-CPPC	PD-1	Cancer	2	02; 15; 17
Sarilumab	Kevzara	IgG1	IL-6R	Rheumatoid Arthritis	9	10
Tocilizumab	Actemra	Humanized IgG1	IL-6R	Rheumatoid Arthritis	2	10
Ustekinumab	Stelara	Human IgG1	IL-12/IL-23	Plaque Psoriasis	6	10
Vedolizumab	Entyvio	Humanized IgG1	Integrin α4β7	Ulcerative colitis and Crohn's disease	6	11

3.3. Statistical Characterization of the Assay

We investigated the stability of the assay across independent screens and the potential influence of confounding experimental factors using a variance component analysis on the full data set, which included the controls as well as the marketed compounds.

The data reported in this study were recorded in 24 different screens over several years. Hence, data replication occurred on various levels (i.e., bevacizumab and KLH were measured in all screens, several compounds were repeatedly measured in some screens, while subsets of compounds were tested on all donors within each screen), we could estimate the variance contributions of the treatments, the donor, the treatment-donor interaction, and the screen. As a typical screen is done in 3–4 batches, running a given subset of donors on all compounds in each batch, we can also assess the batch effect that is nested in the screen. During the course of the study, healthy donors that had given their blood could visit the blood donation center again, and the derived cells were used in two (or more) screens (i.e., same donor, same treatment but different screens). The results of this analysis are summarized in Figure 3a. Treatment-related effects (the expected effect from a compound in the assay, here driven primarily by the large number of strong KLH responses) accounted for 54% of the total variance; in contrast, the contribution of purely experimental factors was quite small (screen-to-screen variability: 0.5%, batch-to-batch variability within a screen: 2.0%). The donor factor (i.e., a factor accounting for a generally higher or lower donor-specific IFN-γ release independently of the treatment) accounted for 6.9% of the total variability; a similar proportion of the variance (5.4%) was attributed to the donor-treatment interaction (i.e., a factor taking account of a subject-specific response to a given treatment). A relatively high proportion of the total variance (23.6%) could not be readily accounted for by the known experimental factors. This could be due, for example, to the unavoidable technical variability in the protocol used to carry out the assay, or to heterogeneities unaccounted for when collecting sample material from a given donor at different times. In general, it would be very difficult to single out these technical and biological sources of variance and to investigate their relative impact on the assay reproducibility without some very cumbersome additional quality control processes.

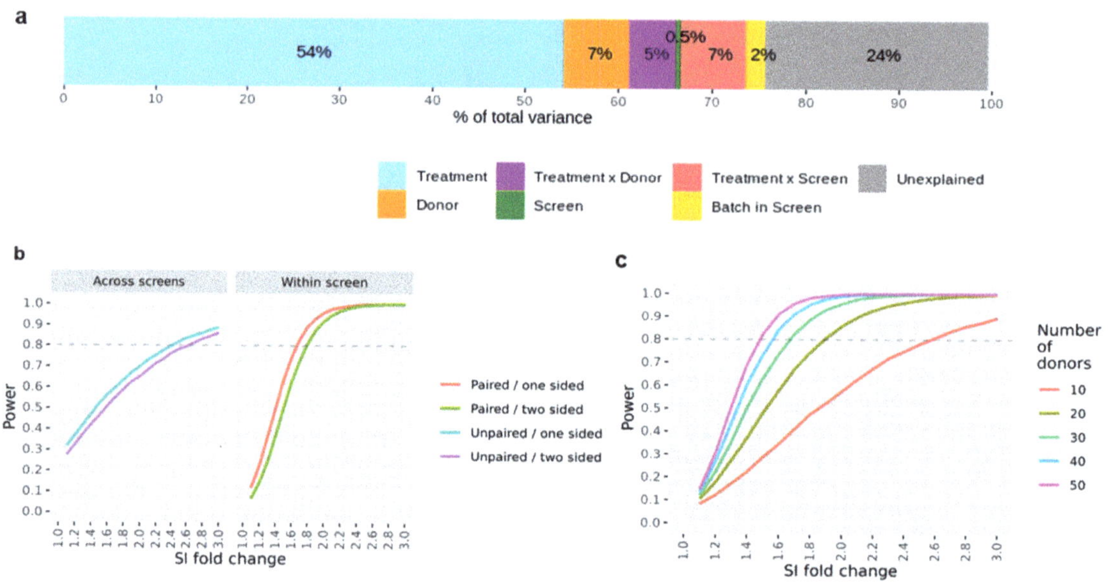

Figure 3. Variance and assay power. (**a**) Main factors contributing to assay variability estimated by a variance component analysis. The fitted model is the following: log2 (SI) ~ (Compound × DonorID) + Screen/Batch. There is a relatively low relative impact of assay batching variables (Screen, Batch within a screen) in comparison to the compound component. (**b**) DC:CD4+ T cell restimulation assay power curves for compound comparison. (**c**) Assay power curves showing the statistical power to detect a treatment effect by comparing a compound with a comparator treatment. A one-sided paired test within-study ($\alpha = 0.05$) according to various donor cohort sizes has been used.

The breakdown of the SI readouts into individual variance components enables us to simulate data sets with specified effect sizes for hypothetical treatments. Hence, we can estimate the statistical power (i.e., the probability of detecting a true compound effect) in a wide range of conditions. For example, Figure 3b shows the resulting statistical power when comparing a compound SI fold-change response with the one of a reference or comparator treatment; here, we differentiate the case where both compounds of interest were assessed in the same screen, in contrast to a comparison that was conducted across different screens. A major advantage of a within-screen comparison is that one could apply paired testing (i.e., using 'donor' as a covariate) to yield higher statistical power because the donor-to-donor variability would be partially accounted for in this approach. This is, in our opinion, the recommended setting for a compound ranking study. Moreover, depending on the hypothesis of interest, some additional statistical power may be gained by using a one-sided testing approach. This is legitimate when only a higher (or lower) compound response is of interest as compared to a reference treatment, which, in fact, could be the most relevant scenario. As a rule of thumb, we expect that SI differences of about 75% on a linear scale (i.e., a SI fold-change of 1.75 or 0.8 log2 units) can be detected with a statistical power of 80%, assuming one-sided testing within the same screen, alpha = 0.05, and n = 30 donors.

Statistical power is also a function of the sample size (here, number of donors per screen); we next examined this dependency and the impact of this variable in the interpretation of our assay results (Figure 3c). We observe a considerable gain in statistical power for studies including up to 30 donors per study. Increasing the number of donors beyond this point leads to noticeably smaller gains in statistical power at the cost of a considerable increase in effort and expenses, which is associated with larger experiments. In our experience, a standard study size of 30 donors per screen strikes the right balance, both for the experimental and statistical angles. In the case that enhanced statistical power is desired, we believe that a reduction of the residual assay variance by experimental protocol refinements could be a more promising approach than merely increasing the donor count.

3.4. Qualification of the Assay Threshold

An essential aspect of the study was to investigate the assay's ability to predict the potential for unwanted immune responses in line with FDA labels [13]. Accordingly, we characterized our assay in terms of accuracy (overall rate of correct predictions on compound level), sensitivity (probability to detect an immunogenic treatment), specificity (probability of correctly identifying a non-immunogenic treatment), and Positive/Negative Predictive Value (confidence in assigning either label correctly). To this end, we tested the aforementioned 24 molecules for which clinical data were available; however, since ADA responses in a limited number of patients would not necessarily be considered a relevant risk, we divided the tested molecules into two categories: high risk (≥20% reported ADA rate) and low risk (<20% reported ADA rate) for immunogenicity according to the reported data upon treatment. This classification was correlated to the proportion of donors for which a given biopharmaceutical triggered a CD4+ T cell-driven IFN-γ production in the assay: a positive assay readout was set to generate a SI statistically significant above 2 compared to the blank control, while a negative assay readout would not.

Using 10% as an optimal threshold (>3/30 positive donors according to our criteria), the assay reported 4 true positives (TP) and 16 true negatives (TN) for a total of 24 tested biopharmaceuticals (6 categorized as high risk, 18 labeled as low risk). It categorized 2 antibodies (daratumumab and pembrolizumab) at high risk of immunogenicity, even though their clinical ADA rates were below 20% (false positives, FP), while brentuximab and atezolizumab are categorized as low risk of immunogenicity, even though their clinical ADA rate were above 20% (false negatives, FN). The accuracy is the sum of true positives and true negatives over the total of tested compounds, yielding an estimated assay accuracy of 83% (20/24). The sensitivity, TP/(TP + FN), and specificity, TN/(TN + FP), are two additional important estimators, which represent the two types of possible errors. At this threshold, the DC:CD4+ T cell restimulation assay provides a 67% sensitivity at 89% specificity, with a 67% (4/6) and 89% (16/18) Positive and Negative Predictive Value, respectively.

3.5. Case Studies in Pre-Clinical Research

An important motivation of running a DC:CD4+ T cell restimulation assay in a pre-clinical setting is to derive information on whether compounds in development might be at risk of inducing an immunogenic response in treated patients. In this context, it is important to reduce false positive compound categorization, even at the expense of a higher false negative rate (i.e., over-classifying new molecules in the high immunogenicity risk category). As part of an integrated immunogenicity risk assessment, other risk factors (e.g., peptide presentation, mode of action, etc.) should also be taken into consideration. Our analysis demonstrates that a direct comparison of the responder rates in the DC:CD4+ T cell restimulation assay with the proportion of ADA-positive patients for a given treatment may not provide the best context of use for this assay. Our proposed strategy is to apply a given threshold to interpret results, essentially reducing the assay output to a binary outcome for biotherapeutics immunogenicity hazard identification. This enables us to retain the essential information on compound risk categorization, while minimizing the impact of noise in the data. Our data suggest that a selected threshold of 10% positive responders to classify a molecule as bearing a higher potential for immunogenicity is the optimal cutoff to flag compounds with high immunogenic potentials, while limiting the number of false negatives at an early stage of preclinical development.

To illustrate the strategy delineated above, we provide here a case study derived from one of our internal programs where seven potential clinical candidates from the same project, which differ from their primary sequence, have been tested in the assay (Figure 4a). The results showed that compounds A, B, D, and G were above the threshold, whereas variants C, E, and F were below the threshold and, therefore, associated with a lower risk of immunogenicity.

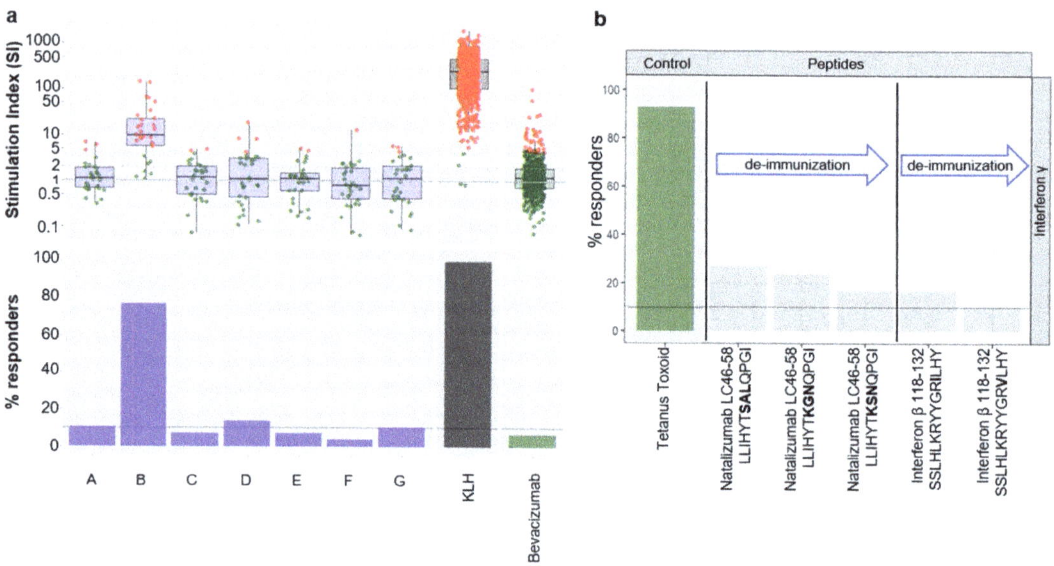

Figure 4. Case studies in pre-clinical research. (**a**) Stimulation Index (SI) obtained for 7 different candidate compounds of the same project (named A to G). The change in color, from green to red, depicts the positivity of the donor within the screen. The lower panel represents the proportion of positive donors. The threshold derived from the validation study is set at 10% positive donors. (**b**) DC:CD4+ re-stimulation results obtained for a selection of known T cell epitopes derived from biotherapeutics and their "de-immunized" counterparts.

Furthermore, we demonstrated that this assay was also suitable for testing whether peptides could trigger a CD4+ T cell response. Hence, we tested known T cell epitopes from Natalizumab and Interferon β, as well as potential deimmunized versions [15,16] (Figure 4b). Peptides were tested at 2 ug/mL and followed the same experimental procedure as described in the Material and Methods section. Results from the assay demonstrate that minor changes in the amino acid sequence of the T cell epitopes could reduce the onset of a CD4+ T cell response, thus confirming the published findings, but also that this assay can accommodate peptides (e.g., peptide based biotherapeutics or T cell epitopes).

4. Discussion

The multifactorial nature of immunogenicity requires that an integrated preclinical risk assessment should be a key element of biotherapeutics development. As T cell-dependent responses are major drivers of immunogenicity, in vitro T cell assays are frequently used as tools to identify and measure CD4+ T cell-dependent responses to biotherapeutics. The DC:CD4+ T cell restimulation assay described here assesses the propensity of a biotherapeutic to trigger a CD4+ T cell response that may result in B cell activation and ADA production. This assay plays a key role in our integrated approach to therapeutic protein immunogenicity risk estimation, which could accelerate drug development.

While a number of assays probing T cell activation in the context of immunogenicity have been published in recent years [6], we believe that the DC:CD4+ T cell restimulation assay described here provides a more comprehensive insight into the role of dendritic cells (taken here as the archetypal APC) in the context of their activation of T cells [17]. The immune response follows a three-signal rule for activation (TCR:MHC/peptide interactions, costimulatory interactions such as via CD28, and cytokine production); the assay published herein captures the interplay of all three signals [18]. In addition, the number of preexisting T cells specific to biotherapeutics is very low, ranging between 1 out of 10^8 (e.g., trastuzumab and etanercept) and 1 out of 10^7 T cells (e.g., rituximab) [9], but the assay format of the DC:CD4+ T cell restimulation assay allows screening of more CD4+ T cells than in a classical PBMC-based assay. We also believe that the re-stimulation step increases the likelihood of capturing a sustained T cell response [19].

A key part of validating the DC:CD4+ T cell restimulation assay was assessing repeatability and reliability, and its potential to categorize biotherapeutics according to their risk of inducing an unwanted immune response in the clinic. To this end, we carried out a detailed analysis of 24 biotherapeutics with various levels of clinically-detected ADA rates as a proxy for immunogenicity risk. Our robust assessment comprised several levels of repetitions, including repeated assessment of some compounds in multiple assay screens, to provide insights into both the potential and inevitable limitations of the assay. Variance component analysis showed that the primary factors governing the experimental setup, i.e., the screen and the batched donor processing within each screen, did not have a major systematic impact on the readouts. Notable variability arose, however, when compounds were re-tested in another screen, presumably related to subtle variation in compound preparation or the used production batches. Nonetheless, the compound batch effect is not specific to this assay. It is likely that the handling and storage of the sample plays a role here, influencing post-translational modifications and aggregation. Additionally, non-product related factors (e.g., DNA and host cell protein contaminations) have an impact on the risk of immunogenicity and could also influence the assay readout. General donor specific inducibility and the donor specific response to individual compounds also explained parts of the signal variability. However, there was a rather substantial residual unexplained variance, which should caution the user with regard to overinterpretation of individual readouts. Notably, quantification of SI changes in a strict sense were not directly informative, as even clinically-tested compounds with low immunogenic risks (for example, bevacizumab) resulted in a few positive readouts. We presume that additional insights might be gained by fundamental and costly changes of the lab protocol, i.e., performing replicated measurements in different experimental batches for every condition. In our experience, donor cohorts of 30 individuals per screen offer a reasonable tradeoff between the cost, timelines, and statistical power of the assay. While not specifically discussed in the manuscript, we found it important to test all compounds using a panel of donors that showed an HLA-DRB1 allele frequency that broadly reflects the world population. It has been demonstrated that certain HLA alleles were associated with an increased immunogenic response towards certain biopharmaceuticals [20–25]. Nevertheless, in the context of use of this preclinical assay carried out in 30 donors, we primarily investigated whether compounds may be at risk of inducing an enhanced immunogenic response in a general population. As each screen usually comprised different sets of donors, an arbitrary selection of pre-typed donors with respect to their allelic HLA-DRB1 composition enabled a higher comparability of the data in the long term.

While tempting, it is problematic to compare SI values measured in a DC:CD4+ T cell restimulation assay with actual ADA rates in the clinic, although it is one of the few available measures directly related to clinical immunogenicity. Assays used to measure ADA in clinics are based on different methodologies sensitive to sample handling, the timing of sample collection, concomitant medications utilized in the study, and the underlying nature of the treated disease [26]. Furthermore, while we believe we have used the most recent information available on the FDA database, most labels may not be updated on a regular basis: in a recent review, Borrega et al. showed that 57% (39/69) of the biological drugs authorized before 2012 did not have updated summaries of product characteristics, especially in the immunogenicity section [27]. In our study, we collected the ADA rates of the 24 assessed marketed compounds as a starting point to build a database to benchmark newly developed immunogenicity estimation methods and to have a retrospective and comprehensive overview of the immunogenicity of marketed antibodies. We used the available data to create two categories of compounds, at high ($\geq$20% reported ADA rate) and low (<20% ADA rate) risk for immunogenicity, on which we calibrated the assay's linear mixed model. Thus, this binary high/low risk paradigm is the most reasonable for implementation in preclinical risk evaluation for therapeutics. To facilitate this process, we found it essential to add in the panel a few standard compounds (at minimum, a negative control, such as bevacizumab, and a positive control, such as KLH; any additional comparators also provide useful comparisons) to help set precise boundaries of low and high risk of immunogenicity while mitigating intrinsic donor variability. Accordingly, in this context of use, one of the most useful applications of the DC: CD4+ T cell restimulation assay is to provide a relative ranking for compounds with similar amino acid sequences and mode of action, or compounds that have different formulation or have been produced in different batches.

While assays measuring T cell activation in response to novel biopharmaceuticals are not yet required by regulatory agencies, there is added value in presenting the results of such assays as part of the risk assessment submitted in the Integrated Summary of Immunogenicity [28]. A current challenge is that none of the published assays is considered to be fully validated. We here propose a new assay format that captures the interaction between DCs and CD4+ T cells by monitoring the production of IFN-γ by CD4+ T cells in response to biotherapeutics processed by DCs. We tested the predictive power of this assay vs. clinical ADA rate by assessing 24 marketed antibodies, which resulted in 83% accuracy. Predicting the actual rate of ADA-positive patients in a clinical setting with a single in vitro assay is unlikely to be possible, given the myriad contributing factors. However, the DC:CD4+ T cell restimulation assay can help flag potentially immunogenic biopharmaceuticals in preclinical drug development, allowing for selection or de-immunization before a clinical trial starts, improving both patient safety and the cost of pharmaceuticals. Implementation of this assay as part of a comprehensive risk assessment has the potential to provide a more robust and informative immunogenicity risk assessment in early drug development.

Supplementary Materials: The following supporting information can be downloaded at: https://www.mdpi.com/article/10.3390/pharmaceutics14122672/s1, Figure S1: Heatmap showing the relative distribution of the HLA-DRB1 supertype frequencies among the donor cohorts per screen of the DC:CD4+ T cell restimulation assay.

Author Contributions: Conceptualization, M.S., G.S., L.C.F., K.B.-F., T.P.H., N.S. and C.M.-D.; Formal analysis, M.S., G.S., L.C.F., A.D. and C.M.-D.; Investigation, M.S., G.S., L.C.F., F.C., J.H., K.H., C.M.L., N.S. and C.M.-D.; Writing—original draft, M.S., G.S., L.C.F., A.D. and C.M.-D.; Writing—review & editing, F.C., J.H., K.H., C.M.L., K.B.-F., O.R., T.P.H. and N.S. All authors have read and agreed to the published version of the manuscript.

Funding: This research received no external funding.

Institutional Review Board Statement: The study was conducted in accordance with the Declaration of Helsinki, and approved by the local Research Ethics Committee (reference number: 21/LO/0474).

Informed Consent Statement: Informed consent was obtained from all subjects involved in the study.

Data Availability Statement: Data are contained within the article or Supplementary Materials.

Conflicts of Interest: M.S., G.S., L.C.F., K.H., C.M.L., A.D., K.B.F., T.P.H. and C.M.D. are employees of F. Hoffmann. La Roche (Switzerland); F.C., J.H. and N.S. are employees of Lonza Biologics (United Kingdom). The company had no role in the design of the study; in the collection, analyses, or interpretation of data; in the writing of the manuscript; or in the decision to publish the results.

References

1. Brummelen, E.M.J.; Ros, W.; Wolbink, G.; Beijnen, J.H.; Schellens, J.H.M. Antidrug Antibody Formation in Oncology: Clinical Relevance and Challenges. *Oncologist* **2016**, *21*, 1260–1268. [CrossRef]
2. Crommelin, D.J.A.; Mastrobattista, E.; Hawe, A.; Hoogendoorn, K.H.; Jiskoot, W. Shifting Paradigms Revisited: Biotechnology and the Pharmaceutical Sciences. *J. Pharm. Sci.* **2019**, *109*, 30–43. [CrossRef] [PubMed]
3. Harding, F.A.; Stickler, M.M.; Razo, J.; DuBridge, R. The Immunogenicity of Humanized and Fully Human Antibodies: Residual Immunogenicity Resides in the CDR Regions. *Mabs* **2010**, *2*, 256–265. [CrossRef] [PubMed]
4. Kuriakose, A.; Chirmule, N.; Nair, P. Immunogenicity of Biotherapeutics: Causes and Association with Posttranslational Modifications. *J. Immunol. Res.* **2016**, *2016*, 1298473. [CrossRef] [PubMed]
5. Bray-French, K.; Hartman, K.; Steiner, G.; Marban-Doran, C.; Bessa, J.; Campbell, N.; Martin-Facklam, M.; Stubenrauch, K.-G.; Solier, C.; Singer, T.; et al. Managing the Impact of Immunogenicity in an Era of Immunotherapy: From Bench to Bedside. *J. Pharm. Sci.* **2021**, *110*, 2575–2584. [CrossRef]
6. Ducret, A.; Ackaert, C.; Bessa, J.; Bunce, C.; Hickling, T.; Jawa, V.; Kroenke, M.A.; Lamberth, K.; Manin, A.; Penny, H.L.; et al. Assay Format Diversity in Pre-Clinical Immunogenicity Risk Assessment: Toward a Possible Harmonization of Antigenicity Assays. *Mabs* **2021**, *14*, 1993522. [CrossRef] [PubMed]
7. Duke, B.R.; Mitra-Kaushik, S. Current In Vitro Assays for Prediction of T Cell Mediated Immunogenicity of Biotherapeutics and Manufacturing Impurities. *J. Pharm. Innov.* **2020**, *15*, 202–218. [CrossRef]
8. Hammond, S.; Thomson, P.; Meng, X.; Naisbitt, D. In-Vitro Approaches to Predict and Study T-Cell Mediated Hypersensitivity to Drugs. *Front. Immunol.* **2021**, *12*, 630530. [CrossRef]
9. Delluc, S.; Ravot, G.; Maillere, B. Quantitative Analysis of the CD4 T-cell Repertoire Specific to Therapeutic Antibodies in Healthy Donors. *FASEB J.* **2011**, *25*, 2040–2048. [CrossRef]
10. Reis e Sousa, C. Dendritic Cells in a Mature Age. *Nat. Rev. Immunol.* **2006**, *6*, 476–483. [CrossRef]
11. Engering, A.; Geijtenbeek, T.B.H.; van Vliet, S.J.; Wijers, M.; van Liempt, E.; Demaurex, N.; Lanzavecchia, A.; Fransen, J.; Figdor, C.G.; Piguet, V.; et al. The Dendritic Cell-Specific Adhesion Receptor DC-SIGN Internalizes Antigen for Presentation to T Cells. *J. Immunol.* **2002**, *168*, 2118–2126. [CrossRef] [PubMed]
12. Presicce, P.; Taddeo, A.; Conti, A.; Villa, M.L.; Bella, S.D. Keyhole Limpet Hemocyanin Induces the Activation and Maturation of Human Dendritic Cells through the Involvement of Mannose Receptor. *Mol. Immunol.* **2008**, *45*, 1136–1145. [CrossRef] [PubMed]
13. FDA Label Search. Available online: https://labels.fda.gov/ (accessed on 24 October 2022).
14. Manda, K.; Glasow, A.; Paape, D.; Hildebrandt, G. Effects of Ionizing Radiation on the Immune System with Special Emphasis on the Interaction of Dendritic and T Cells. *Front. Oncol.* **2012**, *2*, 102. [CrossRef] [PubMed]
15. Cassotta, A.; Mikol, V.; Bertrand, T.; Pouzieux, S.; Parc, J.L.; Ferrari, P.; Dumas, J.; Auer, M.; Deisenhammer, F.; Gastaldi, M.; et al. A Single T Cell Epitope Drives the Neutralizing Anti-Drug Antibody Response to Natalizumab in Multiple Sclerosis Patients. *Nat. Med.* **2019**, *25*, 1402–1407. [CrossRef] [PubMed]
16. Yeung, V.P.; Chang, J.; Miller, J.; Barnett, C.; Stickler, M.; Harding, F.A. Elimination of an Immunodominant CD4+ T Cell Epitope in Human IFN-β Does Not Result in an In Vivo Response Directed at the Subdominant Epitope. *J. Immunol.* **2004**, *172*, 6658–6665. [CrossRef] [PubMed]

17. Xue, L.; Hickling, T.; Song, R.; Nowak, J.; Rup, B. Contribution of Enhanced Engagement of Antigen Presentation Machinery to the Clinical Immunogenicity of a Human Interleukin (IL)-21 Receptor-blocking Therapeutic Antibody. *Clin. Exp. Immunol.* **2016**, *183*, 102–113. [CrossRef]
18. Goral, S. The Three-signal Hypothesis of Lymphocyte Activation/Targets for Immunosuppression. *Dial. Transpl.* **2011**, *40*, 14–16. [CrossRef]
19. Pennock, N.D.; White, J.T.; Cross, E.W.; Cheney, E.E.; Tamburini, B.A.; Kedl, R.M. T Cell Responses: Naïve to Memory and Everything in Between. *Adv. Physiol. Educ.* **2013**, *37*, 273–283. [CrossRef]
20. Diego, V.P.; Luu, B.W.; Hofmann, M.; Dinh, L.V.; Almeida, M.; Powell, J.S.; Rajalingam, R.; Peralta, J.M.; Kumar, S.; Curran, J.E.; et al. Quantitative HLA-class-II/Factor VIII (FVIII) Peptidomic Variation in Dendritic Cells Correlates with the Immunogenic Potential of Therapeutic FVIII Proteins in Hemophilia A. *J. Thromb. Haemost.* **2020**, *18*, 201–216. [CrossRef]
21. Sazonovs, A.; Kennedy, N.A.; Moutsianas, L.; Heap, G.A.; Rice, D.L.; Reppell, M.; Bewshea, C.M.; Chanchlani, N.; Walker, G.J.; Perry, M.H.; et al. HLA-DQA1*05 Carriage Associated With Development of Anti-Drug Antibodies to Infliximab and Adalimumab in Patients With Crohn's Disease. *Gastroenterology* **2020**, *158*, 189–199. [CrossRef] [PubMed]
22. Benucci, M.; Damiani, A.; Gobbi, F.L.; Bandinelli, F.; Infantino, M.; Grossi, V.; Manfredi, M.; Noguier, G.; Meacci, F. Correlation between HLA Haplotypes and the Development of Antidrug Antibodies in a Cohort of Patients with Rheumatic Diseases. *Biol. Targets Ther.* **2018**, *12*, 37–41. [CrossRef] [PubMed]
23. Quarmby, V.; Phung, Q.T.; Lill, J.R. MAPPs for the Identification of Immunogenic Hotspots of Biotherapeutics; an Overview of the Technology and Its Application to the Biopharmaceutical Arena. *Expert Rev. Proteom.* **2018**, *15*, 733–748. [CrossRef] [PubMed]
24. Lagassé, H.A.D.; McCormick, Q.; Sauna, Z.E. Secondary Failure: Immune Responses to Approved Protein Therapeutics. *Trends Mol. Med.* **2021**, *27*, 1074–1083. [CrossRef] [PubMed]
25. McMaster, M.; Mohr, K.; Page, A.; Closmore, A.; Towne, F.; Brooks, B.D. Epitope Characterization of Anti-Drug Antibodies—A Tool for Discovery and Health: An Overview of the Necessity of Early Epitope Characterization to Avoid Anti-Drug Antibodies and Promote Patient Health. *Expert Opin. Biol. Ther.* **2021**, *21*, 705–715. [CrossRef]
26. Gorovits, B.; Peng, K.; Kromminga, A. Current Considerations on Characterization of Immune Response to Multi-Domain Biotherapeutics. *Biodrugs* **2020**, *34*, 39–54. [CrossRef]
27. Borrega, R.; Cruz, J.P.; Taylor, P.; Goncalves, J. Analysis of Immunogenicity Data in the Product Information of Biological Drugs: A Need to Report Immunogenicity Data Systematically. *Biodrugs* **2019**, *33*, 683–691. [CrossRef]
28. Chamberlain, P. Effective Presentation of Immunogenicity Risk Assessments and Related Data in Regulatory Dossiers. *Bioanalysis* **2019**, *11*, 1581–1592. [CrossRef]

pharmaceutics

MDPI

Article

Determination of Mutational Timing of Colistin-Resistance Genes through *Klebsiella pneumoniae* Evolution

Jenna M. Kuhn and Yuanpu Peter Di *

Department of Environmental and Occupational Health, School of Public Health, University of Pittsburgh, Pittsburgh, PA 15261, USA
* Correspondence: peterdi@pitt.edu; Tel.: +1-(412)-624-8718

Abstract: The emergence and dissemination of carbapenem-resistant *Klebsiella pneumoniae* (KP), one of the carbapenem-resistant *Enterobacteriaceae* (CRE), is now an emerging cause of antibiotic-resistant nosocomial infections associated with high rates of morbidity and mortality. Colistin, or polymyxin E, is a last-resort peptide antibiotic used to treat multidrug-resistant (MDR) Gram-negative bacterial infections including KP. Unfortunately, resistance to colistin is rising with increasing use in the clinical setting. Although clinical evidence links certain mutations to colistin resistance (COL-R) in KP, the origination and association of the mutations remain unclear. We hypothesize that the timing of COL-R mutations influences the development and progression of KP resistance to colistin. We performed planktonic and biofilm in vitro experimental evolutions of KP strain ATCC 43816 under increasing colistin concentrations to characterize the temporal regulation of critical COL-R mutations throughout COL-R progression. The resistance generation and mutation profiles of independently evolved bacterial populations with different lifestyles were compared. Genes with various functions theorize the timeline in which key mutations are generated and their roles in the progression of COL-R. Our results aim to advance the research and development of effective therapeutics to treat MDR bacterial infection as the dissemination of CRE continues to be a severe public health threat.

Keywords: *Klebsiella pneumoniae*; colistin; antimicrobial resistance; mutation timing; evolution

Citation: Kuhn, J.M.; Di, Y.P. Determination of Mutational Timing of Colistin-Resistance Genes through *Klebsiella pneumoniae* Evolution. *Pharmaceutics* **2023**, *15*, 270. https://doi.org/10.3390/pharmaceutics15010270

Academic Editor: Ivana Cacciatore

Received: 16 November 2022
Revised: 16 December 2022
Accepted: 4 January 2023
Published: 12 January 2023

1. Introduction

The emergence of carbapenem-resistant Enterobacterales (CRE) has become a global public health threat with high mortality rates of infection and limited available antimicrobial treatment options [1]. According to the US Centers for Disease Control and Prevention 2019 Antibiotic Resistance Threats Report, CRE infections led to approximately 13,100 hospital incidents and 1100 infection-related deaths, with average case numbers remaining steady between 2012 and 2017 [2]. Additionally, the worldwide dissemination of extended-spectrum β-lactamases (ESBLs) in species of *Enterobacteriaceae* and *P. aeruginosa* has facilitated resistance to a number of β-lactam antibiotics, including carbapenems, cephalosporins, monobactams, and penicillins, as well as recent detection of resistance to aztreonam and oxyimino-cephalosporins [3,4]. A pathogen of urgent concern of the *Enterobacteriaceae* family, *Klebsiella pneumoniae* (KP), is a one of six nosocomial ESKAPE pathogens (*Enterococcus faecium*, *Staphylococcus aureus*, *Klebsiella pneumoniae*, *Acinetobacter baumannii*, *Pseudomonas aeruginosa*, and *Enterobacter* species), noted for its virulence and high potential for multidrug resistance (MDR) [5]. KP is characterized as an encapsulated, Gram-negative, nonmotile, facultative anaerobic pathogen acquired in community or healthcare settings that may cause pneumonia, urinary tract infection, soft-tissue infection, bacteremia, and meningitis, especially in immunocompromised individuals [6]. The recent emergence and rising prevalence of carbapenem-resistant hypervirulent KP (CR-hvKP) has several concerning clinical impacts due to high resistance and pathogenicity, high mortality, production of multiple carbapenemases, and gut colonization, facilitating further resistance dissemination [7]. Due to rising numbers of identified ESBLs and the global spread of CR-hvKP, there are limited treatment options available.

Colistin, or polymyxin E, is a last-resort antibiotic reserved for treating complex CR-hvKP infections. Polymyxins are cationic cyclic polypeptides that belong to a family of antimicrobial peptides (AMPs), which are bactericidal components of the host innate immune system present in plant and animal species, as well as some bacteria and fungi. AMPs are usually amphipathic, cationic, and approximately 15–30 amino acids in length; they impose their activity at the cell membrane of bacteria [8]. The two dominant polymyxins used clinically to treat Gram-negative bacterial infections are polymyxin B and polymyxin E (Figure 1) [9,10]. AMPs challenge the development of drug resistance due to their diverse killing mechanisms and low specificity for a given target in host cells [11].

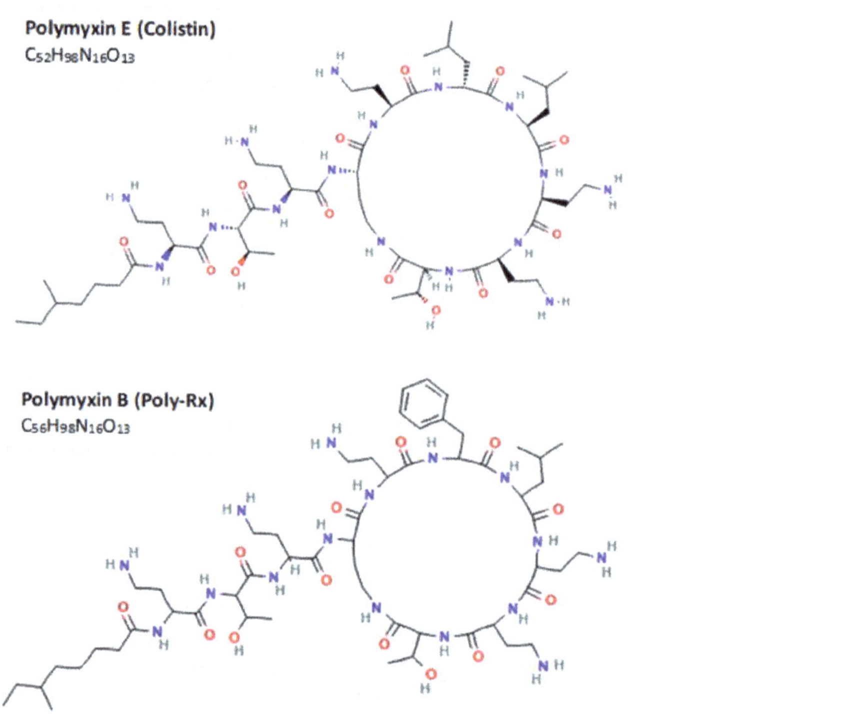

Figure 1. Chemical structure of polymyxins B and E. The primary polymyxins used clinically to treat Gram-negative infections are the cyclic polypeptides Polymyxin B and E. Both polymyxins enact their activity through binding to negatively-charged lipopolysaccharide and disruption of outer membrane permeability. The two-dimensional chemical structures and formulas for polymyxin B and polymyxin E displayed in this figure were obtained from PubChem, with CIDs 4868 and 5311054, respectively.

Colistin is a cationic polypeptide that exerts its activity by binding to anionic regions of lipopolysaccharide (LPS), leading to permeabilization of the cell envelope, cell leakage, and cell death [12]. Since its displacement by other antimicrobials four decades prior, colistin has shown significant activity against KP, *P. aeruginosa*, *A. baumannii*, and other Gram-negative pathogens with low initial resistance levels [13]. However, with increased use in the clinical space, rates of colistin resistance have been on the rise through several main mechanisms, including chromosomal mutations in genes responsible for disrupting the cationic charge of LPS (PhoP/PhoQ and PmrA/PmrB two-component regulatory systems) and MgrB, a transmembrane regulator of PhoP/PhoQ signaling, as well as plasmid-mediated colistin resistance (*mcr-1-10*) [14,15]. It has also been shown that increased production of capsular polysaccharide hinders interactions between colistin with the cell membrane of bacteria, facilitating resistance [14]. While the primary mechanisms of COL-R have been identified, the relationship among resistance generation timing, population frequency, and interactions of mutations facilitating the onset and progression of COL-R has not been thoroughly investigated. The demand for further development of effective antimicrobials against the growing public health threat of MDR infections continues to grow. Now, even last-resort treatments are showing high rates of resistance.

In particular, biofilm-associated infections comprise ~65% of all bacterial infections in the clinical setting and present a serious challenge to the healthcare community due to their diversity and innate defense mechanism to protect against antimicrobials [16,17]. There are several mechanisms via which biofilms may resist antimicrobial agents. The exopolysaccharide matrix of biofilm environments may hinder penetration of negatively charged antimicrobials such as aminoglycosides [17]. Furthermore, microcolonization within the biofilm leads to waste accumulation and fluctuation of pH and CO_2 and O_2 partial pressures, consequently disrupting antimicrobial activity [17]. Additionally, regulation of efflux pumps, expression of antimicrobial chelating enzymes, and quorum sensing are other biofilm-associated resistance mechanisms [17]. KP biofilms may lead to invasive, chronic infections in the urinary, gastrointestinal, or respiratory tracts through cell adhesion to a surface, colony formation, biofilm maturation, and cell detachment [18]. Here, we consider both planktonic and biofilm KP lifestyles in our experimental evolution methods to better understand the timing of critical resistance mutations and their likely impact on the progression of COL-R.

2. Materials and Methods

2.1. Bacteria Strain and Culture Media

Klebsiella pneumoniae American Type Culture Collection (ATCC) strain 43816 was first grown on cation-adjusted Mueller–Hinton Broth 2 (MHB2) agar medium (Sigma-Aldrich, St. Louis, MO, USA) overnight, and a single colony was transferred into liquid M9 medium supplemented with $\frac{1}{2}$MIC level of colistin sulfate treatment (Research Products International, Mount Prospect, IL, USA). M9 minimal growth medium was designed using M9 salts (6 g/L Na_2HPO_4, 3 g/L KH_2PO_4, 0.5 g/L NaCl, and 1.0 g/L NH_4Cl) supplemented with 0.1 mM $CaCl_2$, 2 mM L-Glutamine, 2 mM $MgSO_4$, and 4% glucose to reflect the essential components required by mammalian cells and biofilm formation to prevent additional nutrient selection effects [19,20].

2.2. Bacterial Transfer and Resistance Selection

To investigate the timing and relevance of resistance genes concerning MIC increase, we performed bacterial experimental evolution studies using a serial passage of KP planktonic culture and a single colonized biofilm bead under colistin selection pressure (Supplementary Figure S1). Three replicate populations were subjected to colistin selection according to bacteria lifestyle. For planktonic populations, 1/100 or approximately 2.5×10^6 colony forming units (CFU)/mL of total overnight culture was transferred into fresh M9 media supplemented with colistin. For biofilm populations, a bead transfer-based biofilm evolution system was used according to a method described previously [21–23]. Briefly, a single colonized 7 mm polystyrene bead, approximately 2.5×10^5 CFU/mL, was transferred into M9 medium supplemented with colistin and two sterile polystyrene beads. The sterile beads facilitate the processes of bacteria attachment, biofilm formation, and dispersal. The concentration of colistin treatment was doubled every three days to steadily enhance bacteria resistance selection. Evolution under antibiotic selection continued for 36 days without complete inhibition of growth after each doubling of treatment concentration (Figure 2A).

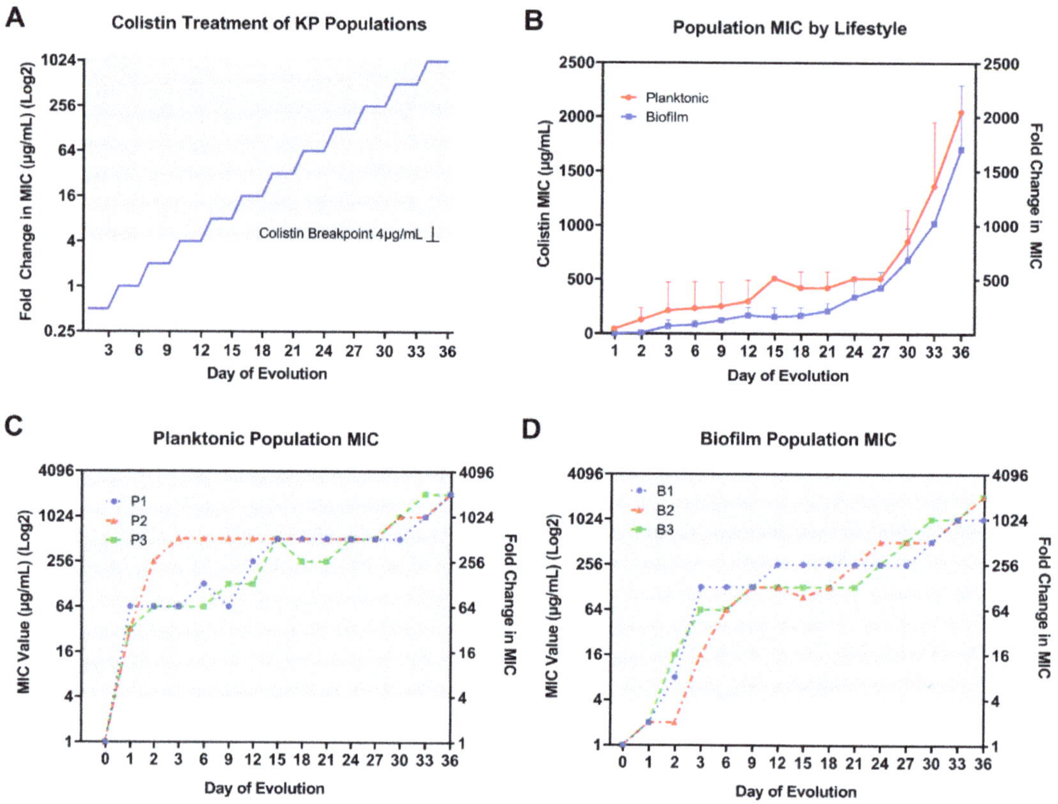

Figure 2. Colistin selection and population MIC result by bacterial lifestyle. (**A**) Colistin treatment for KP planktonic and biofilm populations started at $\frac{1}{2}$ MIC; the selection concentrations were doubled every three days for 36 days. (**B**) Colistin population MIC results averaged among three replicate colistin-treated KP populations by bacteria lifestyle. Rapid resistance to colistin was observed for (**C**) planktonic and (**D**) biofilm individually evolved populations after 36 days of colistin selection pressure. By day 36, there was a 1024–2048-fold increase in colistin MIC compared to baseline level. The resistance is distinguished in gray background when MIC is above the CLSI-determined clinical breakpoint at 4 µg/mL for colistin.

2.3. MIC Measurement

Colistin MIC was determined using the broth microdilution method established by the Clinical and Laboratory Standards Institute (CLSI) [24]. Whole-population MIC was determined for each KP population replicate at the end of every 3 days of serial passage and prior to transfer into the subsequent treatment concentration. The range of antibiotic selection began at a baseline level of $\frac{1}{2}$MIC or 0.5 µg/mL for the ancestor KP clone on days 1–3 and was subsequently increased to a level of 1024 × MIC or 1024 µg/mL for the final days 33–36. Cation-adjusted Mueller–Hinton Broth 2 (MHB2) (Sigma-Aldrich, MO, USA) was used to inoculate each planktonic KP population for overnight incubation at 37 °C prior to MIC testing to increase bacterial cell count. KP biofilm populations were isolated by transferring a single colonized polystyrene bead into a 15 mL conical tube containing 2 mL of phosphate-buffered saline for ultrasonic homogenization (DPS-20 model, PRO Scientific Inc., Oxford, CT, USA). Prior to MIC testing, planktonic and homogenized biofilm populations were cultured in MHB2 medium for 24 h at 37 °C in a shaking incubator (Corning LSE 71L model, Corning, NY, USA). MIC testing was performed using sterile 96-well, microplates (Greiner, Frickenhausen, Germany). Bacterial concentration was adjusted to approximately 5 × 10^5 CFU/mL with PBS prior to plating. MIC values were determined by measuring turbidity via optical density readings at 570 nm wavelength using a Gen 5 Microplate Reader and Imaging software (BioTek Instruments, Winooski, VT, USA, Version 3.04). Samples selected for MIC testing were preserved in 8% dimethyl sulfoxide for further genomic analysis.

2.4. Whole-Genome DNA Sequencing

KP (whole population or single clone) genomic DNA was extracted using a DNeasy Blood and Tissue Kit (Qiagen, Hilden, Germany). Biofilm attached to beads was dissociated by sonication in sterile PBS before DNA extraction. Planktonic cultures were centrifuged and pelleted before DNA extraction. Whole-genome DNA sequencing was performed using an Illumina NextSeq 2000 platform with a paired-end mode of 2 × 150 base pairs and sequencing depth coverage of 200 Mbp for individual clones and 650 Mbp for whole-population bacterial genomes (SeqCenter, Pittsburgh, PA, USA).

2.5. Comprehensive Mutation Analysis

Data preprocessing for raw sequencing data included gentle quality trimming using trimmomatic (version 0.38, parameter: PE -phred33 LEADING:20 TRAILING:20 SLIDING-WINDOW:4:20 MINLEN:70) to filter out low-quality and unpaired reads before computational mutation analysis [25]. Bowtie2 (version 2.4.1), as part of the breseq workflow, was used to build an index of the reference genome and align reads to the index reference genome *K. pneumoniae* strain ATCC 43816 (serotype O1:K2) [26,27]. The reference genome was downloaded from NCBI GenBank with accession number SRR13008124. Breseq was used to detect mutations relative to the reference genome in consensus mode and polymorphism mode for clonal and mixed-population samples, respectively [28]. Both mixed-population and clonal samples were analyzed to identify and confirm the most integral COL-R mutations generated with increased selection. Breseq gdtools COMPARE subcommand was used to create side-by-side mutation comparison tables for individual populations to show how genetic variants and their frequencies change over time.

2.6. Mutation Selection Criteria

A series of criteria were distinguished to select relevant genes and their corresponding mutations of interest. Mutation prediction using breseq and mutation comparison tables were designed and analyzed to identify alterations in genes that were observed in independently evolved populations at greater than 20% frequency. Most of these mutations increased in frequency with enhanced selection pressure or became fixed at 100% frequency in these populations. We studied the function of genes for which these mutations were observed to speculate their potential role in and importance for the resistance mechanism. Using a thorough review of the current literature, we noted mutations that have been previously identified as critical for colistin-resistance development. In addition, we included mutations that have not been studied extensively in *K. pneumoniae* to theorize how they may be influential in the resistance pathway according to their timing and associated gene functions.

2.7. String Test for Hypervirulence

The string test was performed to assess changes in hypervirulence following $\frac{1}{2}$MIC colistin treatment for planktonic KP populations. Tryptic soy agar plates containing 40 mg/L Congo Red (Sigma, MO, USA) and 20 mg/L Coomassie Brilliant Blue (Bio-Rad, Watford, UK) dyes were prepared and used to plate various bacterial dilutions to achieve comparable single-colony counts between each treated population and the evolution ancestor clone. A sterile inoculation loop was used to touch the surface of each single colony and measure the length of mucoid string produced. A positive test was determined for a mucoid string of 5 mm or greater in length, a feature observed clinically to define the hypervirulent phenotype [29]. Statistical comparisons between treatment populations and the original clone were made using one-way ANOVA. A p-value < 0.05 was considered to be statistically significant.

3. Results

3.1. Rapid MIC Increase in Colistin-Treated KP Populations

The baseline MIC for the colistin-susceptible ancestor KP clone (ATCC 43816) was determined to be 1 μg/mL. Rapid increases in MIC were observed for individually evolved KP populations through experimental evolution under colistin selection with increasing concentrations (Figure 2A) independent of bacteria lifestyle (Figure 2B). However, there was a slower rate of MIC increase within the first 3 days and for the extent of evolution for biofilm- compared to planktonic-evolved KP. For planktonic-evolved KP populations, the most substantial jumps in MIC occurred during the first 3 days of colistin selection. More gradual and consistent twofold increases occurred every 3 days afterward and for the remaining days of evolution. Compared to the planktonic lifestyle, biofilm populations showed modest (twofold) increases from baseline MIC after 24 h of incubation at $\frac{1}{2}$MIC level treatment. After 24 h of antibiotic selection, colistin MIC increased over 32-fold for planktonic populations and twofold for biofilm populations (Figures 1D and 2C). By day 3, planktonic populations 1–3 showed 64-, 512-, and 64-fold increases in MIC compared to 128-, 16-, and 64-fold increases for biofilm populations 1–3. Similar to planktonic populations, after the first 3 days of selection, biofilm populations showed consistent 2–4-fold increases in MIC for the remainder of the 36 day evolution experiment. The CLSI breakpoint for colistin of 4 μg/mL was surpassed after only 24 and 48 h evolution at $\frac{1}{2}$MIC level of selection for planktonic and biofilm lifestyles, respectively. After day 27, KP showed further twofold increases in MIC with no hindrance in growth under selection pressure for both lifestyles. Following 36 days of selection, 2048-fold increases in colistin MIC were observed for all three planktonic populations (Figure 2C) as well as 1024-, 2048-, and 2048-fold increases in biofilm populations 1–3, respectively (Figure 2D).

3.2. Temporal Regulation of COL-R Mutations in Planktonic KP

After 24 h of $\frac{1}{2}$MIC level colistin selection, all three replicate planktonic KP populations showed an insertion in mucoid phenotype A regulator RmpA, as well as single-nucleotide polymorphisms (SNPs) in quinolinate synthase NadA and large-conductance mechanosensitive channel protein MscL, at nearly 100% frequency (Table 1, Figure 3A–C). These mutations coincided with a substantial rise in colistin MIC equivalent to 64-, 32-, and 32-fold increases compared to baseline MIC level (1 μg/mL) for populations 1–3, respectively. By day 6 of selection, KP population 1 showed an SNP in two-component system sensor histidine kinase PhoQ at 93% frequency, which aligned with a 128-fold increase in colistin MIC compared to baseline. In addition, this mutation increased to 100% frequency in population 1 in accordance with a 512-fold increase in colistin MIC. By day 36, a SNP at a new position in *phoQ* was acquired at 100% frequency as well as an SNP in two-component system sensor histidine kinase PmrB at high frequency, in line with a 2048-fold increase in colistin MIC compared to baseline (Figure 3A). Despite detection at different positions, mutations in *phoQ* were persistently observed from day 6 onward until day 36, which may contribute to the continued enhanced colistin resistance (Table 1, Figure 3A).

For planktonic population 2, two different SNPs in *phoQ* acquired by day 3 of selection corresponded with a 512-fold increase in colistin MIC. This dramatic rise in MIC during the first 3 days of selection was prominent in population 2, compared to populations 1 and 3, which showed more gradual increases over selection time (Figure 3A–C). Population 2 showed a similar pattern of *phoQ* SNPs generated at different positions through the course of evolution to that of population 1 (Table 1). Colistin MIC level remained consistent from days 3 to 27, in alignment with the duration of a unique SNP in *phoQ* and an SNP in thioredoxin-dependent thiol peroxidase BCP, observed on days 15–27. The final jump in MIC (2048-fold) was linked to mutations at unique positions of UDP-3-

O-(3-hydroxymyristoyl)glucosamine N-acyltransferase LpxD, *phoQ*, and *bcp*, all at 100% frequency. Interestingly, no *prmB* mutations were observed for population 2 for the extent of colistin resistance evolution (Table 1, Figure 3B).

Table 1. Timepoint and MIC of acquired COL-R mutations in planktonic evolved KP populations. COL-R mutations are represented for each individually evolved KP population with colistin MIC (μg/mL) and population frequency. Mutations with shared gene positions between populations are shaded gray. Shared mutations in genes that are acquired in independently evolved populations are color-coded for comparison. Mutations unique to individual populations are in white background. The shading is darker for mutations with higher population frequency.

	Position	Mutation	Gene	Day 1: 64	Day 2: 64	Day 3: 64	Day 6: 128	Day 15: 512	Day 27: 512	Day 36: 2048
Population 1	796,797	SNP	*mscL*	94.20%	100%	100%	100%	100%	100%	100%
	2,898,145	SNP	*nadA*	94.50%	100%	100%	100%	100%	100%	100%
	4,782,985:1	INS	*rmpA*	100%	100%	100%	100%	100%	100%	100%
	3,366,007	SNP	*phoQ*		26.30%	9.80%	92.50%	100%	100%	
	5,228,037	SNP	*bcp*				24.80%	100%	100%	
	2,287,052	SNP	*lpxD*					15.00%	100%	
	4,585,314	SNP	*mgrB*						70.80%	
	3,366,111	SNP	*phoQ*							100%
	2,949,548	SNP	*pmrB*							68.60%
	363,465	SNP	*mdtO*	12.20%						
	Position	**Mutation**	**Gene**	**Day 1: 32**	**Day 2: 256**	**Day 3: 512**	**Day 6: 512**	**Day 15: 512**	**Day 27: 512**	**Day 36: 2048**
Population 2	796,797	SNP	*mscL*	94.20%	100%	100%	100%	100%	100%	100%
	2,898,145	SNP	*nadA*	94.50%	100%	100%	100%	100%	100%	100%
	4,782,985:1	INS	*rmpA*	100%	100%	100%	100%	100%	100%	100%
	3,367,207	SNP	*phoQ*			12.60%	79.30%	100%	100%	
	3,366,550	SNP	*phoQ*			25.10%	14.70%			
	5,228,086	SNP	*bcp*					85.00%	85.50%	
	2,287,052	SNP	*lpxD*						10.40%	
	2,287,023	SNP	*lpxD*							100%
	3,366,222	SNP	*phoQ*							100%
	5,228,407	DEL	*bcp*							100%
	Position	**Mutation**	**Gene**	**Day 1: 32**	**Day 2: 64**	**Day 3: 64**	**Day 6: 64**	**Day 15: 512**	**Day 27: 512**	**Day 36: 2048**
Population 3	796,797	SNP	*mscL*	94.20%	100%	100%	100%	100%	100%	100%
	2,898,145	SNP	*nadA*	94.50%	100%	100%	100%	100%	100%	100%
	4,782,985:1	INS	*rmpA*	100%	100%	100%	100%	100%	100%	100%
	2,949,368	SNP	*pmrB*				12.60%	86.30%	63.60%	
	3,367,778	SNP	*phoP*						33.00%	
	3,366,214	SNP	*phoQ*						50.30%	
	2,287,022	SNP	*lpxD*						37.60%	
	3,367,218	SNP	*phoQ*						22.40%	
	3,366,222	SNP	*phoQ*							72.90%
	796,974	SNP	*mscL*							21.00%
	3,366,013	SNP	*phoQ*							15.90%

Abbreviations: single-nucleotide polymorphism—SNP; insertion—INS; deletion—DEL.

Planktonic population 3 showed a unique pattern of SNPs in *phoQ* and *pmrB* with enhanced selection. On day 6, a SNP in *pmrB* was acquired and increased to 86.3% frequency by day 15. On day 15, a substantial rise in MIC was seen (512-fold) along with an additionally acquired SNP in two-component system response regulator PhoP that exists in the population until day 27 (Table 1, Figure 3C). On day 27, three unique mutations in *phoQ* (two positions) and *lpxD* were observed for which the MIC was maintained at 512-fold compared to baseline. The *phoQ* mutation seen at the same position as population 2 occurred at 73% frequency in population 3 on day 36, consistent with a 2048-fold increase in colistin MIC. The shifts in *phoQ* SNP positions with increased selection pressure likely contributed to the enhanced evolution of colistin resistance. The temporal regulation of COL-R mutations can be seen via the generation of key mutations sequentially, in alignment with considerable increases in MIC, with many alterations rising in population frequency or becoming fixed with enhanced selection pressure (Table 1, Figure 3A–C).

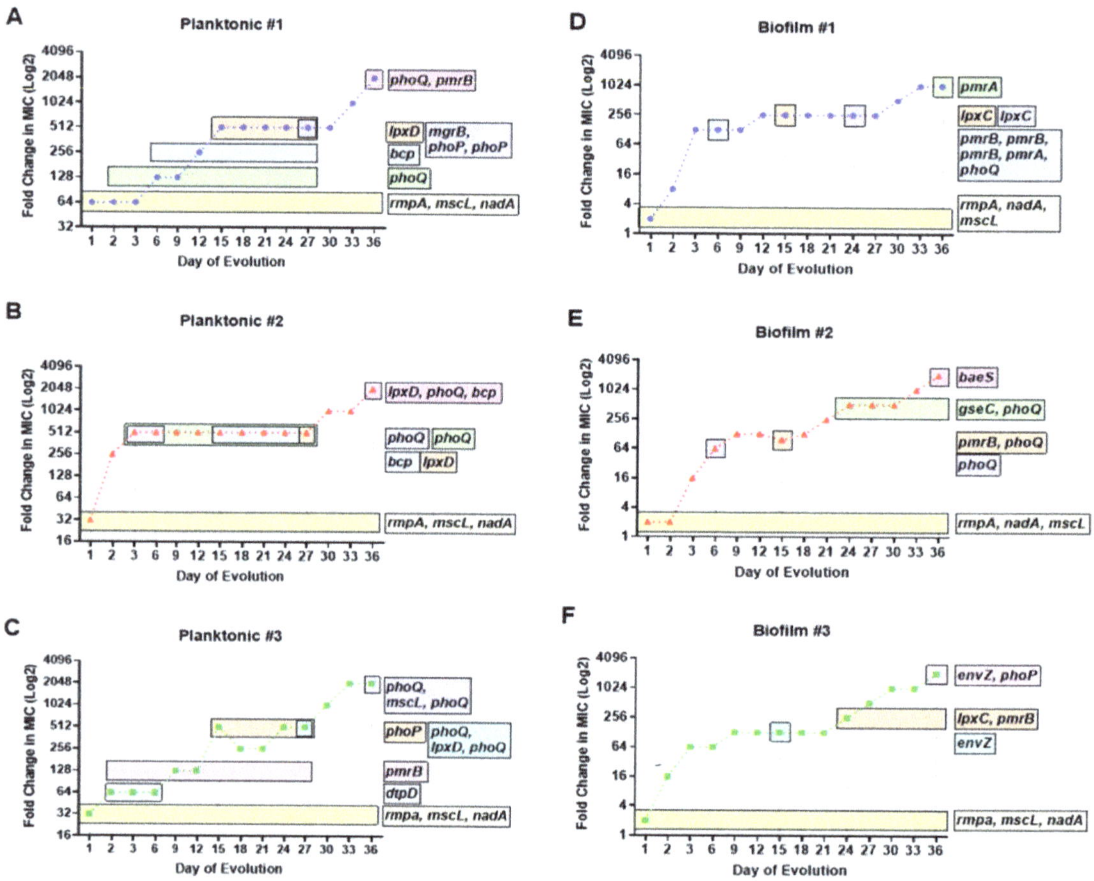

Figure 3. Temporal regulation of colistin-resistant mutations in planktonic and biofilm KP populations. Timelines of acquired mutations having mutation frequencies greater than 10% in known and theorized COL-R genes are represented by horizontal bars throughout a 36 day evolution. The duration of evolution in which the COL-R mutations exist within each planktonic (**A**–**C**) and biofilm (**D**–**F**) population is compared against colistin MIC. For both lifestyles, mutations in genes *rmpA*, *nadA*, and *mscL* were generated after 24 h at $\frac{1}{2}$ MIC colistin selection and were subsequently fixed in each population. Planktonic populations 1 (**A**) and 2 (**B**) acquired a mutation in *bcp* around midpoint of evolution followed by mutations in *lpxD*. Biofilm populations 1 (**D**) and 3 (**F**) acquired mutations in *lpxC* around midpoint of evolution. Mutations in two component system sensor regulator genes *phoQ*/*phoP* and histidine kinase genes *pmrB*/*pmrA* appeared sporadically, in relation to notable (at least twofold) increases in colistin MIC.

3.3. Temporal Regulation of COL-R Mutations in Biofilm KP

Similar to the "first wave" of colistin resistance acquired in planktonic KP, after 24 h selection, all three replicate KP biofilm populations showed mutations in *rmpA*, *nadA*, and *mscL* at 100% frequency (Table 2, Figure 3D–F). Insertions in *rmpA*, and SNPs in *nadA* and *mscL* coincided with 128-, 16-, and 64-fold increases in colistin MIC by day 3 of selection for biofilm populations 1–3, respectively. After 6 days of selection, biofilm population 1 acquired three SNPs in *pmrB* at unique positions, as well as an SNP in *phoQ*, consistent with a 128-fold increase in MIC. These mutations generated by day 6 were lost. Subsequent SNPs in *phoQ* and UDP-3-O-acyl-N-acetylglucosamine deacetylase LpxC were later acquired by day 24, which likely contributed to additional MIC increase (Figure 3D). In addition, a SNP in *lpxC* by day 15 aligned with another twofold increase in MIC, equivalent to a 256-fold increase from baseline. The final day of selection showed a 1024-fold increase in MIC, which overlapped with the timing of a deletion in PhoP/PhoQ regulator and DNA-binding transcriptional repressor MgrB–KdgR and a SNP in two-component system response regulator PmrA (Table 2, Figure 3D).

Table 2. Timepoint and MIC of acquired COL-R mutations in biofilm evolved KP populations. COL-R mutations are represented for each individually evolved KP population with colistin MIC (µg/mL) and population frequency. Mutations with shared gene positions between populations are shaded gray. Shared mutations in genes that are acquired in independently evolved populations are color-coded for comparison. Mutations unique to individual populations are in white background. The shading is darker for mutations with higher population frequency.

	Position	Mutation	Gene	Day 1: 2	Day 2: 8	Day 3: 128	Day 6: 128	Day 15: 256	Day 24: 256	Day 30: 512	Day 36: 1024
Population 1	796,797	SNP	*mscL*	100%	100%	100%	100%	100%	100%	100%	100%
	2,898,145	SNP	*nadA*	100%	100%	100%	100%	100%	100%	100%	100%
	4,782,985:1	INS	*rmpA*	100%	100%	100%	100%	100%	100%	100%	100%
	2,949,706	SNP	*pmrB*				47.60%				
	2,949,547	SNP	*pmrB*				40.70%				
	2,949,548	SNP	*pmrB*				38.60%				
	2,950,483	SNP	*pmrA*				36.40%				
	3,367,219	SNP	*phoQ*				13.10%		52.20%		
	2,178,392	SNP	*lpxC*					100%			
	2,178,666	SNP	*lpxC*						36.70%		
	4,585,299	DEL	*[mgrB]–[kdgR]*							100%	100%
	2,950,399	SNP	*pmrA*								100%

	Position	Mutation	Gene	Day 1: 2	Day 2: 2	Day 3: 16	Day 6: 64	Day 15: 96	Day 24: 512	Day 30: 512	Day 36: 2048
Population 2	796,797	SNP	*mscL*	100%	100%	100%	100%	100%	100%	100%	100%
	2,898,145	SNP	*nadA*	100%	100%	100%	100%	100%	100%	100%	100%
	4,782,985:1	INS	*rmpA*	100%	100%	100%	100%		100%	100%	100%
	3,366,550	SNP	*phoQ*				74.60%				
	2,949,916	SNP	*pmrB*					75.90%			
	3,366,423	SNP	*phoQ*					33.90%			
	554,800	DEL	*qseC*						100%	100%	100%
	3,365,950	SNP	*phoQ*						100%	100%	100%
	4,903,958	SNP	*baeS*								100%

	Position	Mutation	Gene	Day 1: 2	Day 2: 16	Day 3: 64	Day 6: 64	Day 15: 128	Day 24: 256	Day 30: 1024	Day 36: 2048
Population 3	796,797	SNP	*mscL*	100%	100%	100%	100%	100%	100%	100%	100%
	2,898,145	SNP	*nadA*	100%	100%	100%	100%	100%	100%	100%	100%
	4,782,985:1	INS	*rmpA*	100%	100%	100%	100%	100%	100%	100%	100%
	864,664	SNP	*envZ*					100%			
	2,178,740	SNP	*lpxC*						100%	100%	100%
	2,949,703	SNP	*pmrB*						100%	100%	100%
	864,325	SNP	*envZ*								61.80%
	3,367,440	SNP	*phoP*								45.40%

Abbreviations: single-nucleotide polymorphism—SNP; insertion—INS; deletion—DEL.

For biofilm population 2, a similar timeframe for generating two-component system sensor histidine kinase mutations was seen to that of population 1 (Table 1, Figure 3E). By day 6, an SNP of 75% frequency was acquired in *phoQ*, consistent with a 64-fold increase in MIC from baseline. Two unique SNPs in *pmrB* and *phoQ* were generated by day 15 (Figure 3E), with a slight twofold increase in MIC. A 512-fold increase in MIC from baseline occurred by day 24, with the generation of an SNP in *phoQ* and deletion in two-component system sensor histidine kinase QseC, which remain fixed in the population for the extent of the selection. By day 36 of selection, a newly appeared mutation in two-component system sensor histidine kinase BaeS aligned with a 2048-fold increase in colistin MIC from baseline (Table 2, Figure 3E).

For biofilm KP population 3, a delay in the generation of two-component system sensor histidine kinase mutations was observed until day 15 of selection (Table 2, Figure 3F). By day 15, a mutation in two-component system sensor histidine kinase EnvZ was generated, consistent with a 128-fold increase in MIC from baseline. Another twofold increase in MIC was seen by day 24, for which mutations in *lpxC* and *pmrB* were acquired and fixed for the extent of selection. By day 36 of selection, two unique mutations in *envZ* and *phoP* were generated, which aligned with a 2048-fold increase in MIC from baseline (Table 2, Figure 3F). Through monitoring the MIC change via increasing concentrations of colistin selection, key mutations in two-component system sensor and regulators, as well as SNPs in *lpxC*, appear to drive COL-R in biofilm populations. The modification in mutation position for these critical genes and the increase in population frequency over time are also likely contributing factors to enhanced colistin resistance, independent of bacterial lifestyle. Next, we aimed to further characterize COL-R in KP by studying the timing of these mutations in relation to gene function.

3.4. *Timing of Colistin Resistance Mutations by Functional Roles*

Planktonic and biofilm-evolved KP in this study shared many COL-R mutations with various functions in capsule production, cell membrane integrity, energy metabolism, and modification of LPS structure and biosynthesis (Tables 3 and 4). Defense against reactive oxygen species (ROS) by *bcp* was a mutation only observed in planktonic-evolved KP, while mutations in genes regulating fatty-acid biosynthesis (*fadR* and *acpP*), biofilm formation (*qseC*), and peptide transport (*sbmA*) were only observed in biofilm-evolved KP (Tables 3 and 4). A mutation in DUF3413 domain-containing protein was observed in both planktonic and biofilm KP, but remains functionally uncharacterized. However, the generation time of this mutation appears to align with other mutations on genes responsible for the modification of LPS (*lpxC* and *lpxD*) (Figure 4). Next, we assessed the timing of mutations according to their functional roles observed for both planktonic- and biofilm-evolved KP to better understand their influence on the COL-R mechanism.

3.4.1. Capsule Production

Hypervirulent KP has been characterized on the basis of several features, including the virulence gene *rmpA*, a regulator of the mucoid phenotype A, which is an activator for capsular polysaccharide synthesis [30,31]. Low *rmpA* expression has been shown to be correlated with a hypervirulence-negative phenotype in KP [30]. However, less is understood regarding the impact of *rmpA* on COL-R and the potential evolutionary tradeoff between hypervirulence and COL-R. An insertion in *rmpA* was acquired by day 1 of selection in both bacteria lifestyles and was nearly fixed after 24 h. The timing of this insertion aligns with 64-, 32-, and 32-fold increases in MIC for planktonic populations 1–3 (Table 3) and twofold increases in MIC for biofilm populations (Table 4). Notably, these MIC increases were similarly observed for *mscL* and *nadA* mutations, which have functions in cell membrane integrity and energy metabolism, respectively. Our results suggest that capsule production is one of the earliest functional groups modified by colistin selection (Figure 4).

3.4.2. Cell Membrane Integrity

Genes with roles in cell membrane integrity that were altered following colistin selection include *mscL* and *baeS*, which were generated by day 1 and day 36 in planktonic and biofilm populations, respectively (Tables 3 and 4, Figure 4). SNPs in *baeS* were observed in planktonic population 3 and biofilm population 2 on the final day of selection at 75% and 100% frequency, respectively (Tables 3 and 4).

Table 3. Functional roles of COL-R mutations in relation to timing, population frequency (percentage), and colistin MIC (μg/mL) for planktonic-evolved KP. Resistant MICs are shown in red, and mutations independent to planktonic lifestyle are shown in blue.

Planktonic KP Colistin-Resistance Mutations by Functional Role and MIC												
			Population 1			Population 2			Population 3			
Gene	**Description**	**Function**	**Day**	**Freq %**	**MIC μg/mL**	**Day**	**Freq %**	**MIC μg/mL**	**Day**	**Freq %**	**MIC μg/mL**
rmpA	mucoid phenotype A regulator	Capsule production	1–36	100	64–2048	1–36	100	32–2048	1–36	100	32–2048
mscL	large-conductance mechanosensitive channel protein	Cell Membrane Integrity	1–36	94–100	64–2048	1–36	92–100	32–2048	1–36	100	32–2048
baeS	two-component system sensor histidine kinase		-	-	-	-	-	-	36	75	2048
nadA	quinolinate synthase	Energy Metabolism	1–36	95–100	64–2048	1–36	91–100	32–2048	1–36	95–100	32–2048
IT767_20610	XylR family transcriptional regulator		-	-	-	36	100	2048	36	83	2048
mgrB	PhoP/PhoQ regulator	Modification of LPS	27	71	512	-	-	-	-	-	-
phoQ	two-component system sensor histidine kinase		2–27 36	26–100	64-2048	3–6, 3–27 36	13–100	512–2048	27 36	22 73	512 2048
phoP	two-component system response regulator		27	10–15	512	-	-	-	15–27	8–33	512
pmrB	two-component system sensor histidine kinase		36	69	2048	-	-	-	2–27	7–86	64–512
arnC	undecaprenyl-phosphate 4-deoxy-4-formamido-L-arabinose transferase		-	-	-	-	-	-	36	81	2048
lpxD	UDP-3-O-(3-hydroxymyristoyl) glucosamine N-acyltransferase	LPS biosynthesis	15–27	15–100	512–2048	27 36	10 100	512 2048	27	38	512
lpxC	UDP-3-O-acyl-N-acetylglucosamine deacetylase		-	-	-	27	22	512	15	14	512
IT767_01585	DUF3413 domain-containing protein	Unknown	15	16	512	27	30	512	27–36	100	512–2048
bcp	thioredoxin-dependent thiol peroxidase	ROS Defense	6–27	25–100	128–512	15–27 36	8–100	512–2048	-	-	-

Abbreviations: mutation frequency—Freq; hypermucoviscous—HMV; nicotinamide adenine dinucleotide—NAD; lipopolysaccharide—LPS; reactive oxygen species—ROS.

3.4.3. Energy Metabolism

Quinolinate synthase NadA, XylR family transcriptional regulator, and pyrroloquinoline–quinone synthase PqqC are associated with energy metabolism and were modified through colistin selection. SNPs in *nadA* were generated by day 1 and persisted throughout selection. Through carbon catabolite repression, bacteria may activate transcription factor XylR to regulate the metabolism of L-arabinose and D-xylose in place of glucose [32]. Mutations in XylR family transcriptional regulator were only acquired in planktonic populations 2 and 3 on day 36 (Table 3). *PqqC* expression is required for the biosynthesis of pyrroloquinoline quinone, a vitamin and redox cofactor of bacterial dehydrogenases, important for cell growth and metabolic reactions [33]. A mutation in *pqqC* was generated only in biofilm population 3 on day 24 but persisted until the end of selection at 100% frequency (Table 4). While mutations in *nadA* were fixed early in selection, mutations in *xylR* and *pqqC* were required at later timepoints for planktonic and biofilm populations, respectively (Figure 4). It is likely that alterations in bacterial metabolism are direct responses to the environmental stress posed by increasing colistin selection pressure.

Table 4. Functional roles of COL-R mutations in relation to timing, population frequency (percentage), and colistin MIC (μg/mL) for biofilm-evolved KP. Resistant MICs are shown in red, sensitive MICs are shown in green, and mutations independent to biofilm lifestyle are shown in green.

Biofilm KP Colistin-Resistance Mutations by Functional Role and MIC											
			Population 1			Population 2			Population 3		
Gene	Description	Function	Day	Freq %	MIC μg/mL	Day	Freq %	MIC μg/mL	Day	Freq %	MIC μg/mL
rmpA	mucoid phenotype A regulator	Capsule production	1–36	100	2–1024	1–36	100	2–2048	1–36	100	2–2048
mscL	large-conductance mechanosensitive channel protein	Cell Membrane Integrity	1–36	100	2–1024	1–36	100	2–2048	1–36	100	2–2048
baeS	two-component system sensor histidine kinase		-	-	-	36	100	2048	-	-	-
nadA	quinolinate synthase	Energy Metabolism	1–36	100	2–1024	1–36	100	2–2048	1–36	100	2–2048
pqqC	pyrroloquinoline-quinone synthase		-	-	-	-	-	-	24–36	100	256–2048
[mgrB]–[kdgR]	PhoP/PhoQ regulator - DNA-binding transcriptional repressor	Modification of LPS	30–36	100	512–1024	-	-	-	-	-	-
phoQ	two-component system sensor histidine kinase		6, 24	13, 52	128, 256	6, 15, 24–36	34–100	64–2048	-	-	-
phoP	two-component system response regulator		-	-	-	-	-	-	36	45	2048
pmrB	two-component system sensor histidine kinase		6, 6, 6	39–48	128	15	76	96	24–36	100	256–2048
lpxC	UDP-3-O-acyl-N-acetylglucosamine deacetylase	LPS Biosynthesis	15, 24	100, 37	256	-	-	-	24–36	100	256–2048
IT767_01585	DUF3413 domain-containing protein	Unknown	-	-	-	15, 24–36	63, 100	96, 512–2048	15	94	128
fadR	fatty acid metabolism transcriptional regulator	Fatty Acid Biosynthesis	24, 36	34, 100	256, 1024	-	-	-	-	-	-
acpP	acyl carrier protein		24–36	46–100	256–1024	-	-	-	-	-	-
qseC	two-component system sensor histidine kinase	Biofilm Formation	-	-	-	24–36	100	512–2048	-	-	-
sbmA	peptide antibiotic transporter	Peptide Transport	-	-	-	-	-	-	24–30	52–70	256–1024

Abbreviations: mutation frequency—Freq; hypermucoviscous—HMV; nicotinamide adenine dinucleotide—NAD; lipopolysaccharide—LPS; antimicrobial peptide—AMP.

3.4.4. Modification of LPS

The disruption of polymyxin interactions with negatively charged phosphate groups of lipid A of LPS is a commonly observed mechanism of COL-R [34]. Several genes involved in the modification of LPS were observed for both lifestyles through selection, including *mgrB*, *phoQ*, *phoP*, and *pmrB*. Two-component transduction systems PmrAB and PhoPQ regulate the modifications of LPS in response to colistin or a decline or rise in Mg^{2+} and Fe^{3+} levels, respectively [35]. Activation of the PhoPQ signaling system leads to the synthesis of small regulatory transmembrane protein MgrB, which functions as a negative feedback regulator of PhoPQ systems [36]. SNPs in *phoQ* were acquired early in planktonic KP, by days 2, 3, and 27 for populations 1–3. A SNP in *pmrB* was acquired by day 2 and persisted until day 27 in population 3, while another SNP in *pmrB* was generated by day 36 in population 1. Mutations in *phoP* were generated by days 27 and 15 for populations 1 and 3. A single SNP in *mgrB* was observed on day 27 for population 1 at 71% frequency (Table 3). *ArnC*, which encodes undecaprenyl-phosphate 4-deoxy-4-formamido-L-arabinose transferase, was also modified in planktonic population 3 by day 36 (81%). ArnC is one of several enzymes involved in adding an amino sugar L-Ara4N to lipid A, which disrupts the interaction of cationic peptides with LPS, leading to rapid colistin resistance [37].

Timing of COL-R mutations in Planktonic KP

Gene	Timing	Functional Role
rmpA	1-36	Capsule Production
mscL	1-36	Cell Membrane Integrity
nadA	1-36	Energy Metabolism
phoQ	2-36	Modification of LPS
pmrB	2-36	Modification of LPS
bcP	6-27	ROS Defense
phoP	15-27	Modification of LPS
lpxC	15, 27	LPS Biosynthesis
lpxD	15-36	LPS Biosynthesis
IT767_01585	15-36	Uncharacterized Function
mgrB	27	Modification of LPS
arnC	36	Modification of LPS
baeS	36	Cell Membrane Integrity
IT767_20610	36	Energy Metabolism

Timing of COL-R mutations in Biofilm KP

Gene	Timing	Functional Role
rmpA	1-36	Capsule Production
mscL	1-36	Cell Membrane Integrity
nadA	1-36	Energy Metabolism
phoQ	6-36	Modification of LPS
pmrB	6-36	Modification of LPS
lpxC	15-36	LPS Biosynthesis
IT767_01585	15-36	Uncharacterized Function
sbmA	24-30	Peptide Transport
fadR	24, 36	Fatty Acid Biosynthesis
acpP	24-36	Fatty Acid Biosynthesis
pqqC	24-36	Energy Metabolism
qseC	24-36	Biofilm Formation
[mgrB]-[kdgR]	30-36	Modification of LPS
phoP	36	Modification of LPS
baeS	36	Cell Membrane Integrity

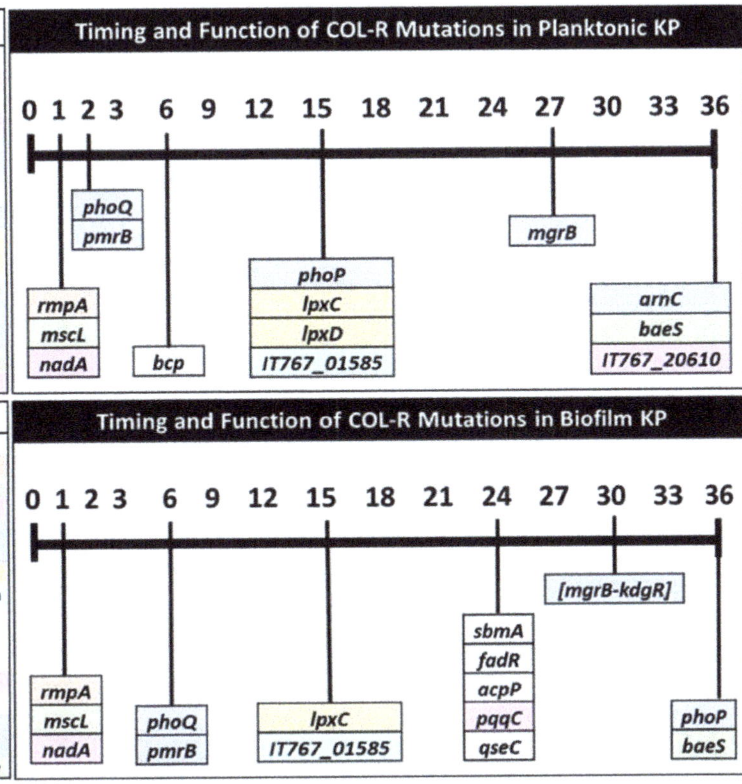

Figure 4. Timing of colistin-resistant mutations by functional role for independent KP lifestyles. Mutations are listed in relation to the first appeared time during evolution and color-coded by functional role to compare COL-R resistance patterns in planktonic and biofilm lifestyles. The timing in which shared mutations are generated and their corresponding functions can be compared between planktonic and biofilm-evolved KP. The colors in the bars are used to identify the individual or combinations of mutations generated at a similar timeframe throughout colistin selection. There is no connection between the color scheme used here and in other figures in this manuscript.

For biofilm populations, SNPs in *phoQ* and *pmrB* were generated by day 6 in populations 1 and 3, while a single SNP in *phoP* was generated on day 36 in population 3. Multiple SNPs in *phoQ* were generated in biofilm population 2 at unique positions, with each additional mutation aligning with a substantial rise in MIC (Table 4). A single deletion in *mgrB–kdgR* was generated on day 30 and was subsequently fixed. Overall, it appears that mutations involved in LPS modifications are observed early on (by days 2–6) in selection and reappear at later timepoints (days 27–36) for both planktonic and biofilm populations. Additionally, mutations involved in LPS biosynthesis are generated around or shortly following modifications in LPS (Figure 4).

3.4.5. LPS Biosynthesis

In our experimental evolution, we observed mutations in *lpxc* and *lpxD* at later timepoints of selection in both planktonic and biofilm KP (Figure 4). SNPs in *lpxD* were acquired by days 15–27 for all planktonic populations, while SNPs in *lpxC* were observed on days

27 and 15 for populations 2 and 3, corresponding to a high level of COL-R (512 μg/mL) (Table 3). For biofilm populations 1 and 3, SNPs in *lpxC* were observed by day 15 and 24 at 100% frequency.

3.4.6. ROS Defense

Mutations in bacterioferritin comigratory protein (BCP) were solely observed in planktonic populations 1 and 2 through selection. A single mutation in *bcp* was acquired on day 6 that persisted until day 27 in population 1, which aligned with a 128–512-fold increase in MIC from baseline. Two mutations in *bcp* were observed in population 2 on days 15–27 and day 36, consistent with 512–2048-fold increases in MIC from baseline (Table 3). BCP has roles in bacteria's defense against environmental ROS such as hydrogen peroxide [38]. It is likely that mutations in *bcp* are responses to oxidative stress posed by colistin treatment on bacterial cells.

3.4.7. Peptide Transport

A single mutation in peptide antibiotic transport gene *sbmA* was observed on days 24–30 for biofilm population 3, with a 256–1024-fold increase in MIC (Table 4). SbmA is an inner membrane transporter that facilitates the transport of antimicrobial peptides, especially those that are proline-rich, into the cell [39]. It has recently been observed that mutations in *sbmA* confer resistance to certain peptide conjugates [40]. Biofilm KP mutations in *sbmA* occurred following LPS biosynthesis modifications, in accordance with two mutations having roles in fatty-acid biosynthesis (Figure 4).

3.4.8. Fatty-Acid Biosynthesis

Two mutations were observed in biofilm population 1 with roles in fatty-acid biosynthesis. A mutation in fatty-acid metabolism transcriptional regulator FadR was observed on days 24 and 36, which corresponded to 256- and 1024-fold increases in MIC from baseline. This gene activates fatty-acid synthesis while repressing fatty-acid degradation in response to environmental fatty-acid levels and may influence bacterial cell size [41]. A mutation in acyl carrier protein ACP was generated and fixed after day 24 and aligned with a 246–1024-fold increase in MIC (Table 4). ACP is a highly conserved transport protein for acyl intermediates and is necessary for fatty-acid biosynthesis [42]. Mutations in *fadR* and *acpP* aligned with the timing of a mutation in *pqqC*, which affects energy metabolism, and a mutation in *qseC*, which has roles in biofilm formation (Figure 4).

3.4.9. Biofilm Formation

A single mutation in a two-component system sensor histidine kinase QseC was generated in biofilm population 2 on day 24 and became fixed. This mutation coincided with a 512–2048-fold increase in MIC from baseline (Table 4). *QseC* is an integral gene of the QseBC two-component system of quorum-sensing with roles in modulating biofilm formation and potentially regulating virulence in KP [43].

3.4.10. Mutations with Uncharacterized Function

A mutation in a DUF3413 domain-containing protein was observed following alterations in LPS synthesis genes *lpxC* and *lpxD*, independent of bacterial lifestyle. While the function of this gene is uncharacterized, it was generated by days 15–27 in planktonic and biofilm populations. The mutations in this gene aligned with a 512-fold increase in MIC from baseline for planktonic populations (Table 3). Biofilm population 2 acquired two independent mutations in this gene; one mutation on day 15 corresponded to a 96-fold increase in MIC from baseline, while another mutation that was fixed after day 24 aligned with 512–2048-fold increases in MIC from baseline. A single mutation was observed on day 15 for population 3, with a 128-fold increase in MIC compared to the baseline (Table 4).

3.5. Theoretical Pathways of Colistin Resistance in *K. pneumoniae*

On the basis of the similarities in mutations observed with increasing colistin MIC by bacterial lifestyle, we theorize a pathway of colistin resistance, including the timing of affected gene functions and their corresponding mutations (Figure 5). We predict that an initial mutation in *rmpA* leads to loss of capsule polysaccharide (cps) synthesis, followed by regulation in osmotic pressure and efflux (mutations in *mscL* and *baeS*). Next, one could observe changes in energy metabolism through a modification in *nadA*. Additionally, mutations in *mgrB*, *phoQ/phoP*, and *pmrB* led to the addition of 4-amino-4-deoxy-L-arabinose and/or the transfer of phosphoethanolamine (pEth) by enzyme phosphoethanolamine transferase. Lastly, mutations in *lpxA* and *lpxD* critical for lipid A synthesis contribute to LPS loss. These mutations allow for the development and progression of colistin resistance in hypervirulent *K. pneumoniae*.

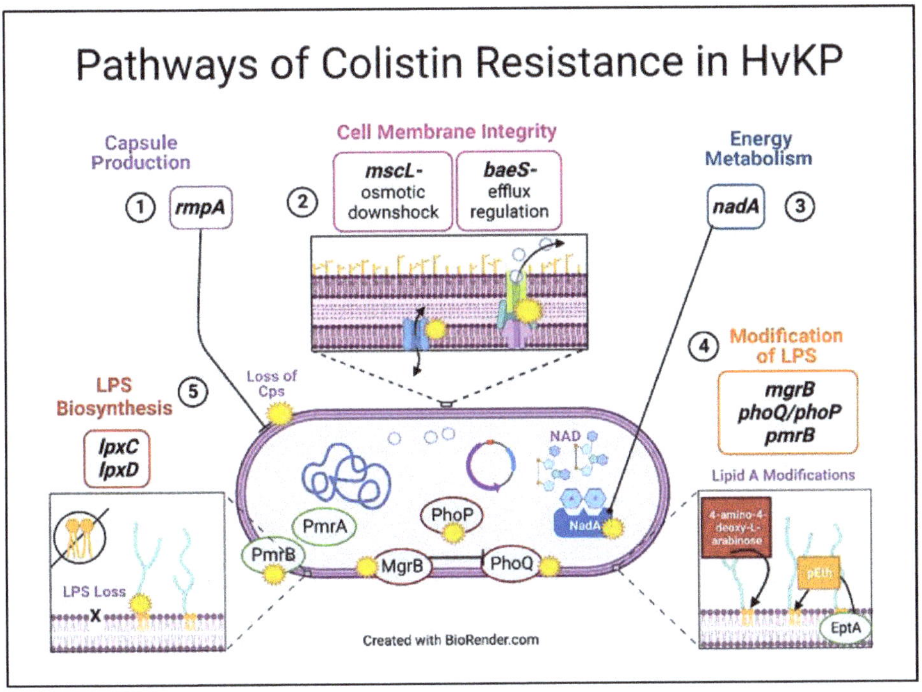

Figure 5. Theoretical pathways of colistin resistance in hypervirulent-*K. pneumoniae*. Colistin selection led to changes in five major gene functional groups, possibly in a sequential manner in both bacterial lifestyles: capsule production, cell membrane integrity, energy metabolism, modifications of LPS, and LPS biosynthesis loss. We posit that the mutations related to these functional groups facilitate and allow for enhanced colistin resistance in *K. pneumoniae*.

3.6. COL-R Isolates Remain Susceptible to Dual-Inhibitor Antibiotics

Recently, physicians have started using the newly approved β-lactam/β-lactamase dual-inhibitor antibiotics to treat clinical infections that are resistant to colistin treatment. We investigated the potential for cross-resistance of these COL-R isolates to a novel dual-inhibitor antibiotic, ceftazidime–avibactam (CAZ/AVI). Throughout the 36 day experimental evolution under colistin selection, we also tested the MIC of CAZ/AVI for both planktonic and biofilm populations every three days. We found that both planktonic and biofilm lifestyles showed sustained susceptibility to CAZ/AVI, throughout the 36 days of selection, even with substantial COL-R progression (Figure 6).

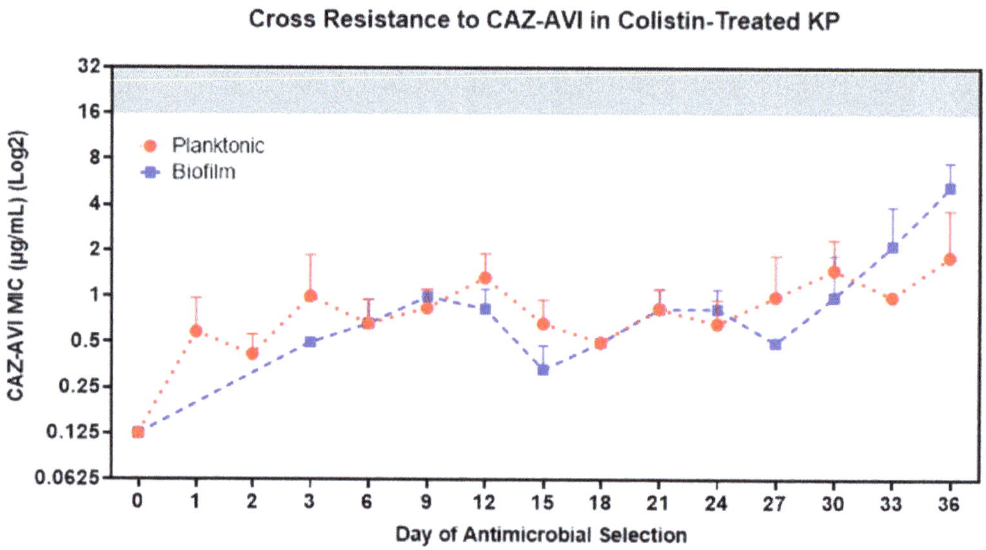

Figure 6. COL-R isolates do not generate cross-resistance to CAZ/AVI. Following 36 days of experimental evolutions of KP ATCC 43816 under colistin selection, cross-resistance to CAZ/AVI was assessed for COL-R planktonic (red) and biofilm (blue) isolates from all populations in all timepoints. The clinical breakpoint for CAZ/AVI is 16 µg/mL, and the resistance MIC area is shown in a gray background. The data points reflect the mean MIC to CAZ/AVI of three replicate populations by lifestyles and colistin treatment conditions. Data are representative of three independent experiments.

3.7. Loss of Hypermucoviscous Phenotype with Colistin Selection

Our results indicated that rapid COL-R was initially dependent on mutations in mucoid phenotype A regulator gene *rmpA*, leading to changes in capsular polysaccharide synthesis and the hypermucoviscous (HMV) phenotype, independent of bacterial lifestyle. To demonstrate the *rmpA* mutations and alterations in HMV, we conducted the string test for hypervirulence on 1 day evolved (under $\frac{1}{2}$ MIC colistin treatment) planktonic population clones compared to the evolutionary ancestor clone. We found a significant decrease in the percentage of colonies that passed string tests following 1 day of colistin treatment for all three planktonic populations (Supplementary Figure S2). These results suggest that there may be an evolutionary tradeoff between hypervirulence and progressive resistance to colistin.

4. Discussion

Colistin resistance has been extensively studied with meaningful mutations and their roles in the resistance mechanism identified. Modifications in lipid A moiety of LPS, overexpression of two-component regulatory systems PhoPQ and PmrAB, plasmid-mediated transfer of mobilized colistin resistance genes *mcr-1* to *mcr-8*, and the inactivation of the PhoQ/PhoP signaling regulator MgrB are some of the most understood mechanisms of colistin resistance in Gram-negative bacteria [14,37,44–50]. In this study, we narrow the knowledge gap in understanding the importance of mutation timing in the progression of COL-R by gene function for both planktonic and biofilm KP lifestyles. We show a similar pattern of resistance mutation timing through colistin selection between bacteria lifestyles (Figure 4). The "first wave" of early resistance was consistent between lifestyles, with the generation of mutations relevant to capsule production, cell membrane integrity, and energy metabolism. Mutations in *mscL* and *nadA*, regulating osmotic stress and energy metabolism, respectively, were likely immediate responses to environmental stress posed by colistin treatment. NadA is a catalyst in the biosynthesis of nicotinamide adenine dinucleotide (NAD^+), an essential cofactor, signaling molecule, and coenzyme for redox reactions of energy metabolism [51,52]. MscL is a pore-forming membrane protein that protects the cell from osmotic downshock, by promoting the efflux of various molecules such as potassium, glutamate, and proline, from the cytosol [53]. MscL is considered a potential drug target by acting as an emergency release valve, allowing for the uptake of extracellular molecules, including antibiotics [54]. These mutations may promote the development of further resistance mutations with roles in LPS modification and synthesis.

Following the "first wave", sporadic mutations affecting LPS structure were observed after day 2 in planktonic populations and occurred later on, by day 6, in biofilm populations. It appears that the positioning of the two-component system sensor and response regulator mutations responsible for LPS modifications may impact COL-R progression. For instance, for planktonic and biofilm populations, mutations at unique gene positions were generated in *phoQ* and *pmrB*. It is possible that mutation positions facilitating optimized COL-R became fixed in each population through enhanced selection pressure. Interestingly, mutations in LPS biosynthesis, likely leading to LPS loss, were generated following modifications in lipid A. For both lifestyles, lipid A modifications preceded mutations in *lpxC* and *lpxD* genes responsible for LPS biosynthesis. This suggests that LPS production may have been altered through colistin selection. Complete loss of LPS is a known mechanism of COL-R in *A. baumannii*, through modifications in *lpxA*, *lpxC*, and *lpxD*, the primary genes involved in lipid A synthesis, leading to a dramatic increase in colistin MIC of greater than 256 μg/mL [49]. Loss of LPS through deletions in *lpx* genes associated with COL-R has also been observed in *Escherichia coli* [55]. However, less is understood regarding the potential loss of LPS in KP, leading to rapid COL-R. Decreased LPS and modifications in LPS are likely the dominant mechanisms of COL-R observed in both planktonic and biofilm evolutions in our study. In addition, an uncharacterized gene mutation (*IT767_0158*) was detected in both lifestyles following mutations *lpxC* and/or *lpxD*, suggesting that this gene may have roles in regulating LPS. Further investigation is necessary to understand the role of these gene mutations and whether their function is related to LPS biosynthesis.

We also found that bacterial evolution to resistance is likely specialized according to bacterial lifestyle in terms of mutation timing and MIC increase. The biofilm environment itself is designed to protect and respond in defense to environmental stressors, shown by a lag time in COL-R development compared with planktonic-evolved populations. In addition, biofilm development requires more time and metabolic demand for cell attachment, colonization, and maturation, compared to free-living cell propagation. Several mutations were theoretically specialized for biofilm-evolved COL-R, including mutations with roles in peptide transport, fatty-acid biosynthesis, and biofilm formation (*sbmA*, *fadR*, *acpP*, and *qseC*). The sensory kinase QseC is part of the two-component-based quorum sensing system (QseBC), which responds to environmental stress, including changes in osmotic pressure, heat shock, and oxidative stress [56]. Further investigation is required to understand the role of the quorum sensing system QseBC in COL-R development. Mutations generated at later timepoints (on or past day 27) had roles in regulating LPS modifications (*mgrB*, *phoP*, and *arnC*), as well as the envelope stress response and efflux pump expression (*baeS*). Mutations in *baeS*, generated by day 36 for both lifestyles, were likely attributed to bacterial defense against interactions of colistin at the bacterial cell membrane. The membrane-bound sensor histidine kinase BaeS is part of a two-component system involved in envelope stress response that responds to environmental stressors and regulates the expression of different efflux pumps [57]. The pattern of COL-R mutation timing between bacteria lifestyles appears to be both similar and specialized due to growing metabolic demands and likely differences in bacteria susceptibility to colistin treatment.

Through our long-term 36 day experimental evolution, we were able to monitor changes in mutation position and frequency and addition of newly acquired mutations, with the intent to better understand how KP continues to adapt to increasing colistin pressure. These methods allowed us to assess the temporal regulation and interactions of COL-R mutations for over 374 generations (36 days × ~10.2 generations per day). Additionally, we were able to study the timing of various impacted COL-R genes and their functions by bacterial lifestyle. By performing the experimental evolutions using both planktonic and biofilm growth, we were able to account for the effects and demands of bacteria lifestyle as a part of the COL-R mechanism and compare/contrast the most beneficial mutations that led to substantial colistin MIC increase with enhanced selection pressure.

A limitation of in vitro experimental evolution studies is the inability to capture the resistance mechanisms associated with plasmid-mediated transfer in addition to somatic gene mutations. As expected, we did not observe any horizontal transfer of plasmid-borne COL-R genes such as the *mcr* variants, which have been rising in prevalence in nature and human patients [15]. Nonetheless, the system of in vitro experimental evolutions under a controlled environment allows us to unmistakably identify the most beneficial chromosomal alterations leading to progressively enhanced colistin resistance.

Considering the rapid resistance to colistin detected for KP in both lifestyles, it is crucial to determine if the COL-R isolates that could exist in clinical settings will remain susceptible to the newly approved β-lactam/β-lactamase dual-inhibitor antibiotics. CAZ/AVI has been approved for administration in the United States since 2015 to combat ESBL and carbapenemase-producing infections. Avibactam prevents ceftazidime hydrolysis by carbapenemases (KPCs) and ESBLs, while the third-generation cephalosporin ceftazidime exhibits bactericidal activity by inhibiting cell-wall synthesis [58]. Our results demonstrate the diversity in bacterial resistance mechanisms to various antimicrobials, including AMPs and conventional dual-inhibitor antibiotic treatments designed to prevent resistance development. These results support that CAZ/AVI remains a critical treatment option for colistin-resistant infections, which is increasingly vital with the rise in high mortality and hypervirulent MDR bacterial infections in the clinical setting.

In summary, this study presents a clearer understanding of the timing and significance of COL-R mutations in KP to ultimately aid in the development of new clinical treatments that successfully eradicate and/or prevent CR-KP infections in patients. Experimental approaches using in vitro and in vivo models of antibiotic selection similar to laboratory systems described here and isolates sampled from patients before and after treatment [59,60] would facilitate the identification of potentially critical resistance mutations. In addition, we can study the influence of epigenetic factors such as DNA methylation of bacteria, which regulate the expression of genes [61]. With further gain- or loss-of-function studies to investigate the impact of individual and combinations of mutations on antimicrobial resistance [62–64], these data are useful for the clinical situation by elucidating the critical targets that are utilized by bacteria for adaptive resistance to antimicrobials. Along with understanding the influence of timing of critical mutations and timing of affected genes by their functional roles for cell survival through adaptive evolution to resistance, studies like this will be important in enhancing drug development to prevent or lessen the rapid resistance to antimicrobials.

Supplementary Materials: The following supporting information can be downloaded at https://www.mdpi.com/article/10.3390/pharmaceutics15010270/s1: Figure S1. Schematic of planktonic and biofilm experimental evolution workflow; Figure S2. Evolved KP isolates quickly lost their hypervirulence.

Author Contributions: Conceptualization, J.M.K. and Y.P.D.; formal analysis, J.M.K. and Y.P.D.; investigation, J.M.K. and Y.P.D.; funding acquisition Y.P.D.; resources, Y.P.D.; supervision, Y.P.D.; writing—original draft, J.M.K.; writing—review and editing, Y.P.D. All authors have read and agreed to the published version of the manuscript.

Funding: This work was supported by the National Institutes of Health (R01 AI-133351 to Y.P.D.). The funding agencies had no role in the study design, data collection, and analysis, the decision to publish, or the preparation of the manuscript.

Institutional Review Board Statement: Not applicable.

Informed Consent Statement: Not applicable.

Data Availability Statement: Data are contained within the article or Supplementary Materials. The sequencing data have been deposited with links to BioProject accession number PRJNA922161 in the NCBI BioProject database (https://www.ncbi.nlm.nih.gov/bioproject/PRJNA922161).

Acknowledgments: The authors thank Jaydep Halder and Pranav Kumar Kaliperumal for their assistance in carrying out string test experiments.

Conflicts of Interest: The authors declare no conflict of interest. The funders had no role in the design of the study; in the collection, analyses, or interpretation of data; in the writing of the manuscript; or in the decision to publish the results.

References

1. Jean, S.-S.; Harnod, D.; Hsueh, P.-R. Global Threat of Carbapenem-Resistant Gram-Negative Bacteria. *Front. Cell. Infect. Microbiol.* **2022**, *12*, 823684. [CrossRef] [PubMed]
2. Centers for Disease Control and Prevention. *Antibiotic Resistance Threats in the United States, 2019*; Centers for Disease Control and Prevention: Atlanta, GA, USA, 2019.
3. Bradford, P.A. Extended-Spectrum β-Lactamases in the 21st Century: Characterization, Epidemiology, and Detection of This Important Resistance Threat. *Clin. Microbiol. Rev.* **2001**, *14*, 933–951. [CrossRef]
4. De Angelis, G.; Del Giacomo, P.; Posteraro, B.; Sanguinetti, M.; Tumbarello, M. Molecular Mechanisms, Epidemiology, and Clinical Importance of β-Lactam Resistance in *Enterobacteriaceae*. *Int. J. Mol. Sci.* **2020**, *21*, 5090. [CrossRef] [PubMed]
5. Rice, L.B. Federal Funding for the Study of Antimicrobial Resistance in Nosocomial Pathogens: No ESKAPE. *J. Infect. Dis.* **2008**, *197*, 1079–1081. [CrossRef] [PubMed]
6. Podschun, R.; Ullmann, U. *Klebsiella* spp. as Nosocomial Pathogens: Epidemiology, Taxonomy, Typing Methods, and Pathogenicity Factors. *Clin. Microbiol. Rev.* **1998**, *11*, 589–603. [CrossRef]
7. Lan, P.; Jiang, Y.; Zhou, J.; Yu, Y. A global perspective on the convergence of hypervirulence and carbapenem resistance in *Klebsiella pneumoniae*. *J. Glob. Antimicrob. Resist.* **2021**, *25*, 26–34. [CrossRef]
8. Jorge, P.; Pérez-Pérez, M.; Pérez-Rodríguez, G.; Pereira, M.O.; Lourenço, A. A network perspective on antimicrobial peptide combination therapies: The potential of colistin, polymyxin B and nisin. *Int. J. Antimicrob. Agents* **2017**, *49*, 668–676. [CrossRef]
9. National Center for Biotechnology Information. PubChem Compound Summary for CID 5311054, Polymyxin E. 2022. Available online: https://pubchem.ncbi.nlm.nih.gov/compound/5311054 (accessed on 12 December 2022).
10. National Center for Biotechnology Information. PubChem Compound Summary for CID 4868, Polymyxin b. 2022. Available online: https://pubchem.ncbi.nlm.nih.gov/compound/4868 (accessed on 12 December 2022).
11. Jorge, P.; Lourenço, A.; Pereira, M.O. New trends in peptide-based anti-biofilm strategies: A review of recent achievements and bioinformatic approaches. *Biofouling* **2012**, *28*, 1033–1061. [CrossRef]
12. Falagas, M.E.; Kasiakou, S.K.; Saravolatz, L.D. Colistin: The Revival of Polymyxins for the Management of Multidrug-Resistant Gram-Negative Bacterial Infections. *Clin. Infect. Dis.* **2005**, *40*, 1333–1341. [CrossRef]
13. Li, J.; Nation, R.L.; Milne, R.W.; Turnidge, J.D.; Coulthard, K. Evaluation of colistin as an agent against multi-resistant Gram-negative bacteria. *Int. J. Antimicrob. Agents* **2005**, *25*, 11–25. [CrossRef]
14. Poirel, L.; Jayol, A.; Nordmann, P. Polymyxins: Antibacterial Activity, Susceptibility Testing, and Resistance Mechanisms Encoded by Plasmids or Chromosomes. *Clin. Microbiol. Rev.* **2017**, *30*, 557–596. [CrossRef]
15. Liu, Y.-Y.; Wang, Y.; Walsh, T.R.; Yi, L.-X.; Zhang, R.; Spencer, J.; Doi, Y.; Tian, G.; Dong, B.; Huang, X.; et al. Emergence of plasmid-mediated colistin resistance mechanism MCR-1 in animals and human beings in China: A microbiological and molecular biological study. *Lancet Infect. Dis.* **2016**, *16*, 161–168. [CrossRef] [PubMed]
16. Lewis, K. Riddle of biofilm resistance. *Antimicrob. Agents Chemother.* **2001**, *45*, 999–1007. [CrossRef] [PubMed]
17. del Pozo, J.L.; Patel, R. The Challenge of Treating Biofilm-associated Bacterial Infections. *Clin. Pharmacol. Ther.* **2007**, *82*, 204–209. [CrossRef] [PubMed]
18. Wang, G.; Zhao, G.; Chao, X.; Xie, L.; Wang, H. The Characteristic of Virulence, Biofilm and Antibiotic Resistance of *Klebsiella pneumoniae*. *Int. J. Environ. Res. Public Health* **2020**, *17*, 6278. [CrossRef] [PubMed]
19. Kimura, T.; Kobayashi, K. Role of Glutamate Synthase in Biofilm Formation by *Bacillus subtilis*. *J. Bacteriol.* **2020**, *202*, e00120-20. [CrossRef]
20. Palwe, S.; Bakthavatchalam, Y.D.; Khobragadea, K.; Kharat, A.S.; Walia, K.; Veeraraghavan, B. In-Vitro Selection of Ceftazidime/Avibactam Resistance in OXA-48-Like-Expressing *Klebsiella pneumoniae*: In-Vitro and In-Vivo Fitness, Genetic Basis and Activities of β-Lactam Plus Novel β-Lactamase Inhibitor or β-Lactam Enhancer Combinations. *Antibiotics* **2021**, *10*, 1318. [CrossRef]
21. O'Rourke, D.; FitzGerald, C.E.; Traverse, C.C.; Cooper, V.S. There and back again: Consequences of biofilm specialization under selection for dispersal. *Front. Genet.* **2015**, *6*, 18. [CrossRef]
22. Cooper, V.S. Experimental Evolution as a High-Throughput Screen for Genetic Adaptations. *mSphere* **2018**, *3*, e00121-18. [CrossRef]
23. Lin, Q.; Pilewski, J.M.; Di, Y.P. Acidic Microenvironment Determines Antibiotic Susceptibility and Biofilm Formation of *Pseudomonas aeruginosa*. *Front. Microbiol.* **2021**, *12*, 747834. [CrossRef]

24. Clinical and Laboratory Standards Institute. *CLSI Document M07-A10: Methods for Dilution of Antimicrobial Susceptibility Tests for Bacteria That Grow Aerobically; Approved Standard—10th Edition*; Clinical and Laboratory Standards Institute: Wayne, PA, USA, 2015.
25. Bolger, A.M.; Lohse, M.; Usadel, B. Trimmomatic: A flexible trimmer for Illumina sequence data. *Bioinformatics* **2014**, *30*, 2114–2120. [CrossRef] [PubMed]
26. Langmead, B.; Trapnell, C.; Pop, M.; Salzberg, S.L. Ultrafast and memory-efficient alignment of short DNA sequences to the human genome. *Genome Biol.* **2009**, *10*, R25. [CrossRef]
27. Budnick, J.A.; Bina, X.R.; Bina, J.E. Complete Genome Sequence of *Klebsiella pneumoniae* Strain ATCC 43816. *Microbiol. Resour. Announc.* **2021**, *10*, e01441-20. [CrossRef] [PubMed]
28. Deatherage, D.E.; Barrick, J.E. Identification of Mutations in Laboratory-Evolved Microbes from Next-Generation Sequencing Data Using *breseq*. In *Engineering and Analyzing Multicellular Systems*; Sun, L., Shou, W., Eds.; Methods in Molecular Biology; Springer: New York, NY, USA, 2014; Volume 1151, pp. 165–188, ISBN 9781493905539.
29. Fang, C.-T.; Chuang, Y.-P.; Shun, C.-T.; Chang, S.-C.; Wang, J.-T. A Novel Virulence Gene in *Klebsiella pneumoniae* Strains Causing Primary Liver Abscess and Septic Metastatic Complications. *J. Exp. Med.* **2004**, *199*, 697–705. [CrossRef]
30. Lin, Z.-W.; Zheng, J.-X.; Bai, B.; Xu, G.-J.; Lin, F.-J.; Chen, Z.; Sun, X.; Qu, D.; Yu, Z.-J.; Deng, Q.-W. Characteristics of Hypervirulent *Klebsiella pneumoniae*: Does Low Expression of *rmpA* Contribute to the Absence of Hypervirulence? *Front. Microbiol.* **2020**, *11*, 436. [CrossRef]
31. Lai, Y.-C.; Peng, H.-L.; Chang, H.-Y. RmpA2, an Activator of Capsule Biosynthesis in *Klebsiella pneumoniae* CG43, Regulates K2 *cps* Gene Expression at the Transcriptional Level. *J. Bacteriol.* **2003**, *185*, 788–800. [CrossRef]
32. Ni, L.; Tonthat, N.K.; Chinnam, N.; Schumacher, M.A. Structures of the *Escherichia coli* transcription activator and regulator of diauxie, XylR: An AraC DNA-binding family member with a LacI/GalR ligand-binding domain. *Nucleic Acids Res.* **2013**, *41*, 1998–2008. [CrossRef]
33. Magnusson, O.T.; Toyama, H.; Saeki, M.; Rojas, A.; Reed, J.C.; Liddington, R.C.; Klinman, J.P.; Schwarzenbacher, R. Quinone biogenesis: Structure and mechanism of PqqC, the final catalyst in the production of pyrroloquinoline quinone. *Proc. Natl. Acad. Sci. USA* **2004**, *101*, 7913–7918. [CrossRef]
34. Herrera, C.M.; Hankins, J.V.; Trent, M.S. Activation of PmrA inhibits LpxT-dependent phosphorylation of lipid A promoting resistance to antimicrobial peptides. *Mol. Microbiol.* **2010**, *76*, 1444–1460. [CrossRef]
35. Leung, L.M.; Cooper, V.S.; Rasko, D.A.; Guo, Q.; Pacey, M.P.; McElheny, C.L.; Mettus, R.T.; Yoon, S.H.; Goodlett, D.R.; Ernst, R.K.; et al. Structural modification of LPS in colistin-resistant, KPC-producing *Klebsiella pneumoniae*. *J. Antimicrob. Chemother.* **2017**, *72*, 3035–3042. [CrossRef]
36. Poirel, L.; Jayol, A.; Bontron, S.; Villegas, M.-V.; Ozdamar, M.; Türkoglu, S.; Nordmann, P. The mgrB gene as a key target for acquired resistance to colistin in *Klebsiella pneumoniae*. *J. Antimicrob. Chemother.* **2015**, *70*, 75–80. [CrossRef]
37. Gunn, J.S.; Lim, K.B.; Krueger, J.; Kim, K.; Guo, L.; Hackett, M.; Miller, S.I. PmrA–PmrB-regulated genes necessary for 4-aminoarabinose lipid A modification and polymyxin resistance. *Mol. Microbiol.* **1998**, *27*, 1171–1182. [CrossRef] [PubMed]
38. Jeong, W.; Cha, M.-K.; Kim, I.-H. Thioredoxin-dependent Hydroperoxide Peroxidase Activity of Bacterioferritin Comigratory Protein (BCP) as a New Member of the Thiol-specific Antioxidant Protein (TSA)/Alkyl Hydroperoxide Peroxidase C (AhpC) Family. *J. Biol. Chem.* **2000**, *275*, 2924–2930. [CrossRef] [PubMed]
39. Ghilarov, D.; Inaba-Inoue, S.; Stepien, P.; Qu, F.; Michalczyk, E.; Pakosz, Z.; Nomura, N.; Ogasawara, S.; Walker, G.C.; Rebuffat, S.; et al. Molecular mechanism of SbmA, a promiscuous transporter exploited by antimicrobial peptides. *Sci. Adv.* **2021**, *7*, eabj5363. [CrossRef]
40. Ghosal, A.; Vitali, A.; Stach, J.E.M.; Nielsen, P.E. Role of SbmA in the Uptake of Peptide Nucleic Acid (PNA)-Peptide Conjugates in *E. coli*. *ACS Chem. Biol.* **2013**, *8*, 360–367. [CrossRef]
41. Cronan, J.E. The *Escherichia coli* FadR transcription factor: Too much of a good thing? *Mol. Microbiol.* **2021**, *115*, 1080–1085. [CrossRef]
42. Byers, D.M.; Gong, H. Acyl carrier protein: Structure-function relationships in a conserved multifunctional protein family. *Biochem. Cell Biol.* **2007**, *85*, 649–662. [CrossRef]
43. Lv, J.; Zhu, J.; Wang, T.; Xie, X.; Wang, T.; Zhu, Z.; Chen, L.; Zhong, F.; Du, H. The Role of the Two-Component QseBC Signaling System in Biofilm Formation and Virulence of Hypervirulent *Klebsiella pneumoniae* ATCC43816. *Front. Microbiol.* **2022**, *13*, 817974. [CrossRef]
44. Boinett, C.J.; Cain, A.K.; Hawkey, J.; Do Hoang, N.T.; Khanh, N.N.T.; Thanh, D.P.; Dordel, J.; Campbell, J.I.; Lan, N.P.H.; Mayho, M.; et al. Clinical and laboratory-induced colistin-resistance mechanisms in *Acinetobacter baumannii*. *Microb. Genom.* **2019**, *5*, e000246. [CrossRef]
45. Masood, K.I.; Umar, S.; Hasan, Z.; Farooqi, J.; Razzak, S.A.; Jabeen, N.; Rao, J.; Shakoor, S.; Hasan, R. Lipid A-Ara4N as an alternate pathway for (colistin) resistance in *Klebsiella pneumonia* isolates in Pakistan. *BMC Res. Notes* **2021**, *14*, 449. [CrossRef]
46. Nummila, K.; Kilpeläinen, I.; Zähringer, U.; Vaara, M.; Helander, I.M. Lipopolysaccharides of polymyxin B-resistant mutants of *Escherichia coii* are extensively substituted by 2-aminoethyl pyrophosphate and contain aminoarabinose in lipid A. *Mol. Microbiol.* **1995**, *16*, 271–278. [CrossRef]

47. Moskowitz, S.M.; Brannon, M.K.; Dasgupta, N.; Pier, M.; Sgambati, N.; Miller, A.K.; Selgrade, S.E.; Miller, S.I.; Denton, M.; Conway, S.P.; et al. PmrB Mutations Promote Polymyxin Resistance of *Pseudomonas aeruginosa* Isolated from Colistin-Treated Cystic Fibrosis Patients. *Antimicrob. Agents Chemother.* **2012**, *56*, 1019–1030. [CrossRef]
48. Jin, X.; Chen, Q.; Shen, F.; Jiang, Y.; Wu, X.; Hua, X.; Fu, Y.; Yu, Y. Resistance evolution of hypervirulent carbapenem-resistant *Klebsiella pneumoniae* ST11 during treatment with tigecycline and polymyxin. *Emerg. Microbes Infect.* **2021**, *10*, 1129–1136. [CrossRef]
49. Moffatt, J.H.; Harper, M.; Harrison, P.; Hale, J.D.F.; Vinogradov, E.; Seemann, T.; Henry, R.; Crane, B.; St. Michael, F.; Cox, A.D.; et al. Colistin Resistance in *Acinetobacter baumannii* Is Mediated by Complete Loss of Lipopolysaccharide Production. *Antimicrob. Agents Chemother.* **2010**, *54*, 4971–4977. [CrossRef]
50. Quan, J.; Li, X.; Chen, Y.; Jiang, Y.; Zhou, Z.; Zhang, H.; Sun, L.; Ruan, Z.; Feng, Y.; Akova, M.; et al. Prevalence of *mcr-1* in *Escherichia coli* and *Klebsiella pneumoniae* recovered from bloodstream infections in China: A multicentre longitudinal study. *Lancet Infect. Dis.* **2017**, *17*, 400–410. [CrossRef]
51. Covarrubias, A.J.; Perrone, R.; Grozio, A.; Verdin, E. NAD^+ metabolism and its roles in cellular processes during ageing. *Nat. Rev. Mol. Cell Biol.* **2021**, *22*, 119–141. [CrossRef]
52. Cantó, C.; Menzies, K.J.; Auwerx, J. NAD^+ Metabolism and the Control of Energy Homeostasis: A Balancing Act between Mitochondria and the Nucleus. *Cell Metab.* **2015**, *22*, 31–53. [CrossRef]
53. Booth, I.R.; Blount, P. The MscS and MscL Families of Mechanosensitive Channels Act as Microbial Emergency Release Valves. *J. Bacteriol.* **2012**, *194*, 4802–4809. [CrossRef]
54. Wray, R.; Wang, J.; Iscla, I.; Blount, P. Novel MscL agonists that allow multiple antibiotics cytoplasmic access activate the channel through a common binding site. *PLoS ONE* **2020**, *15*, e0228153. [CrossRef]
55. Moosavian, M.; Emam, N.; Pletzer, D.; Savari, M. Rough-type and loss of the LPS due to *lpx* genes deletions are associated with colistin resistance in multidrug-resistant clinical *Escherichia coli* isolates not harbouring *mcr* genes. *PLoS ONE* **2020**, *15*, e0233518. [CrossRef]
56. He, L.; Dai, K.; Wen, X.; Ding, L.; Cao, S.; Huang, X.; Wu, R.; Zhao, Q.; Huang, Y.; Yan, Q.; et al. QseC Mediates Osmotic Stress Resistance and Biofilm Formation in *Haemophilus parasuis*. *Front. Microbiol.* **2018**, *9*, 212. [CrossRef]
57. Leblanc, S.K.D.; Oates, C.W.; Raivio, T.L. Characterization of the Induction and Cellular Role of the BaeSR Two-Component Envelope Stress Response of *Escherichia coli*. *J. Bacteriol.* **2011**, *193*, 3367–3375. [CrossRef]
58. Ehmann, D.E.; Jahić, H.; Ross, P.L.; Gu, R.-F.; Hu, J.; Durand-Réville, T.F.; Lahiri, S.; Thresher, J.; Livchak, S.; Gao, N.; et al. Kinetics of Avibactam Inhibition against Class A, C, and D β-Lactamases. *J. Biol. Chem.* **2013**, *288*, 27960–27971. [CrossRef]
59. Ku, Y.-H.; Lee, M.-F.; Chuang, Y.-C.; Chen, C.-C.; Yu, W.-L. In vitro activity of colistin sulfate against *Enterobacteriaceae* producing extended-spectrum β-lactamases. *J. Microbiol. Immunol. Infect.* **2015**, *48*, 699–702. [CrossRef]
60. Zhang, Y.; Wang, X.; Wang, S.; Sun, S.; Li, H.; Chen, H.; Wang, Q.; Wang, H. Emergence of Colistin Resistance in Carbapenem-Resistant Hypervirulent *Klebsiella pneumoniae* Under the Pressure of Tigecycline. *Front. Microbiol.* **2021**, *12*, 756580. [CrossRef]
61. Muhammad, J.S.; Khan, N.A.; Maciver, S.K.; Alharbi, A.M.; Alfahemi, H.; Siddiqui, R. Epigenetic-Mediated Antimicrobial Resistance: Host versus Pathogen Epigenetic Alterations. *Antibiotics* **2022**, *11*, 809. [CrossRef]
62. McConville, T.H.; Giddins, M.J.; Uhlemann, A.-C. An efficient and versatile CRISPR-Cas9 system for genetic manipulation of multi-drug resistant *Klebsiella pneumoniae*. *STAR Protoc.* **2021**, *2*, 100373. [CrossRef]
63. Wang, Y.; Wang, S.; Chen, W.; Song, L.; Zhang, Y.; Shen, Z.; Yu, F.; Li, M.; Ji, Q. CRISPR-Cas9 and CRISPR-Assisted Cytidine Deaminase Enable Precise and Efficient Genome Editing in *Klebsiella pneumoniae*. *Appl. Environ. Microbiol.* **2018**, *84*, e01834-18. [CrossRef]
64. Sun, Q.; Wang, Y.; Dong, N.; Shen, L.; Zhou, H.; Hu, Y.; Gu, D.; Chen, S.; Zhang, R.; Ji, Q. Application of CRISPR/Cas9-Based Genome Editing in Studying the Mechanism of Pandrug Resistance in *Klebsiella pneumoniae*. *Antimicrob. Agents Chemother.* **2019**, *63*, e00113-19. [CrossRef] [PubMed]

MDPI
St. Alban-Anlage 66
4052 Basel
Switzerland
www.mdpi.com

Pharmaceutics Editorial Office
E-mail: pharmaceutics@mdpi.com
www.mdpi.com/journal/pharmaceutics